D0875422

SENSORY PROCESSES

The New Psychophysics

SENSORY PROCESSES

The New Psychophysics

LAWRENCE E. MARKS

*John B. Pierce Foundation Laboratory
and Department of Psychology
Yale University
New Haven, Connecticut*

ACADEMIC PRESS   New York and London   1974
A Subsidiary of Harcourt Brace Jovanovich, Publishers

ACADEMIC PRESS, INC.
111 Fifth Avenue, New York, New York 10003

United Kingdom Edition published by
ACADEMIC PRESS, INC. (LONDON) LTD.
24/28 Oval Road, London NW1

Library of Congress Cataloging in Publication Data

Marks, Lawrence E
        Sensory processes: the new psychophysics.

        Bibliography: p.
        1.   Senses and sensation.   2.   Psychometrics.
I.   Title.   [DNLM: 1.   Psychophysics.   2.   Sensation.
BF237 M346s 1974]
BF237.M33   152.1   73-2070
ISBN 0–12–472950–9

TO JOYA

CONTENTS

PREFACE

In books on sensory psychology it is not uncommon to find prefaces that begin by pointing out the unique role that the senses play in human behavior and the acquisition of knowledge: To wit, the senses are portals to the mind, the senses detect and organize information about the world, and so on. In fact, such introductory remarks are so frequent that I shall dispense with them.

Psychology, philosophy, and physiology all converge on the study of the senses. Of these the psychological approaches—phenomenological and psychophysical—are propaedeutic. First comes the phenomenological examination of sensory experience, of the attributes and qualities of sensory perceptions. This leads almost naturally to the psychophysical examination of sensory experience: How do the perceptual attributes of sensations relate to the physical stimuli that produce them?

The psychophysics of human sensory processes has made startling advances over the past two decades. In large measure, these advances are the outcome of application of direct scaling procedures—what the major exponent, S. S. Stevens, has called "the new psychophysics"—to the study of sensation. Through the treatment of the intensities of sensations as measurable quantities, capable of direct assessment on the part of human subjects, sensory psychologists have been able not only to evaluate sensory magnitudes, but perhaps more importantly, to utilize such evaluations in order to elucidate the nature of fundamental sensory processes.

This book summarizes, describes, and theorizes on the application of the new psychophysics to the study of sensory processes. Its aim is to deal with substantive issues in sensory psychology, primarily by treating sensory dimensions and attributes as measurable quantities. A particular goal is to integrate results obtained via procedures that employ direct scaling with results obtained via older, more traditional procedures such as intrasensory matching. The emphasis, however, is clearly on the new psychophysics.

Like most books on sensory psychology, the present volume does deal extensively with visual and auditory processes, but the psychophysics of olfaction, taste, and the skin senses is far from neglected. I attempted in particular to emphasize studies of these three "lower senses," in large part because these sense departments have benefited proportionately the most from the new psychophysics. Many of the recent advances in knowledge and understanding of olfactory, gustatory, and somesthetic processes derive from application of direct scaling procedures.

Sensory Processes will be found useful by researchers and graduate students in sensory psychology, but will also be useful as a reference text for advanced undergraduate students interested in the senses. Its core consists of four central chapters that deal with the ways that stimulus variables—intensity, composition, duration, spatial distribution—influence and determine the magnitudes and qualities of our sensations. Since the approach taken is topical, that is, it deals with topics such as temporal summation and spatial inhibition, it was possible to point out, when applicable, those principles of sensory functioning that pertain to several senses. The chapters that straddle this core provide a brief history and introduction to the methods and basic results of sensory scaling, a detailed example of how scaling was used to examine one particular substantive issue in the domain of sensory processes, and some theoretical issues that arise in the realm of sensory scaling.

There are many people to whom I am indebted. Foremost are S. S. Stevens and Joseph C. Stevens, who have served in several roles: teachers, critics, and models. Special thanks go to Joseph C. Stevens, William S. Cain, Marc H. Bornstein, and Linda M. Bartoshuk for reading and evaluating the mansucript. Mrs. Carol Mikalavicius performed the unenviable task of typing the manuscript amidst a plethora of other duties. Mrs. Fenna Bouhuys prepared the graphs.

The last word is reserved for my wife, who persevered with me through the many years of work that led to the book's completion—not just through the last few years of work on the manuscript itself, but also through the formative years.

SENSORY PROCESSES

The New Psychophysics

Chapter 1

INTRODUCTION TO SENSORY SCALING

The author's favorite lunchtime restaurant is quite dimly lit: Upon emerging from it, objects on the street, illuminated by the afternoon sun, are almost painfully bright.

One of the author's friends complains that various substances—orange juice, cigarettes—remain unpleasant tasting for a period of time after using toothpaste.

A finger or elbow dipped in a heated bath feels only slightly warm in comparison to the massive thermal sensation experienced when the entire body is immersed.

Certain odors linger in a room almost independently of the amount of fresh air that is introduced, whereas other odors disappear rapidly when so diluted in concentration.

Such a list could be collected and added to almost indefinitely. Most or all of the phenomena just described (or variants of them) should be familiar to the reader. They are examples of sensory phenomena, phenomena that reflect the ways the sense organs operate on signals that impinge upon them. In fact, all of these phenomena are reflections or examples of basic, fundamental properties of human sensory systems. A primary concern of this book is to elucidate such fundamental properties as temporal adaptation and spatial summation.

How does brightness decline with time after initial exposure to light? How does warmth increase as more and more fingers are immersed into heated water? These are *psychophysical* questions. That is, they deal with the relations between psychological or sensory quantities on the one hand and physical or stimulus quantities on the other: brightness versus time,

1

warmth versus area, and so on. Our major concern is with what has been dubbed the "new psychophysics." What is the new psychophysics? Quite simply, it consists of attempts to answer *directly* the aforestated questions! Thus, an answer to the first question might be that brightness declines exponentially with duration of the stimulus. In order to answer psychophysical questions in this manner, it is necessary to quantify both the stimulus variables and the response variables. In order to say that magnitude of warmth sensation is proportional to area, for instance, not only must we be able to measure area (in square centimeters, say), but also to measure warmth (in subjective units).

New versus Old Psychophysics

New Psychophysics

Fundamental to the new psychophysics is the view that human subjects can make meaningful evaluations of the magnitudes of their sensory experiences, at least under certain conditions. The primary developer of methods for evaluating sensory magnitudes is S. S. Stevens, a man whose assiduous interest in measurement and sensory processes dates back to the early 1930s. Stevens has developed and elaborated methods—often called direct scaling methods—for determining magnitudes of sensations. Whereas measurement of stimulus magnitudes may be difficult at times, the principles of measurement have been well worked out. Unfortunately, the same cannot be said for measurement of sensations. But the last two decades have seen great strides made in the quantification of sensory responses, e.g., to saying how many times one sensation is as great as another. This is the domain of the new psychophysics.

Old Psychophysics

Attempts to measure sensation and to relate sensory to stimulus magnitudes have at least two centuries of history. The first quantitative statement (to the best of the author's knowledge) appeared in the eighteenth century, when Krüger (1743) proposed that sensation grows proportionally in its strength with increases in stimulus strength. And the first empirically based "law"—Fechner's logarithmic rule—appeared more than a century ago. Nevertheless, for many years the main thrust of attempts to measure (quantify) sensations and to relate sensations to stimuli fell under the rubric of what may be termed the "old psychophysics." The viewpoint of the old psychophysics was that it is meaningless, difficult, or unimportant to measure sensation directly. So, instead, we measure the stimulus. Take, for example, the question, "How does brightness decrease with time after initial exposure?" One way to assess the decline is to present the test (adapting) light to one eye and a comparison light to the other. The comparison light flashes on briefly, every 5 sec, say, and the subject's task is to adjust its intensity to match the brightness of the test light that is seen continuously. At onset, prior to any adaptation, the comparison will be adjusted to be equal in intensity to the test light. But as time goes on, and the brightness of the test light declines, the intensity of the comparison, adjusted so as to equal the brightness of the test light, must also diminish. The extent of diminution gives a measure of "brightness adaptation."

This sort of measurement—done in accordance with the old psychophysics—for decades dominated the study of the senses. In fact, it still holds sway, providing a lion's share of the information about sensory functioning. But, as this book will attempt to document, more and more research in recent years has shown the limitations imposed by the very nature of the old psychophysics, and, concomitantly, the advantages that often accrue to the use of methods of the new psychophysics.

Obviously, in the sort of measurement of brightness adaptation just described, we do not really have a measure of the magnitude of brightness sensation. All we can say is, for example, that after 2 min of viewing a continuous stimulus, its brightness appears the same as the brightness of a flash of light that is one-eighth (or whatever) as intense. The old psychophysics relied on the method of *intrasensory matching*. What the new psychophysics adds to our knowledge is the relation between brightness and intensity. Thus if we know that an eightfold change in intensity corresponds to a twofold change in brightness, we can conclude that after 2 min of viewing, the brightness fell to a level one-half the initial brightness value.

There is another important approach under the rubric of the old psychophysics, one that deserves special mention. That approach consists of the determination of *sensory invariances*. In the example given above, the time course of adaptation is determined by varying the intensity of a comparison stimulus in order to follow the changing level of brightness. To determine sensory invariance or equivalence, it is necessary to find out how two variables interact to maintain a *constant sensory experience*. For example, adaptation might be studied by maintaining constant the intensity of the comparison light, but adjusting the intensity of the test light over time in order to keep its brightness at the same constant level as that of the comparison. This latter sort of measurement employs the subject as a "null instrument." One purpose of this approach is to avoid some of the problems that arise from nonlinear transformation between stimulus and sensation magnitudes. As we just saw, with the previous procedure (tracking the course of brightness over time), it is necessary to know the psychophysical relation between brightness and intensity of flashes of light in order to understand how brightness diminishes; with the null method, that problem is avoided, since brightness does not change. Thus, the null approach may be considered as the opposite end of the spectrum from the approach of the new psychophysics.

PSYCHOPHYSICAL, SENSORY–PHYSICAL, AND PSYCHOSENSORY MEASUREMENT

Perhaps most important to keep in mind is that all of these types of "sensory measurement"—of the old and of the new psychophysics—are mutually complementary. "Psychophysics" as a term encompasses all of these, as well as several other, approaches to the study of sensation. That use of the term psychophysics is a broad one. We may speak of psychophysics in the narrow sense, however, as the study of *stimulus* and *sensation*. Then, the term *psychophysics* in the narrow sense refers only to what we have called the new psychophysics, i.e., to studies of the relations between sensation and stimulus when both are measured as quantities. It becomes convenient in that case to look for another term to describe the old psychophysics. The term to be employed here is *sensory physics*. Sensory physics is defined here to refer to the evaluation of sensory responses purely in terms of measurements of variations in the corresponding physical stimuli. There exists a third category of relationships under the broad term, psychophysics; this third category also arose as a result of the new psychophysics. It consists of *psychosensory* measurement, the determination of interrelations among sensory variables without regard to how they may depend on the stimulus. We have few examples of psychosensory data, however, so the concept is at present of more formal than practical interest. Some examples appear in Chapter 5 (binaural additivity of loudness) and in Chapter 6 (relations among the auditory attributes of loudness, volume, and density).

Early Attempts to Quantify Sensation

As was already mentioned, interest in the quantitative relations between sensation and stimulus well predates the new psychophysics. The greatest impetus to that interest was the publication in 1860 of G. T. Fechner's *Elemente der Psychophysik*, which propounded the notions that sensory intensities could be measured and that the fundamental relation between sensation and stimulus intensities is logarithmic. Unfortunately, attempts at quantification of sensory magnitudes have often led to much rancor and interminable "philosophical" types of dispute, the outcome of which was usually little or no gain in understanding how sensory systems operate. At the point in his textbook where consideration of Fechner's work became appropriate, William James (1892) commented, "Fechner's psycho-physic formula, as he called it, has been attacked on every hand; and as absolutely nothing practical has come of it, it need receive no farther notice here [p. 22]."

It is worthwhile mentioning that psychologists have not been alone in their interest in the measurement of sensory magnitudes. Workers concerned with the application of knowledge in sensory psychophysics, such as acoustical engineers, have often found themselves looking for answers to psychophysical questions of this nature. In fact, it is probably not unfair to state that, to a large extent, the continued interest in the relation between sensory magnitudes and stimulus intensities through the early decades of this century was the result of dissatisfaction among acoustical workers with Fechner's logarithmic psychophysical law.

Among acoustical engineers it had become popular to use decibel notation. Since the decibel is a logarithmic measure of relative energy flow, it had become common, because of Fechner's logarithmic law, to treat the decibel scale of sound energy as a scale of sensory intensity, or loudness. Fletcher for a period of time used the decibel as a unit of loudness (see Fletcher & Munson, 1933). Starting at absolute threshold, constant increases in decibels were assumed to correspond to constant increases in loudness. But the logarithmic (decibel) law was clearly unsatisfactory. Churcher wrote in 1935,

> In common with others engaged on noise problems, the experience of the author and his colleagues over many years is that the numbers assigned by the decibel scale to represent sensation magnitudes are not acceptable to introspection as indicating their relative magnitudes. Two instances will suffice. The loudness of the noise of a motor assessed at 80 dB above threshold in terms of a pure reference tone is, to introspection, enormously greater than twice that of a motor assessed at 40 dB. In some experiments on the reduction of the noise of a geared turbo generator, successive measured values of 104 and 100 dB were obtained or what would ordinarily be taken to indicate a 4 percent reduction in loudness. However, to the ear the loudness reduction appeared much greater than 4 percent and an onlooker who knew nothing of the measurements volunteered the opinion that the reduction was about 20 percent. On other evidence we know that it was actually about 27 percent [p. 217].

Had Fechner's logarithmic law been less egregiously at variance with direct experience, no doubt there would have been much less an impetus to discover the correct nature of the psychophysical law, and in particular the psychophysical function for loudness.

FECHNER'S PSYCHOPHYSICAL LAW

It appears appropriate at this point to review briefly some of the history of attempts to generate scales of sensory magnitude, and, by relating these scales to scales of corresponding physical intensities, thereby determine the nature of the psychophysical function for sensation magnitudes. Mention has already been made of Fechner (1860), who was one of the first scientists with audacity enough to suggest a mathematical relation between sensation and stimulus magnitudes. Fechner's logarithmic law may be expressed

$$\psi = k \log(\phi/b) \tag{1.1}$$

where ψ stands for the sensation magnitude, ϕ stands for the stimulus magnitude, b stands for the stimulus magnitude at the absolute threshold, and k is a constant of proportionality. Equation (1.1) was called by Fechner the *Massformel* or *measurement formula*. Fechner's law can be stated as, "Equal stimulus ratios produce equal sensation intervals." Or, "As stimulus intensity increases geometrically, sensation intensity increases arithmetically." When the stimulus is measured in units of the absolute threshold (b), so that the sensation aroused by intensity b is zero, then the sensation magnitude aroused by stimulus intensity $10b$ is k, that aroused by stimulus $100b$ is $2k$, that aroused by $1000b$ is $3k$, etc.

Fechner came to base his law on groundwork done by E. H. Weber (1834). Weber established (approximately) that discrimination is relative: the amount by which the intensity of a stimulus must be increased or decreased $(\Delta\phi)$ in order for a sensory change to be detected is a constant fraction of the original stimulus intensity (ϕ), or

$$\Delta\phi = c\phi \tag{1.2}$$

the expression known as Weber's law.

Weber's law is a general statement concerned with the discriminability of stimuli. Fechner's idea was to make detected changes (often called "just noticeable differences") into units by which sensory magnitude could be measured. The central assumption was that each just noticeable difference (JND) can be considered a minimal increment in sensation, and, furthermore, that all JNDs are equal in subjective magnitude. One method for computing sensory magnitude consists of counting up JNDs starting at the absolute threshold. For example, if 10 JNDs occur between threshold and stimulus intensity z, then the sensation magnitude at z is 10; if 50 JNDs occur between threshold and stimulus intensity y, the sensation magnitude at y is 50. [It is of some historical interest that a scheme basically the same as Fechner's was devised independently some years later by W. F. Tyler (1904, 1907). Tyler, who apparently was totally unaware of Fechner's work, described a procedure for generating scales of sensory intensity by counting off JNDs. Tyler's primary interest was climate; Titchener (1909) reviewed the work of Tyler and several other early investigators of the psychophysics of climate.]

Fechner proposed an alternative method for evaluation of sensory magnitudes, one that is more formal. He assumed that since a JND corresponds to a minimal sensory change, then it can be reduced further and expressed as a mathematical differential, namely $\delta\psi$. Fechner further assumed that Weber's law can be written in terms of mathematical differentials; so, because of the assumed constancy in the subjective size of the JND, Fechner wrote

$$\delta\psi = c \left(\frac{\delta\phi}{\phi} \right) \tag{1.3}$$

Equation (1.3) was termed by Fechner the *Fundamentalformel*, or fundamental formula. Integration of Eq. (1.3) gives Eq. (1.1), that is, Fechner's logarithmic law.

CRITIQUES AND EXTENSIONS

We shall make no attempt to discuss the long and often tedious controversy that followed publication of Fechner's *Elemente*. We may, however, take the opportunity to describe briefly some of the critical issues and ideas. There were three major objections to Fechner's arguments and conclusions. The first objection has been called the "quantity objection." Simply stated, it says that sensations do not have magnitudes, so, obviously, their magnitudes cannot be measured. A major proponent of this viewpoint was James (1890), who quoted Stumpf: "One sensation cannot be a multiple of another. If it could, we ought to be able to subtract the one from the other, and to feel the remainder by itself. Every sensation presents itself as an indivisible unit [p. 547]." [As a reply to this type of critique, Richardson and Ross (1930) constructed the parody: "One mountain cannot be twice as high as another. If it could, we ought to be able to subtract the one from the other and to climb up the remainder by itself. Every mountain presents itself as an indivisible lump (p. 301)."]

The quantity objection had an impact. In particular, Delboeuf and Titchener rejected the idea that sensations per se can be measured, but substituted the notion that sense distances can be measured. Whereas, according to this viewpoint, loudness of tones cannot be measured, the loudness intervals between tones can. Thus, for example, Delboeuf (1873) modified Fechner's law in order to make it refer to sense distances (or, as Delboeuf termed them, *contrastes sensibles*).

A second objection to Fechner's conclusion was that Weber's law held only as an approximation. At low stimulus intensities in particular there often appear clear departures from simple proportionality between $\Delta\phi$ and ϕ. This objection was relevant only to the formal derivation of the logarithmic law, but not to the graphical derivation of psychophysical relations. Psychophysical functions can be obtained graphically—by summing JNDs—regardless of the form of the discrimination function. Formal derivation of Fechner's logarithmic law, however, relies on the validity of Weber's law.

The third major objection, the one that has carried into the twentieth century, concerned Fechner's assumption that all JNDs are equal in subjective magnitude. Again quoting James (1892): "The many pounds which form the just perceptible addition to a hundredweight feel bigger when added than the few ounces which form the just perceptible addition to a pound. Fechner ignored this fact [p. 21]."

One specific alternative to the hypothesis that all JNDs are subjectively equal is that they grow in size as sensory magnitude increases. Brentano (1874) suggested that a rule like Weber's law might also hold for sensation. If there were so then, following the Fechnerian argument, we could write

$$\Delta\psi = c'\psi \tag{1.4}$$

and, in differential form

$$\delta\psi = c'\psi \tag{1.5}$$

Combining with Weber's law and integrating, we obtain

$$\frac{\delta\psi}{\psi} = \frac{c'}{c}\left(\frac{\delta\phi}{\phi}\right)$$

$$\log \psi = (c'/c) \log \phi + k'$$

or

$$\psi = k\phi^{c'/c} \tag{1.6}$$

Thus, we would conclude that the psychophysical relation is a power function.

This third objection has perhaps been the most important for subsequent research. For if sensation magnitudes, or at least sense distances, can be measured, why not attempt to assess sensory magnitudes or distances directly? If direct measurements agree with JND scales, then the subjective equality of JNDs would be verified. In fact, Plateau had already attempted to equate intervals of lightness in experiments conducted in 1852 (and therefore prior to publication of Fechner's *Elemente*), although the results were not published until 1872. Incidentally, Plateau concluded that his data were not consistent with a logarithmic psychophysical law, but rather with a power law. Along similar lines Breton (1887) found a square-root relation between lightness and light intensity. On the other hand, Delboeuf's (1873) results were much more consistent than was either of those with a logarithmic law.

During the late nineteenth century it was considered of critical importance that results obtained from the determination of equal sense distances corroborate Fechner's law if that law were to be accepted. It is important to note that in passing from Fechner's approach via discrimination to the later approach of equating sense distances, we have made a major advance in extending our notion of what operations an experimental subject can be expected to perform with some reliability and validity. Fechner did not believe that sensations could be measured directly; sensations could be measured only indirectly, i.e., by means of discriminability [a notion that traces back to Körber (1746)]. The equation of sense distances, on the other hand, is a procedure for directly evaluating sensations (on an interval scale) or sensation intervals (on a ratio scale). At least it is assumed that the experimental subject is capable of making valid judgments of the equality of sensory intervals.

Titchener (1905) described in great detail the methods, results, and critiques involved in the study of sense distances. Notable is the discussion of the results obtained by Merkel (1888, 1889, 1894), which were consistent in their deviation from predictions made by Fechner's law. If three stimuli are used to mark off two adjacent sense intervals, Fechner's law predicts that the intensity of the middle stimulus (the "bisection point") will equal the geometric mean of the extreme stimuli. Merkel's data fell consistently closer to the arithmetic mean, and Merkel interpreted those results to invalidate Fechner's law. Merkel in fact believed that psychophysical relation to be a linear one, i.e., sensation is proportional to stimulus—Krüger's law (1743). Persuasive enough were the data in showing deviations away from Fechner's prediction and closer to Merkel's (although rarely, if ever did his "bisection points" actually fall at the arithmetic mean, as a linear law predicts), that several psychologists, including eventually Wundt (1902), concluded that there are two psychophysical laws: one law holding for minimal sense distances, i.e., JNDs, and being approximately logarithmic; the other law holding for large sense distances and being approximately linear.

ALTERNATIVES TO FECHNER'S LAW

Over the years that followed Fechner's proclamation of a logarithmic psychophysical law, various modifications and alternatives appeared. For the value of historical completeness—although at the obvious risk of tedium—some of these alternatives are here recorded. First, of course, it should be pointed out that Weber's law, upon which the logarithmic law rested, is only an approximation. Fechner was aware of the fact that Weber's law is not true at low levels of stimulus intensity (and for a long time it was thought to fail also at high intensities; however, the latter breakdown seems to occur only under special circumstances). It is now clear that for many sensory modalities, at least under certain conditions of measurement, a linear generalization of Weber's law is accurate (Fechner, 1860; Miller, 1947). According to this modification, $\Delta\phi$ is proportional to ϕ plus a constant, rather than proportional to ϕ. Other modifications of Weber's law have been proposed, and several attempts were made to derive psychophysical laws by means of Fechner's approach as outlined above.

Helmholtz (1866), Broca (1894), and Schjelderup (1918) all proposed identical formulas for visual psychophysics. These were based upon functions for luminance discrimination that made $\Delta\phi$ proportional to $(\phi + a)^{-1} - (\phi + b)^{-1}$ (effectively a quadratic function). By integration, it was found that sensation ψ is proportional to

$$\log \frac{\phi + a}{\phi + b}.$$

Most likely, this formula had a special appeal to its discoverers in that it predicts sensory magnitude to asymptote—reach a maximum—at high stimulus intensities. [It was, and still is, a common belief that sensory magnitudes must reach some upper limit. In fact, given steady-state viewing of lights, brightness does reach an upper asymptote. However, there seems to be no firm evidence that a ceiling on sensory magnitude is ever reached—shy of stimuli that might damage the sense organ—by transient stimulation. At least one experiment found subjective magnitude to increase up to the highest stimulus intensity possible (Eisler, 1965). As one possible exception, however, there may be an upper limit to sensation in the case of thermal pain (Hardy, Wolff, & Goodell, 1947).]

Other variants of the equation describing discrimination appeared from time to time. One of these was a power relation between $\Delta\phi$ and ϕ (Guilford, 1932), which leads to a power function between stimulus and response so long as the exponent of the discrimination function is not unity. It has even been suggested that the Weber fraction $\Delta\phi/\phi$ follows a normal distribution, so the psychophysical function becomes, by Fechnerian integration, an integral of a phi–gamma function (Houston, 1932).

One unfortunate aspect of these attempts to determine the mathematical nature of the psychophysical function is that they rest on questionable mathematical premises. Luce and Edwards (1958) demonstrated that Fechner's method for integrating discrimination functions is valid for certain types of functions, e.g., Weber's law and its linear generalization, but not for others. The invalid examples include all of the others just described. By the use of appropriate approximations, it is possible to do formal integrations of discrimination functions that are mathematically valid (Krantz, 1971), but the psychophysical functions that emerge will be somewhat different from those derived by the simple Fechnerian approach.

Pütter (1918) suggested that sensory magnitude grows as an exponential function of stimulus intensity—proportional to $(1 - e^{-b\phi})$; a modification of that formula was considered by Zinner (1930–1931), but finally rejected in favor of the proposal that sensation is the exponential of the tangent of intensity $(e^{b\tan\phi})$. Again, both of these equations suggested that sensation reaches a ceiling over high intensities.

One of the more interesting suggestions was made by Bénéze (1929), who propounded an analogy between the psychophysical relation and the trigonometry of the viewing of distant objects. According to his analogy, if the height of such an object reflects stimulus intensity, then the angle subtended by the object at the eye corresponds to sensory intensity. It follows by simple trigonometry that sensation increases as the arctangent of intensity. Truly a self-evident argument!

The Psychophysical Law in Regard to Sensory Physics

By this time the reader might wonder, "What difference does it make what the psychophysical law is?" Is it important—does it matter whether the equation is logarithmic, pseudo-logarithmic, power, exponential, tangential, arctangential, or anything else? In particular, does knowing the mathematical nature of the psychophysical law add at all to our understanding of how sensory systems operate? This is a fundamental question that perhaps can be best examined by looking at the relation between the psychophysics and the sensory physics of sensory processes.

Recall that psychophysical laws relate sensory quantities, measured as such, to physical quantities. Often one particular relationship is called *the psychophysical law*. This is the relation between sensory magnitude and stimulus intensity. In a broader sense, however, there can be other psychophysical laws. Additional psychophysical laws might relate sensory magnitude to some other quantitative, physical variables, such as stimulus duration, stimulus size, etc. The important point is this: Whatever the means by which the quantities are measured or determined, psychophysics (in the narrow sense) deals with relations between two domains, one psychological, the other physical.

By way of contrast, sensory physics deals with relations among quantities that remain totally within a single domain, namely the physical. The statement, "stimulus intensity multiplied by stimulus duration is a constant value," is an example of a sensory-physical law. It is applicable to the fact that two brief visual stimuli—the first of which is half the duration of the second, but the second of which is twice the intensity of the first—will appear equally bright. The statement does not say how bright they will look, but just that they will be the same. At first glance, then, psychophysical relations seem independent of sensory-physical ones. The nature of the psychophysical relation does not matter: logarithmic or power relations can both be consistent with sensory-physical functions.

However, the independence of sensory physics from psychophysics may not be complete. In particular, it is likely that quantification of certain sensory phenomena will, via theory, point more directly toward certain psychophysical relations than toward others. An example is the loudness of sounds heard by two ears and by one: A theory of binaural listening, if it is to be a good one, will almost certainly reveal how the loudness of a sound heard with two ears compares to the loudness of the same sound heard with one. Once we know the rule for binaural summation of loudness, it is easy to verify which psychophysical law is correct. A sound heard binaurally appears to be as loud as a sound heard monaurally when the latter is about 10 dB greater in intensity. Should the theory predict binaural loudness to be twice as great as monaural loudness of a sound at the same intensity then the correct psychophysical law for loudness would state that loudness doubles with every dB increase in stimulus intensity. Final verification of psychophysical functions becomes, in this way, an important adjunct to sensory physics, albeit one that is fundamentally theoretical in nature. The theories by means of which psychophysics and sensory physics join together are themselves theories involving sensory processes. It is unlikely that the two can be joined together through theories or experiments that concern themselves either with psychophysics or scaling alone.

Early Development of the New Psychophysics

All of the attempts and procedures to quantify sensation that were Fechnerian in nature (i.e., that rested on the measurement of discrimination) fall under the rubric of *indirect scaling*. They are called indirect because the measure of sensory magnitude comes about second hand, via the measurement of something other than magnitude per se. Other procedures, such as the methods for setting up equal-appearing intervals of sensation, fall under the complementary rubric of *direct scaling* procedures. With these methods, the measurements of sensory magnitude arise directly from the judgments and observations made by the subjects. In spite of the fact that this terminology has not received universal acceptance, it is convenient, unambiguous in the distinctions it makes, and, therefore, will be used throughout this book (for a criticism of the terminology see Zinnes, 1969).

Interval Scaling

It is convenient to distinguish two types of direct scaling procedures. In one type, the subject's task is to set sensory intervals to appear equal. An example is the method of Plateau (1872), who asked artists to paint a gray that appeared to fall midway between a white and a black. That method is often called *bisection*. Bisection was a popular scaling procedure in the late nineteenth and early twentieth centuries (Titchener, 1905). Another important interval-scaling procedure is the method of *category rating*, wherein the task is to assign categories (often integer numbers) to stimuli in such a way that each succeeding category marks off another constant step in sensation. The rating method traces back to the beginning of this century.

RATIO SCALING

Late in the nineteenth century, the first example of a second type of direct scaling procedure appeared. Merkel (1888) introduced a method called the *Methode der doppelten Reize,* or method of doubled stimuli. What Merkel meant, however, was the doubling of sensation: the subject's task was to set a variable stimulus so that the sensation it produced appeared twice as great as that produced by a fixed stimulus. Fullerton and Cattell (1892) used a similar procedure, asking subjects to adjust sensations to be various fractions and multiples of a standard. A major advantage of this sort of procedure lies in the fact that it can lead to a ratio scale, i.e., a scale of sensation that has a real, nonarbitrary zero point. On a ratio scale, the ratio relations among scale values are meaningful. With a method such as bisection, for example, we cannot ever tell without additional information when one sensation is twice another—only intervals, not ratios, are marked off.

Unfortunately, the ratio methods did not catch on in the early years of sensory scaling. Ratio methods did emerge, however, in the 1930s, particularly under the impetus of acoustical engineering. Note has already been made of the unhappiness that existed with regard to use of the decibel as a unit of sensation. A landmark study in the development of the new psychophysics was the experiment of Richardson and Ross, published in 1930. Richardson's earlier studies (1928, 1929) convinced him that sensation could be measured by means of direct, quantitative estimation on the part of subjects. Consequently, and subsequently, Richardson and Ross asked 11 people to assign numbers to stand for perceived levels of loudness of tones (550–1100 Hz). (The unit hertz—Hz—refers to number of cycles per second.) This method was later designated *magnitude estimation* (Stevens, 1956b, 1958.) In this early study, a standard tone (fixed intensity) was presented, and the subject was told to give to its loudness the numerical value 1. The estimation of loudness of other tones was made with respect to the standard. Richardson and Ross concluded that estimated loudness grows as a power function of sound pressure, with an average exponent of the power function approximately equal to .44. In terms of sound energy, the exponent would be .22, since sound pressure and energy are related by a square function (see Appendix).

Subsequent studies by Ham and Parkinson (1932), Laird, Taylor, and Wille (1932), Geiger and Firestone (1933), and Rschevkin and Rabinovitch (1936) gave generally similar results (although those of Laird *et al.* did not agree so well with the others). As a matter of fact, Rschevkin and Rabinovitch found that loudness increased as the .22 power of sound

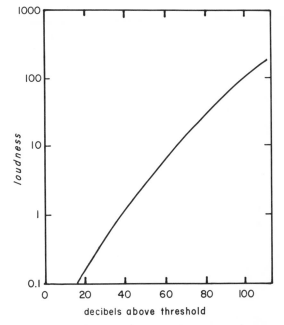

Figure 1.1. The original sone scale of loudness, as a function of the sensation level in decibels of a 1000-Hz tone. [From Stevens (1936). Courtesy of the author.]

energy! All of these experiments employed ratio procedures, but procedures of a type somewhat different from magnitude estimation. In particular, the procedures resembled those of Merkel (and of Fullerton and Cattell) in that they called for the production of loudnesses set to fixed fractions or multiples of a standard. The subject might set one tone to sound one-half, one-fourth, or one-tenth as loud as a standard, or to sound 2, 4, or 10 times as loud. The results from all of the experiments were summarized by Churcher (1935), and Churcher's summary was in turn employed by Stevens (1936) to produce a loudness scale: the original *sone* scale.

Stevens started with the definition of one sone as the loudness of a 1000-Hz tone, heard binaurally, at an intensity level 40 dB above its threshold. He proceeded to construct the psychophysical function shown in Figure 1.1. Although the scale was later revised (Stevens, 1955), a still more recent summary (Stevens, 1972) suggests that the original sone scale comes very close to an accurate portrayal of how the typical subject estimates the loudness of a 1000-Hz tone at different sound pressures.

In general, the two decades that followed the construction of the sone scale saw the development of sensory scales (and concomitantly of psychophysical relations between sensation magnitude and stimulus intensity) for a number of sense attributes and modalities. These included pitch

of pure tones (Stevens, Volkmann, & Newman, 1937; Stevens & Volkmann, 1940), apparent rate of flashes of light (Reese, 1943), taste intensity (Beebe-Center & Waddell, 1948), heaviness of weights (Guilford
& Dingman, 1954; Harper & Stevens, 1948), and brightness of lights
(Hanes, 1949a & b). The method used to determine all of the sensory
scales was *fractionation:* the task of adjusting the intensity of one stimulus to make its sensory magnitude appear some fraction of that of a fixed
stimulus. In a basic way, though, these scales and their psychophysical
functions, remained isolated, unrelated to much of the rest of sensory
psychology, and not clearly related to one another, until the 1950s, when
direct ratio scaling began to be applied extensively to the study of sensory
processes.

DIRECT RATIO SCALING AND THE PSYCHOPHYSICAL LAW

A signal event in contemporary psychophysics was S. S. Stevens's discovery (or rediscovery of Richardson and Ross's discovery) of the method of magnitude estimation. More than any other direct scaling procedure, magnitude estimation has proven versatile and adaptable to many
different sorts of experimental situation, and it has proved the bulwark of
attempts to study sensory processes by means of scaling procedures.
Magnitude estimation has several advantages over the other direct ratio
procedures such as fractionation. For one, fractionation requires the subject to adjust stimulus intensity, and it is not always easy to put control of
the stimulus in the hands of the subject; fractionation is especially difficult in cases where only discrete stimulus values can be produced. Furthermore, magnitude estimation tends to be more rapid: The subject
merely emits a numerical response to every stimulus, so the time for
judgment is a few seconds at most. This contrasts with the great lengths
of time often needed in order to adjust stimuli to effect a fractionation.
Speed and ease of response are more than just matters of convenience.
Because the method of magnitude estimation usually makes it easy to
present and judge relatively large numbers of stimuli in a single experimental session, it is particularly conducive to experiments in which more
than one stimulus parameter is varied. Work by Stevens and his
collaborators on loudness (Stevens, 1955, 1956b), brightness (Stevens &
J. C. Stevens, 1960; J. C. Stevens & Stevens, 1963), and many other sensory dimensions demonstrated the value and utility of the method of
magnitude estimation in the measurement of sensory magnitudes.

Figure 1.2 shows some typical results of 10 experiments on 7 sensory
dimensions. Both coordinates of Figure 1.2 are logarithmic, and all of the

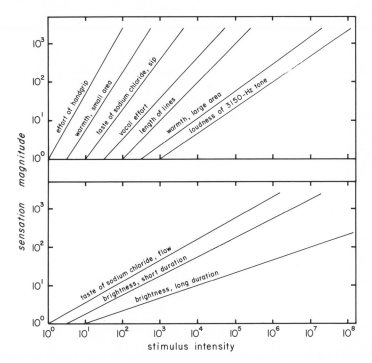

Figure 1.2. Psychophysical functions showing how perceived magnitude increases with stimulus intensity on seven sensory dimensions. Both coordinates are logarithmic. The exponents of the power functions are: brightness (long), .33; brightness (short), .5; sodium chloride (flow), .55; loudness, .67; warmth (large), .7; length, 1.0; vocal effort, 1.2; sodium chloride (sip), 1.3; warmth (small), 1.5; effort, 1.7.

psychophysical functions are straight lines. If ψ represents the (estimated) psychological magnitude and ϕ the stimulus intensity, then straight lines conform to the equation

$$\log \psi = \beta \log \phi + k' \qquad (1.7)$$

Taking antilogarithms we obtain

$$\psi = k\phi^{\beta} \qquad (1.8)$$

where $\log k = k'$.

Equation (1.8) is a simple power function, the function shown by Stevens and others to apply to more than two dozen sensory attributes. A convenience of the double-logarithmic plot is that the size of the expo-

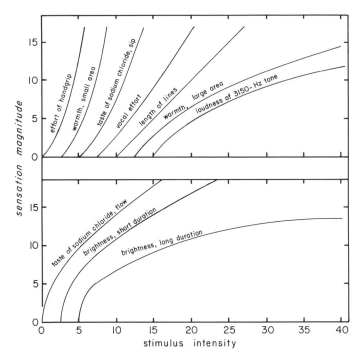

Figure 1.3. Psychophysical functions showing how sensory magnitude increases with stimulus intensity on seven sensory dimensions. Both coordinates are linear.

nent β is given directly by the slope of the straight line. For example, the brightness of a flash of light presented to the dark-adapted eye grows as the cube root of luminance (shallow slope), whereas the apparent effort of muscular contractions increases as the 1.7 power of the force applied (steep slope). Sensory modalities with intermediate-sized exponents have slopes that fall in between. The same psychophysical functions are plotted also in the linear coordinates of Figure 1.3. Here differences in the nature of rate of growth are more apparent. A cube-root function, such as the one governing brightness, is negatively accelerated in linear coordinates: brightness increases less and less rapidly with greater and greater increase in luminance. On the other hand, apparent effort grows more and more rapidly with increasing force: the psychophysical relation is positively accelerated in linear coordinates. Recall that Fechner's logarithmic equation requires that psychophysical functions be negatively accelerated for all sensory dimensions. Two other points are worthy of mention here. First, rate of growth (exponent) sometimes depends strongly on procedure. When sodium chloride is flowed over the tongue, the psychophysical function is negatively accelerated, but when sipped in the mouth, the function is positively accelerated. Second, rate of growth frequently depends on stimulus variables. A brief (10 msec or less) flash of light yields a square-root brightness function, but a longer (1 sec) flash yields a cube-root function.

The psychophysical power Eq. (1.8) has received empirical support time after time, both with repeated experimentation on the same sensory modalities as well as with a large number of different modalities. The support was extensive enough so that Stevens (1957) proposed a power law as a general psychophysical law. The psychophysical power law states that "Equal stimulus ratios produce equal sensation ratios." An 8-to-1 ratio of luminances, for example, may produce a 2-to-1 ratio of brightnesses; this corresponds to a power function with exponent ⅓. But a 10-to-1 ratio of forces produces a 50-to-1 ratio of perceived effort; this corresponds to a power function with exponent 1.7. As the stimulus intensity increases geometrically, so sensation magnitude also increases geometrically. The power law states, therefore, that the relations among sensations follow patterns that are similar to patterns of relations among stimuli. That patterning forms the basis for at least one hypothesis of the origin of the psychophysical power law, namely that it is based on both a neural modeling of the external world and on a principle of perceptual invariance (Yilmaz, 1967): Equal ratios in the sensory domain correspond to equal ratios in the stimulus domain; multiplicative transformation of the stimulus domain leaves the ratios of sensations invariant. In any case, regardless of the appropriateness of Yilmaz's model, it is probably not extravagant to remark that the power law is one of the best-established quantitative statements in psychology.

GENERALITY OF THE POWER LAW

Before accepting Eq. (1.8) as a general psychophysical law, however, there are several questions that require answers. One question is whether the power law applies to all psychophysical relations between sensory and stimulus magnitudes. The answer is, no, there are some sensory attributes that clearly fail to show power–law behavior. An example is the relation of pitch to the frequency of pure tones. The mel scale of pitch (Stevens *et al.*, 1937; Stevens & Volkmann, 1940) is not a power function of frequency. Instead, pitch relates in a complex way to frequency, and the relation seems for the most part determined by the relation to frequency of place of maximal stimulation on the basilar membrane. In fact, there exists a class of sensory continua, sometimes termed *metathetic* (Stevens,

1957; Stevens & Galanter, 1957), that appear often not to display power–law behavior. These metathetic continua contrast with *prothetic* continua, for example, loudness, brightness, and effort, that do seem to follow power functions. The distinction between metathetic and prothetic was made, however, not on the basis of whether the various continua follow the power law, but rather on other psychophysical criteria, such as whether the subjective size of the JND is constant (metathetic) or variable (prothetic). The distinction corresponds roughly to one between qualitative attributes on the one hand and quantitative ones on the other. It may be noted that the distinction has not met with universal acceptance (Piéron, 1959; Warren & Warren, 1963). Sensory modalities or dimensions may even vary in degree of "protheticness" (Eisler, 1963a).

PSYCHOPHYSICAL FUNCTIONS FOR INDIVIDUAL SUBJECTS

Even if we limit ourselves to prothetic continua, some other questions remain to be answered. Many of the earliest studies reported only average estimations made by groups of subjects, so the question naturally arose whether the power law also can describe an individual's estimations. Taking all of the relevant data into consideration, the conclusion must be that there is no evidence for systematic failure of the power law at the level of individual subjects. A few investigators have reported such failure. Pradhan and Hoffman (1963) reported results for lifted weights that suggested the power relation to be an artifact of averaging data across subjects; Luce and Mo (1965) found some systematic departures from power relations for heaviness and loudness in results of some of their subjects. It is not surprising that deviations appear in some individual data. One weakness of the method of magnitude estimation is the tendency on the part of many subjects to prefer whole numbers to fractions or decimals. That sort of response bias can lead to distortions of the psychological scale, although the effects should be primarily local. But such a tendency can be exacerbated by certain other procedural decisions, such as to present the same stimulus a large number of times. This is one possible explanation for results obtained by Luce and Mo (1965). The explanation in terms of response bias is supported by Luce and Mo's finding that distributions of numerical responses were highly peaked.

Nevertheless, individual psychophysical functions have been reported in many sensory modalities, and the overall results strongly support the power law. Evidence includes individual functions for loudness (J. C. Stevens & Guirao, 1964), brightness (Marks & J. C. Stevens, 1966), taste (Ekman & Åkesson, 1965), muscular effort (J. C. Stevens & Mack,

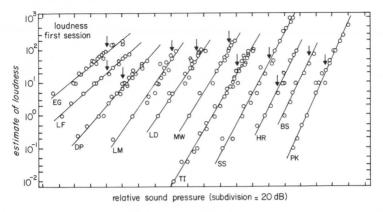

Figure 1.4. Loudness functions (log loudness of noise versus sound pressure in decibels) for 11 individual subjects. The experiment combined features of magnitude production and magnitude estimation, in that each subject both adjusted sound pressure and estimated the loudness. Each point plotted is a single production-estimation. [From J. C. Stevens and M. Guirao, Individual loudness functions. *Journal of the Acoustical Society of America*, 1964, 36, 2210–2213. Courtesy of the authors and *The Journal of the Acoustical Society of America*.]

1959), and heaviness of weights (Ekman, Hosman, Lindman, Ljungberg, & Åkesson, 1968). An example of individual loudness functions is shown in Figure 1.4.

It should be pointed out that individual psychophysical functions often do show a considerable degree of variability in the data. Thus, to say that the results are consistent with the power law is another way of saying only that no large, systematic deviations from power-law behavior usually occur. Detailed studies on individual subjects may well disclose regular departures from power functions. In fact, departures frequently appear in group results pooled over many subjects.

DEVIATIONS FROM THE POWER LAW

This last consideration leads to another question, namely whether there are any systematic deviations from power-law behavior in modalities for which the power equation does provide a good approximation. That is to say, might the power law be only a crude approximation to the relation between sensation and stimulus intensities? Figure 1.5 shows clearly an example of one type of departure from Eq. (1.8). At low levels of luminance, the brightness function deviates from linearity in log–log coordinates. The same type of deviation has been noted in most

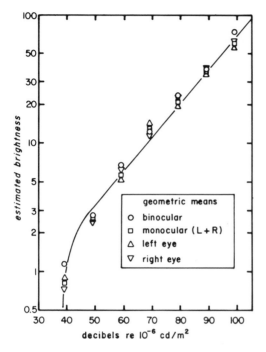

Figure 1.5. Magnitude estimates of brightness, as functions of luminance in decibels *re* 10^{-6} cd/m². The departure from simple power functions is similar for monocular (squares and triangles) and binocular viewing (circles). [From J. C. Stevens (1967a). Courtesy of the author and *Perception & Psychophysics*.]

other modalities, and it seems to reflect a change in the input–output characteristic of sensory systems that occurs when the level of stimulation approaches the absolute threshold.

Several attempts have been made to modify the simple power Eq. (1.8) in order to account for the form of the psychophysical function near threshold. One such modified equation is

$$\psi = k(\phi - \phi_o)^\beta \tag{1.9}$$

Here the constant ϕ_o is subsracted from the stimulus levels (Ekman, 1956; Galanter & Messick, 1961; Luce, 1959; Stevens, 1959b). If ϕ_o is interpreted as a constant that corresponds to absolute threshold, and if the power transformation between stimulus and sensation is the result primarily of neural processes in the periphery of sensory systems, then Eq. (1.9) implies that threshold is determined at a relatively peripheral locus in the sensory system.

Another proposed modification of Eq. (1.8) is

$$\psi = k(\phi^\beta - \phi_o{}^\beta) \tag{1.10}$$

(Ekman, 1961; Fagot, 1963). Equation (1.10) implies that threshold is determined at some locus relatively central to the locus of the power transformation. The subtractive constant ϕ_0 of Eq. (1.10) has also been interpreted in another way, namely as a reflection of response bias, rather than of threshold (Irwin & Corballis, 1968). In either case, note that when the exponent β equals unity, Eqs. (1.9) and (1.10) become identical. Unfortunately, it is often very difficult to decide, given any particular set of data, which of the two equations provides a better description. Furthermore, these alternatives are not exhaustive; other possibilities exist and have been proposed. These include a linear relation between sensation and stimulus at low intensities (Ekman & Gustaffson, 1968; Garner, 1948; Zwislocki, 1965) and a combination of corrections (Borg, 1962; Fagot, 1966). Furthermore, there is little compelling theoretical reason to prefer any one of the formulations. (From a sensory-physical point of view, postulation of a linear relation near threshold has the advantage of providing a simple basis for predicting other quasi-linear phenomena, such as spatial and temporal summation, that take place at low intensities.) Nonetheless, from an empirical point of view the deviations from simple power functions that arise at low intensities are real and significant.

Although the parameter ϕ_0 in Eqs. (1.9) and (1.10) seems to *relate* to absolute threshold, it may be incorrect to consider ϕ_0 to *correspond* to threshold. The concept of absolute threshold has come under attack from several quarters (Corso, 1963; Swets, 1961). One reason is that traditional measures of threshold confound sensory discriminability with response biases; use of more modern procedures, including analysis in terms of the theory of signal detectability, usually fails to produce any evidence for a pure sensory threshold. However, it is nonetheless true that when stimulus intensity is low (a) detectability is usually imperfect, and (b) psychophysical functions depart from a simple power equation. Furthermore, changes in the conditions of stimulation that affect detectability (e.g., changes in stimulus duration or size) also affect the estimated value of ϕ_0. There is no doubt that a connection exists between estimates of ϕ_0 and the concept traditionally referred to as absolute threshold.

There is good reason to believe that some psychophysical relations will fail to follow the power law precisely, even at intensities well above threshold. One of the interesting examples is loudness. Stevens (1972) made extensive examination of cross-frequency loudness-matching data. He concluded that the functions describing how the loudness of white noise and the loudness of pure tones in the vicinity of 1000 Hz depend on sound pressure fail to conform exactly to a power law. In log–log coordinates both of those functions are slightly concave downward (negatively accelerated). It is interesting that we now find Stevens's loudness function for a 1000-Hz tone to be quite similar to the original sone function shown in Figure 1.1.

In the warmth sense, too, there is evidence that the power law does not hold exactly (Marks & J. C. Stevens, in press). The breakdown there may reflect a transition from one type of mechanism or receptor to another. Perhaps the most likely place to find the effect of such a transition is in vision, since brightness is mediated by only one population of receptor (rods) at low stimulus intensities, but by two populations (cones and rods) at higher intensities. Search of the literature has disclosed only one scaling experiment involving brightness and intensity that reveals evidence of such a transition (Marks, 1973), but it does not appear that any scaling studies have been undertaken specifically with that aim. Under careful and appropriate experimental study there is little doubt such a transition could be measured. Brightness-matching data obtained by van den Brink (1962) suggest that sizeable rod–cone breaks occur in brightness functions measured during dark adaptation.

No Single Psychophysical Function Characterizes a Sensory Attribute

Looking again at the individual loudness functions in Figure 1.4, we see that the functions vary noticeably in the sizes of their exponents. The range from smallest to largest is more than 2 to 1, not an uncommon range of variation (e.g., Marks & J. C. Stevens, 1966). The variability of the exponent, and all of the reasons for variability, are matters of concern and importance. However, differences from one subject to another do not constitute the most significant type of variation. Perhaps the most important source of variation in exponent consists of variation in the state of the sensory system and variation in parameters of stimulation. This is to say that psychophysical functions must vary with all of the stimulus variables that determine sensory magnitude. In a fundamental sense, these sorts of variations are the subject matter of much of the remainder of this book. A few examples will have to suffice at this point. The exponent of the psychophysical function for warmth can be varied by a factor of 2 or more simply by varying the areal extent of stimulation. Loudness increases more rapidly with sound pressure (has a larger exponent) when tonal frequency is very low than when frequency is higher. The brightness of a brief flash of light grows more rapidly with luminance than does the brightness of a longer flash. These results and others point to a fundamental concept, namely that *there is no one psychophysical function for any particular sensory attribute.* There is no one psychophysical function for brightness, no one function for loudness, no one function for taste, no one function for smell, no one function for warmth. At best, it is proper to speak only of the psychophysical function for brightness given knowledge of all of the other relevant variables: wavelength composition, duration, stimulus size, state of adaptation of the visual system, presence of other stimuli in the visual field, etc.

Because sensory magnitudes depend on so many variables in addition to stimulus intensity, it is often convenient to specify *standard conditions* of stimulation under which psychophysical relations may be determined. Thus, it becomes possible, for instance, to state that brightness grows as the cube root of luminance, as long as the standard conditions are understood. Here the conditions consist of dark-adapted eye; field size, 1–5°; duration, 1–2 sec. Similarly, the statement that loudness grows as the ⅔ power of sound pressure assumes sound frequency to be greater than about 400 Hz, and listening to be in the quiet. It then becomes relatively easy to specify how changes in conditions effect changes in the psychophysical relations.

VARIATIONS RELATED TO NUMBER BEHAVIOR OF SUBJECTS

There are other sources of variation in the psychophysical function, sources which lie outside the sense organ and the stimulus. One source is the number behavior of subjects, or what Stevens (1961b) termed the "conception of a subjective ratio." Studies by Jones and Marcus (1961), Jones and Woskow (1962), Ekman, Hosman, Lindman, Ljungberg, and Åkesson (1968), and Teghtsoonian and Teghtsoonian (1971) showed that exponents of the power function obtained for any one modality correlate with exponents obtained for other modalities. That is, those subjects who give relatively large exponents for heaviness of lifted weights tend also to give relatively large exponents for loudness, and so on. However, that tendency does not seem to be a permanent characteristic of the number behavior of subjects, since the correlations virtually disappear after passage of very long intervals of time—a year (Teghtsoonian & Teghtsoonian, 1971). Nevertheless, there is little doubt that much of the difference among individual psychophysical functions determined under otherwise identical conditions arise from differences in the subjects' use of numbers.

This is not to deny the possibility that real individual differences may exist with regard to exponent, e.g., differences that are related to the functioning of the sensory system. Ekman *et al.* (1968) found some slight evidence that this may be so, and Luce (1972) has argued that individual differences probably exist. Luce goes on to question whether psychophysical scales can provide extensive measurement of sensation if there are real individual sensory differences of this sort.

There is no doubt that some real differences do exist among psychophysical functions for different people. Certain pathological conditions provide the most clear-cut cases. Foremost is auditory nerve deafness, which results in what is called "recruitment." In recruitment, sensitivity to weak sounds is diminished relative to normal sensitivity, but sensitivity to intense sounds may be virtually unaffected. Thus, the growth of loudness with sound pressure is greater for the recruiting, than for the normal, listener (Stevens, 1966e).

RANGE CONSTRICTION

Let us return to the question of variation in exponent as determined by the number behavior of subjects. It is worthwhile at this point to describe a variable that has a sizeable effect on exponents of psychophysical functions, namely scaling procedure. In particular, the size of the exponent depends systematically on whether the procedure used is magnitude estimation or magnitude production. Magnitude production inverts the roles of stimulus and response, so that the experimenter gives numbers to the subject, who in turn adjusts stimulus intensity so that sensory magnitudes match the numbers. Given the number 60, the subject tries to adjust the sensation to be 6 times that which he set to match 10. Figure 1.6 gives an example of results obtained from the same subjects by magnitude production and estimation. Clearly, the psychophysical function is steeper (exponent is larger) for production. That difference in exponent has

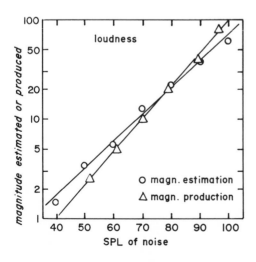

Figure 1.6. Circles: Magnitude estimates of the loudness of a band of noise, as functions of sound pressure levels in decibels (magnitude estimation). Triangles: loudnesses plotted as functions of average sound pressure levels produced (magnitude production). The exponent of the power function (*re* sound pressure) by estimation is .6, the exponent by production is .7. [From Stevens and Greenbaum (1966). Courtesy of the authors and *Perception & Psychophysics*.]

been observed and verified by numerous investigators (Meiselman, 1971; Reynolds & Stevens, 1960; J. C. Stevens & Marks, 1965). Stevens and Greenbaum (1966) summarized the relevant data, and they used the term "regression" to describe the phenomenon. By regression, they meant only that the subject behaves as if to constrict the range of intensities of whatever variable is under his control. When the task is magnitude estimation, he constricts (relatively) the range of numbers; when it is production, he constricts the range of stimulus intensities.

Johnson (1952) first related the phenomenon of range constriction—regression in the psychophysical sense—to the regression—in the statistical sense—that occurs when two variables are not perfectly correlated. He argued that regression occurs whenever the variability for any one variable is large compared to the total variability. However, Johnson's hypothesis remains neither verified nor refuted. For that reason, it would perhaps be preferable to employ a more neutral term, such as "response constriction." Since regression as a term is widely used, though, for the phenomenon, its usage will be continued here. It is important to note that regression is not limited to scaling experiments. It occurs when subjects make intramodality matches as well: it can even occur when a subject is asked to match the loudness of one tone to the loudness of another tone of identical frequency. Hollingworth (1909) called this effect the "central tendency of judgment," and it is a pervasive phenomenon in psychophysics.

One way to look at certain variations that occur in the exponent of the power function is in terms of regression. Stevens and Greenbaum pointed out that the degree of regression seems to depend on difficulty of the task; the more difficult, the greater the constriction of range. Thus, we may expect that difficult tasks will produce lower exponents when the procedure is magnitude estimation, but higher exponents when the procedure is magnitude production. Defining *difficulty* is itself difficult. Stevens and Greenbaum imply that conditions that add "noise," that increase variability, increase regression. Possible examples are long intervals between stimuli (producing a load on memory) and variation of sensations along several perceived attributes (requiring subjects to abstract a single relevant dimension).

A possible solution to the problem raised by regression is to employ both magnitude estimation and production and to take an average of the results. Implicit in that suggestion is the view that such an average can be used to eliminate the biasing effects of regression and to give a "true" estimate of the exponent of the function. On the other hand, since variation in exponent from subject to subject and from experiment to experiment is a fact of psychophysical scaling, we may ask what justification there is for selecting an average as the measure of the true exponent. It is always possible that one task is intrinsically more biased than another. This leads to the fundamental question concerning validity. Is there any way to ascertain what sensory scale is valid, what exponent is the true one?

CROSS-MODALITY MATCHING AND THE INTERNAL CONSISTENCY OF SENSORY SCALES

One method proposed to verify exponents of the psychophysical power function is known as cross-modality matching. Cross-modality matching consists of adjusting stimuli on one dimension so that the sensation magnitudes appear to match those on another sensory continuum. For example, the experimenter might present various levels of luminance of a light, and the subject's task would be to adjust the sound pressure of a tone so as to produce loudnesses that match the brightnesses. One of the earliest attempts at cross-modality matching was that made by Münsterberg in 1890. But it was the experiments conducted by Stevens and his colleagues (J. C. Stevens, Mack, & Stevens, 1960; J. C. Stevens & Marks, 1965; Stevens, 1959a) that brought the method into prominence. The results of a very large number of cross-modality matching studies have been taken to support the psychophysical power law.

How do results of cross-modality matching verify the power law? Assume that the equation governing the psychophysical behavior of one sensory modality is

$$\psi_a = k_a \phi_a{}^{\beta_a} \tag{1.11}$$

and that governing a second modality is

$$\psi_b = k_b \phi_b{}^{\beta_b} \tag{1.12}$$

Cross-modality matching involves setting subjective values equal on the two continua. Thus, $\psi_a = \psi_b$, and

$$k_a \phi_a{}^{\beta_a} = k_b \phi_b{}^{\beta_b}$$

so

$$\phi_a = \left(\frac{k_b}{k_a} \right)^{1/\beta_a} \phi_b{}^{(\beta_b/\beta_a)} \tag{1.13}$$

The predicted relation then between the two stimulus sets is a power function with an exponent equal to the ratio of the exponents of the two

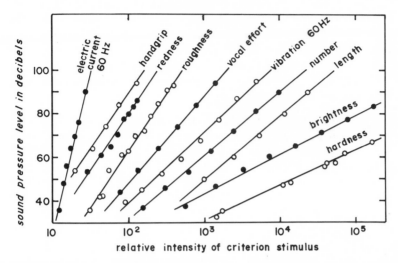

Figure 1.7. Cross-modality matching between loudness and 10 other perceptual dimensions. Plotted are sound pressure levels in decibels against log stimulus intensities of the other continua. Power function exponents range from .8 (loudness versus hardness of rubber) to 7.5 (loudness versus electric shock), when slopes are calculated in terms of sound pressure. [From Stevens (1966c). Courtesy of the author and *Perception & Psychophysics*.]

psychophysical functions. Figure 1.7 gives examples of cross-modality matches of loudness to 10 other continua. Assuming loudness grows as the two-thirds power of sound pressure (the one-third power of sound energy), then if loudness is matched to brightness (exponent of one-third), the matching function between sound energy and light energy should be linear. When matched to handgrip (exponent 1.7), the relation between sound energy and force should have an exponent of 5.1 (equal to 1.7/.33). These predictions have almost invariably been confirmed. Occasional exceptions do occur to that conclusion. Mashhour and Hosman (1968) obtained cross-modality comparisons (among loudness, apparent duration, apparent length, apparent finger span, heaviness, lightness of grays, and apparent visual area) and compared the results to predictions made from magnitude estimations. Predictability of cross-modality matching exponents was poor. However, in the cases of 10 of the 17 comparisons, Mashhour and Hosman employed additive constants in the power equation, solely to improve goodness of fit. Elimination of the use of those additive constants and accounting for regression leads to marked improvement in predictability of exponents (Stevens, 1969a).

Cross-modality matching can be looked upon as an extension of the intrasensory matching procedures that are widely used in sensory psychology. The matching of numbers to sensations (magnitude estimation) or of sensations to numbers (magnitude production) can be subsumed under the general rubric of cross-modality matching. Stevens (1966d) has emphasized this generality to the concept of matching.

THE QUESTION OF VALIDITY

There is little doubt, as cross-modality matching has helped to show, that direct ratio scales of sensory magnitude provide an internally consistent system. However, this internal consistency does not suffice to validate either the individual scales themselves, nor the average exponents of power functions that have been reported (cf., Ekman, 1964; Treisman, 1964; Zinnes, 1969). If the exponents were transformed, for instance, by multiplying all of them by a constant, cross-modality matches would be unaffected. For this reason, still other attempts have been made to verify the results of direct scaling experiments. One such attempt involved asking subjects to judge stimuli whose intensity changes continuously over time. Marks and Slawson (1966) generated 100 pulse/sec buzzes that increased their sound pressure over time. If loudness grows as the two-thirds power of sound pressure, then when sound pressure is increased according to the 1.5 power of time (inverse of two-thirds), the loudness should appear to grow linearly. Of the sample of stimuli used by Marks and Slawson, loudness was judged to increase most linearly when the sound pressure grew as the 1.67 power of time. Similar experiments were carried out by Eskildsen (1966) for brightness, and by Borg, Edström, and Marklund (1970) for perceived muscular effort, and the results have tended to verify both the power law and the associated sensory scales.

Even these results may be questioned with respect to validity. The judgment of linearity merely replaces one set of assumptions (concerning the validity of subjects' conceptions of sensory ratios) with another (concerning the validity of their conceptions of linearity). Further consideration of the question of validity is taken up in Chapter 7. From a realistic point of view, it seems reasonable to accept the approximate validity of sensory scales, at least of those derived under certain conditions. For example, judgments of linear extent (e.g., length of rods) and duration are usually found to be just about proportional to actual length and duration (Stevens & Greenbaum, 1966; Teghtsoonian & Teghtsoonian, 1965). This appears to be a most reasonable outcome, since those two perceptual dimensions are ones with which we have extensive experience in attempting to make veridical judgments; furthermore, we have frequent opportunity to have errors corrected.

A great deal of heat, but little light, has been shed on the question of validity. Much of the speculation has been philosophical in nature, and usually the result has been nil. Perhaps the strongest reason why the question remains not totally resolved is that answers to questions concerning the validity of sensory scales have been sought too often in isolation from studies of sensory processes. Nearly all of the attempts at verification (or "unverification") have involved scaling experiments of one form or another, the aims of which were to discover how subjects go about making psychophysical judgments. It seems likely that in order to understand how sensory magnitudes grow with stimulus intensity, it is necessary at the same time to understand how sensory magnitude depends on other stimulus variables. Scaling cannot stand in isolation from the remainder of sensory psychology. Perhaps the source of greatest potential for validation of sensory scales is the context of experiments on sensory functioning. To say this another way, what is needed are solid theories that can incorporate all of the relevant sensory data—psychophysical, sensory-physical, and perhaps even psychosensory—and thereby delimit the possible psychophysical relations. A start toward such a theory has already been made for visual brightness (Marks, 1972), where the exponents of the brightness function measured under dark adaptation (and, therefore, sensory scales for brightness) appear to be related to other (sensory-physical) properties of brightness vision, properties that need not themselves be studied by scaling procedures.

The approach just outlined parallels the statements made by Anderson (e.g., 1970) about functional measurement, although his applications are aimed primarily at different types of data. Closely related also is conjoint measurement (Luce & Tukey, 1964) in which, as in functional measurement, the goal is to combine scaling with the discovery of behavioral laws. In particular, the aim is to find situations in which two variables add their effects linearly, with no interaction. For example, Levelt, Riemersma, and Bunt (1972) applied conjoint measurement theory to the loudness of binaural tones. Their results were consistent both with binaural additivity and with power functions between loudness and sound pressure (see Chapter 5 for a more detailed account).

The overall purposes of functional measurement, of conjoint measurement, and of the present approach are the same, namely to integrate scaling data with all of the other data that are relevant to the psychological or sensory variables being studied. To that end, the most valuable contribution of direct scaling procedures is to the understanding of how sensory variables depend on all of the relevant parameters of stimulation.

Chapter 2

THE MAGNITUDE ESTIMATION EXPERIMENT

The great value of direct ratio scaling procedures, such as magnitude estimation and magnitude production, derives from their applicability to the study of sensory processes. Direct scaling procedures provide one means of evaluating how sensations depend on a multiplicity of stimulus variables. In this respect, direct scaling procedures resemble the more traditional psychophysical procedures: threshold mapping and intrasensory matching. The knowledge provided by direct scaling procedures is somewhat broader in scope, however; whereas both direct scaling and more traditional procedures can provide us with information about sensory physics, only scaling procedures can provide us with information about the psychophysics of sensory processes. If the application of direct scaling were also limited in scope—for example, limited to the development of sensory (psychophysical) scales only—then the value of direct scaling procedures would indeed be meager. But when applied to substantive problems of sensory functioning, direct scaling procedures become invaluable. They provide a rapid, accurate, and exhaustive means to establish laws of sensory behavior. Examples, some of which we will examine in detail in later chapters, are how loudness depends on the temporal distribution of sound energy, how warmth depends on the spatial distribution of heat, how hue depends on light intensity and wavelength.

There is another application of results of direct ratio scaling. This application is to problems in human sensory engineering. For example, knowing how the average person estimates loudness of sounds is crucial to development of a "sound" program of noise abatement.

To repeat, one of the major advantages of direct scaling procedures stems from their capacity to provide two types of information. The first type is psychophysical information: thus, we can learn how some aspect of our sensory experience covaries with parameters of stimulation. The second type is sensory-physical information: when these procedures are used with multivariate stimuli, we can learn how stimulus parameters must covary to provide a constant sensory experience. These latter types of relation have been the grist and meat of sensory studies for many years, and they provide the most stable body of data in sensory psychology.

Starting with that point of view, some of the direct scaling procedures can be considered to be variants of direct, intrasensory matching. Instead of asking a subject to adjust the energies of two lights of different wavelengths in order to make their brightnesses appear equal, he is asked to judge the brightnesses of lights of different energies and wavelengths, and we can select those pairs of lights that elicit constant judgment of brightness. From the point of view of sensory physics, the numerical judgments of brightness appear as intervening variables. There is often a significant advantage, however, to the use of scaling procedures over the use of sensory matching without numerical intermediary. When scaling procedures are used, it is necessary to present only one stimulus at a time. Experiments can be designed in which it is not necessary to provide stimuli for comparison (matching). This is often an invaluable advantage, especially when there is a strong possibility of interaction between test and comparison stimuli. An example occurs in the sense of taste: in order to match tastes, it is preferable to present stimuli successively. Simultaneous presentation, e.g., to the two sides of the tongue, is not only difficult to implement, but also difficult to judge. Subjects would find it extremely difficult (if not "distasteful") to match for sensory intensity the tastes on the two sides of the tongue. On the other hand, successive matching also has its pitfalls. Usually, it is necessary to provide a distilled water rinse between sapid stimuli. Thus, the large interval of time separating stimuli becomes a problem in matching experiments, where several stimulus presentations may be needed before a match is effected. However, it is no special problem in scaling experiments, where a verbal response or cross-modality match may be emitted after every stimulus presentation.

It should not be difficult to see why a method such as magnitude estimation often turns out to provide as good a means as any to answer questions about sensory physics. In the present chapter we shall look in detail at the way that one direct scaling procedure was used experimentally to answer questions about sensory processes (psychophysics and sensory physics). Although any of the direct ratio scaling methods described in Chapter 1 can be used with success, the most commonly employed procedure (and the one with the greatest versatility and most general application) is *magnitude estimation*. Accordingly, magnitude estimation will be the procedure in this example. We shall examine a large number of aspects of the experiment, from the point of view of learning why the particular version of the procedure was selected, how the data were analyzed, and what sorts of information the data provided. Naturally, many details of the procedure, data analyses, and interpretation are quite specific to the sensory problem investigated. Obviously, experimental design cannot always be the same for studies of vision, warmth, and olfaction. Choices of experimental procedures should quite properly be determined by the nature of the sensory problem under investigation. On the other hand, some decisions can be made on the basis of what is known about scaling procedures and the influences of procedural variables on perceptual judgment. Sometimes consideration of both sensory process and scaling procedure must be taken into account and weighed before a decision can be made as to procedural detail. A major purpose of the remainder of this chapter is to examine the bases upon which such decisions were made in a particular experiment.

The sensory problem of concern in our example is spatial summation in the warmth sense, and the experiment was one conducted by J. C. Stevens and Marks (1971). The psychophysical question that the investigators asked was a rather simple one, namely, how does warmth sensation depend both on the intensity and on the areal extent of thermal stimulation? At the time of the experiment, there was already extensive knowledge about spatial summation at the absolute threshold for warmth. Hardy and Oppel (1937) showed that summation extends over (at least) several hundred square centimeters of body surface, and Kenshalo, Decker, and Hamilton (1967) found that threshold summation was nearly complete. That is, with either radiant intensity or change in skin temperature as a measure of stimulus intensity, Kenshalo *et al.* found that the product of area and intensity was constant at threshold. Take a just-detectable stimulus, then double its area and halve its intensity. The new stimulus will also be just-detectable.

However, as recently as 1971 virtually nothing was known about suprathreshold spatial summation, except that it seemed to exist (Herget, Granath, & Hardy, 1941). But it was not known whether suprathreshold summation of warmth is complete, i.e., whether a stimulus twice as intense, but half as extensive as another, feels just as warm. One aim of the experiment by Stevens and Marks was to see whether and how the sensory physics of spatial summation might vary with level of stimulus intensity. Sensory physics at threshold sometimes does, sometimes does not give a good indication of sensory-physical relations above threshold.

Experimental Apparatus and Arrangement

The experimenters selected two regions of the body on which to examine warmth: the forehead and the back. We shall concern ourselves here with the experiment for the forehead only; results for the back were quite similar. The forehead provides a good surface for the study of warmth since sensitivity to heat is both quite high and quite uniform over the surface from one spot to another.

The stimuli were produced by a 1000-W projector lamp that was positioned to radiate the forehead uniformly. Radiant intensity was changed by varying the voltage to the lamp. As happens when voltage is changed, the spectral distribution of the radiation also changes. That spectral change can create difficulties, because the reflectance of "white" skin is not uniform throughout the spectral range of irradiation. White skin reflects a good deal of the incident radiation, especially radiation in the visible region of the spectrum, but reflects much less infrared radiation. A change in the incident spectrum of heat entails a change in the percentage of reflected and (concomitantly of absorbed) radiation. For that reason, the foreheads of all the subjects were painted black with India ink. Then absorption of the radiation was both uniform across the spectrum and also maximal at all voltages.

The path of the radiation was interrupted by a shutter. The shutter was operated by a solenoid which in turn was connected to a timer. This arrangement permitted control of stimulus duration. Duration was set at 3 sec. Although the choice of stimulus duration was arbitrary, that choice was guided by certain considerations, such as durations used in previous studies. The duration selected was one commonly used in studies of thermal sensations (warmth and pain).

Figure 2.1 Apparatus for thermal irradiation of the forehead. Level of irradiation is controlled by adjusting the voltage to the 1000-W lamp, and areal extent of stimulation is controlled by varying the size of the aperture in the mask. [From J. C. Stevens and Marks (1971). Courtesy of *Perception & Psychophysics.*]

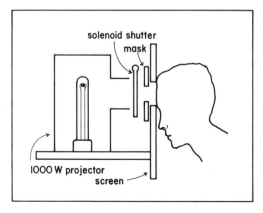

Each subject leaned his forehead against a cork-lined aluminum shield, which had a section cut out. In order to vary the areal extent of stimulation, aluminum masks could be set in the shield, and each mask had a different size of aperture (their areas varied from 2.55–21.8 cm^2). Thus for each presentation of the stimulus the experimenter could vary both stimulus intensity and stimulus size. A schematic diagram of the experimental set-up is shown in Figure 2.1.

Range and Number of Stimuli

One aim of the experiment was to explore as large a region as possible of the psychological range of warmth sensation and to see how stimulus intensity and areal extent would contribute to the magnitude of warmth sensation. Thus, it behooved the experimenters to use stimulus ranges that were as large as possible. With regard to area, the stimuli went up to 21.8 cm^2. Given the variation from subject to subject in size and shape of forehead, that was about as large an area of forehead as could be expected to lie flat against the aperture in the shield. On the other side of the coin, some pilot work had suggested that it is difficult to feel warmth at any but extremely high intensities when stimulus area is very small. The smallest stimulus used was 2.55 cm^2, about one-ninth the size of the largest. A total of six different areas were used, and they were selected so as to be spaced in approximately equal logarithmic steps.

Similar considerations guided the selection of stimulus intensities. At the high end of the range, the limitation was the capacity of the projector lamp and system. In order to provide a relatively uniform radiant field at the forehead, some energy had to be wasted. The highest irradiance available was 209 mW/cm^2. For a 3-sec exposure, that irradiance is sufficient to raise skin temperature about 3.5° C. At the low end of the stimulus range, the limitation was the subject's ability to detect thermal stimulation. Clearly, it does little good to present stimulus levels that cannot be detected on all trials by all subjects. Furthermore, since a significant degree of spatial summation is known to operate at the absolute threshold, the lowest irradiances used were made to vary from one stimulus area to another. The aim was to make the smallest irradiance at each area always detectable, though weak. (As it turned out, the experimenters were not completely successful in this regard: a few stimuli could not be detected on all trials.)

To some extent, the choice of stimulus range can color the results of scaling experiments. What really appears to count in this respect is the range of sensory magnitudes that is produced. Many of the early studies of direct scaling, and some later ones, looked at the way range affects the results. And there is no doubt that range does exert some influence. If all of the stimuli are spread out from one another and cover a wide range, the psychophysical function (in log–log coordinates) is usually flatter than if the stimulus range is smaller (Björkman & Strangert, 1960; Engen, 1956; Engen & Levy, 1958; Ekman, Frankenhaeuser, Levander, & Mellis, 1964; Jones & Woskow, 1966; Künnapas, 1960; Pitz, 1965; Poulton, 1968; Sternbach & Tursky, 1964; Stevens, 1956b; Stevens & J. C. Stevens, 1960; Strangert, 1961). When dealing with a power function, use of a larger stimulus range, therefore, leads to a smaller exponent of the function. It is as if the subject tends to spread out his range of numbers, relatively speaking, to fill up a given stimulus range.

This effect, however, has not been universally observed (Pradhan & Hoffman, 1963) and is not always very large (Engen & Levy, 1958). Subjects do not use a constant range of numbers regardless of range of stimuli, but are somewhat, although not wholly, successful in avoiding the influence of stimulus range. Nevertheless, it is worthwhile to keep this effect in mind when interpreting the results of any particular scaling experiment. Stimulus range, as well as a number of other variables, can influence certain aspects of the outcome of a scaling experiment. It may not be wise to attempt to take as too precise the psychophysical functions obtained from any one scaling experiment. At the same time, effects such as those of stimulus range should have no influence on the sensory-physical relations. These relations will remain invariant as long as the subject consistently assigns a single number to the same sensory magnitude. An analogy to the measurement of threshold sensitivity may be in order: Measurements of threshold sensitivity also depend on choice of experimental procedure; e.g., measured thresholds are different by method of constant stimuli, forced choice, tracking, etc. But what does not seem to change systematically from one procedure to another are sensory-physical relations. The ways that various stimulus parameters (e.g., area and intensity) interact remain invariant across procedure. Thresholds change, but relative sensitivity measured by various procedures remains the same.

Closely related to the question of influence of stimulus range is influence of stimulus spacing. Again, stimulus spacing was a variable examined early in the course of study of direct scaling. Spacing, like range, exerts an effect on the results, but the overall effect is not large (see Beck & Shaw, 1965). What seems to happen is that stimuli closely spaced over a region of the stimulus scale produce a stretching out of numerical responses in that local region of the scale. In terms of the psychophysical function that obtains, a bunching of stimuli often leads to a *local* steepening (greater exponent) of the function, but has little overall effect on the function's slope (J. C. Stevens, 1958; Stevens, 1956b). More sizeable effects of spacing have also been reported, however (Pradhan & Hoffman, 1963). It is interesting that stimulus spacing has a potent effect on categorical judgment (Marks, 1968b; J. C. Stevens, 1958).

Scaling procedures appear to differ with regard to their sensitivity to contextual effects. For example, the method of fractionation seems to be quite sensitive to context; by manipulating the available choices of stimuli it is possible to change drastically the stimulus selected to be half as loud as another (Garner, 1954a). On the other hand, constant-sum judgments (whereby the subject divides up a constant number, such as 100, into portions that represent the sensory magnitudes) seem to be much less sensitive (Engen & Levy, 1958). And even fractionations vary in sensitivity to context depending on the precise way the procedure is used (Engen & Tulunay, 1957).

Of course, in a procedure such as magnitude estimation there is normally no reason to space stimuli in a very irregular manner. In the experiment by Stevens and Marks, an attempt was made to space stimulus intensities in approximately equal logarithmic steps at each stimulus area. Interpretation of the influence of spacing is made more complicated, as it is in the case of stimulus range, by virtue of the fact that it is the spacing of the sensations, not the spacing of the stimuli, that really counts. Since the subjects were, for the most part, quite unable to distinguish stimuli on the basis of size (spatial discrimination of warmth is especially poor), it would be necessary to evaluate the spacing among the sensory magnitudes aroused by the whole complex of stimuli. And since that could not be done prior to the experiment (that is, without first knowing what the answer would be) there was no way a priori to ensure that spacing would be uniform. For example, several different combinations of area and intensity might all be very much alike with regard to the warmth aroused. For that reason, among others, a very large number of stimuli was used—60 in all. Use of such a large number of stimuli accomplishes two things: first, it makes it likely that the magnitudes of warmth sensations will be well distributed over the perceptual range examined, and second, the trend of results will be that more firmly established for a larger than for a smaller set of stimuli.

Temporal Spacing of Stimuli

Another procedural question concerns the rate at which stimuli should be presented. From the subject's point of view, the judgmental task may seem more difficult if long periods of time separate stimuli. The faster they come (within limits!) the better. But the ideal rate of stimulation must be determined primarily by considering such sensory variables as adaptation. In the thermal senses, it is just not feasible to present stimuli (except when their intensities are very low) at a rate exceeding one every 30 sec. That was the rate used by Stevens and Marks. Sometimes, it is necessary to wait several minutes between the end of a very strong thermal stimulus and the onset of the next stimulus (Marks & J. C. Stevens, 1972).

Number of Subjects and Number of Judgments

Because of the large number (60) of different stimuli that were used, it was not feasible to present each stimuli more than once in a given session. Two or more presentations of each stimulus is a common procedure, but in the present case it is likely that the subject's performance near the end of so long a session would have deteriorated. Many subjects find it taxing to make as many as 60 judgments in a single session, unless they come quickly one after another. To make 120 judgments would require exceptional stamina. As an alternative means to obtaining two judgments per stimulus in the session, each subject came back on another day and judged all 60 stimuli again (presented in a different order).

Eighteen men served in the experiment, so a total of 36 different judgments were obtained for each stimulus. It is not possible to state a single fixed number of requisite judgments that will ensure stable results. Indeed, in some experiments, the major concern may be with the application of scaling data obtained for individual subjects (e.g., Ekman & Åkesson, 1965; Marks, 1970). One important variable that determines the number required is the stimulus modality: brightness and loudness, as examples, are usually relatively easy to judge and do not give especially variable results, whereas thermal pain is difficult to judge and appears to be a "noisy" continuum. Another consideration is the number of stimulus parameters that are varied. The larger the number of *perceptible* dimensions of variation, the more difficult may be the subject's task, and the greater may be the variability. (Greater difficulty may also lead to greater degree of range constriction or regression, as discussed in Chapter 1.) If only a small number (say two) of stimulus parameters is being varied, 10 or 12 subjects, and a total of 20 or 24 judgments per stimulus, may suffice. On the other hand, a larger number will be needed for modalities such as temperature and pain that tend to produce more variable results. Part of the greater variation on those continua may be due to greater fluctuation in the magnitude of the proximal stimulus. Just-detectable thermal pain, for instance, is produced when the skin is heated to about 45°C. If the base-line temperature of the skin fluctuates from moment to moment, the effectiveness of a radiant stimulus of constant intensity will vary concomitantly with the fluctuations. It is easy to show to oneself that the same nominal stimulus (intense radiation) just does not produce the same magnitude of sensation from one application to another. With scaling experiments on the temperature senses and on pain, it is probably wise to employ 15–20 subjects, if two judgments per stimulus per subject are obtained.

The Estimation Procedure

We come now to an important aspect of the procedure, namely the instructions to the subjects. Here is an example of one sort of instruction:

> I am going to present a number of heat stimuli to your forehead. Your task is to judge how warm each stimulus feels to you by assigning numbers to stand for the degree of apparent warmth. To the first stimulus, assign whatever number seems to you most appropriate to represent the degree of warmth. Then, for succeeding stimuli, assign other numbers in proportion to warmth. If one stimulus seems three times as warm as another, assign a number three times as great; if it feels one-fifth as warm, assign a number one-fifth as great. Any type of number—whole number, decimal, or fraction—may be used.

These instructions typify the procedure of magnitude estimation—(a) with an *estimation* rather than a *matching* approach, (b) with no fixed *standard* stimulus, and (c) with no *modulus*. Let us look at each of the three points. First, the instructions clearly state that the subject should attempt to assign numbers in proportion to warmth. The subject is asked to estimate warmth numerically. Hellman and Zwislocki (1963, 1964) have proposed another sort of instruction, in which the subject is told simply to match numbers to some aspect of his sensations (in their experiments, to loudness). That procedure has been termed *absolute magnitude estimation*, to signify that subjects are asked to make absolute, not just relative, judgments. If this is so, then it should be possible to compare directly results obtained in sessions run on different days with different sets of stimuli. That will be so if the subjects tend to match the same number to the same sensation on different occasions. Be that as it may, it seems that either the standard estimation instruction or the absolute estimation instruction can be used successfully.

In some of the early experiments that used the method of magnitude estimation, it was common practice to employ a *standard* stimulus, one that might either precede every test stimulus, appear at regular intervals in the course of the experiment, or be available whenever the subject requested it. The reason for the standard was, presumably, to provide a stable base line from which the subject could make his judgments. Typically associated with the standard was a *modulus*, i.e., a number given by the experimenter to be assigned to represent the sensory magnitude of the standard. A very common procedure was to select a standard stimulus from the middle of the range and to give its sensation the value of 10.

It became quite clear in the course of those early investigations that both the stimulus value selected to be the standard and the modulus assigned to the standard acted somewhat to influence the results. The usual effect is as follows. When the standard is either one of the weakest or one of the strongest of the stimuli (near the bottom or top of the stimulus range), the psychophysical function obtained is less steep (has a lower exponent) than it is when the standard is taken from the middle of the range (Beck & Shaw, 1961, 1965; Engen & Levy, 1955; Engen & Lindström, 1963; Engen & Ross, 1966; Hellman & Zwislocki, 1961; J. C. Stevens & Tulving, 1957; Stevens, 1956b). [However, sometimes other directions of influence are found. Ekman (1961) reported that rate of growth increased as level of standard increased. On the other hand, Ross and di Lollo (1968) reported the opposite; they found the exponent to decrease when the standard increased.] Given a weak or a strong standard stimulus, the subject tends to constrict the range of his numerical responses. Position of standard stimulus can also influence the variability of the response. When the standard is low, variability of responses to (magnitude estimates of) intense stimuli is relatively large (Eyman & Kim, 1970). When the standard is near the middle of the range, the variability is more nearly equal across stimuli. Since exponents are usually greatest when the standard is near the middle of the range, there would appear to be a correlation between degree of constancy of variability across stimuli and size of exponent.

In a systematic investigation of scaling of loudness, Hellman and Zwislocki (1961) found that when the standard is weak, stimuli that are even weaker produce a steep psychophysical function, but stimuli that are stronger produce a relatively flat function. When the standard stimulus is strong, stimuli that are even stronger produce a steep function. When the standard is at an intermediate level (in the estimation of loudness of a 1000-Hz tone, when the standard is at about 80 dB SPL), the slope of the loudness function is uniform both above and below the standard.

There seems to be little reason to use a standard stimulus in scaling (magnitude estimation) experiments. Certainly, a standard need not be presented along with every test stimulus. J. C. Stevens and Tulving (1957) found the slope (exponent) of the loudness function obtained by magnitude estimation to be somewhat steeper when a standard of moderate intensity was used than when no standard was used. Engen and Levy (1955) found similar effects on judgments of lightness of grays and heaviness of weights. But even if the subject's range of numbers is larger when a moderate standard is used compared to when no standard is used, that difference in itself is not sufficient reason to require the use of a standard stimulus.

It could be argued that a standard stimulus exists even when the experimenter fails to designate one as such. For example, in the experiment on warmth that is under scrutiny, the first stimulus might have served, from the subject's point of view, as a standard. However, no modulus was assigned to it (the subject selected the number assigned to the first stimulus), and the first stimulus itself varied from subject to subject. J. C. Stevens and Tulving (1957) found that when subjects were free to select the number assigned to the first stimulus, the slope of the loudness function was related to the size of that number (the intensity of the standard stimulus was always the same). When the first number was small, weak stimuli were underestimated (relative to the estimates given by other subjects). Conversely, when the initial number was large, stronger stimuli were underestimated. Systematic effects related to the number assigned to the first, or standard, stimulus have been noted by Stevens (1956b) and by Beck and Shaw (1962, 1965). Hellman and Zwislocki (1961) reported an interaction between standard and modulus. Changes in standard and modulus tend to produce opposing effects. For example, the effect of using a low-intensity stimulus as a standard is similar to that of assigning a large number as modulus. To some extent, the effect of change in one of either the magnitude of standard or modulus can be compensated by an appropriate change in the other. Roughly speaking, the results will be similar given a low standard and small modulus, medium standard and intermediate modulus, and high standard and large modulus. Generally similar interactions were noted by Beck and Shaw (1965) and by Engen and Ross (1966).

Measurements of the influences exerted by standard and by modulus do not exhaust the study of procedural variables. For example, there also exist measurable effects due to choice of the level of the second stimulus —the stimulus after the standard (Beck & Shaw, 1963; Poulton & Simmonds, 1963; Stevens & Poulton, 1956). But in order to maintain perspective on the problem, it is worthwhile to point out, with Poulton

(1968), that effects of choice of standard and modulus themselves tend to be small. When the aim of an experiment is the study of sensory processes, probably the most adequate procedure is to vary the level of the first stimulus from one subject to another (or select it from the middle of the stimulus range) and permit the subject to assign freely a number to represent the subjective magnitude of the first stimulus. This procedure may not be totally satisfactory if the aim is to understand how procedural variables influence numerical judgment; on the other hand, such a procedure—one that minimally constrains the subject—appears the procedure of choice when the aim is to study some sensory process.

As a final note on procedure, it is worth mentioning the question of use of naive versus experienced subjects. Often in psychophysics it is deemed advisable to collect data from well-practiced subjects only. Certainly to the extent that some familiarity with the procedure, apparatus, and general experimental situation puts the subject at ease, practice may reduce the variability in his response; furthermore, it is even possible that practice may exert some effect on the nature of psychophysical functions. However, naive, unpracticed subjects are quite capable of yielding reliable and consistent results (J. C. Stevens & Tulving, 1957; Stevens & Poulton, 1956). There appears to be no evidence that any sizeable degree of practice or training is necessary to obtain data for most tasks involving the direct estimation of sensory magnitudes under multivariate stimulus conditions.

Analysis of the Results of Magnitude Estimation

In some investigations, the nature of the task demands that the results for each subject be analyzed individually. This is particularly true when there is reason to expect sizeable individual sensory differences, e.g., when subjects are sampled from different sensory populations. On the other hand, for many investigations, and for most of those that have used direct scaling procedures, it has been deemed adequate to obtain some sort of average result from a group of subjects. Underlying this approach is the assumption that the significant relations, especially sensory-physical functions, do not differ to any large degree from one normal subject to another. The warmth experiment conducted by Stevens and Marks falls into this category, so we shall examine how data are averaged and analyzed at the group level. Keep in mind, however, that most of the analyses and conclusions could also easily be applied to data for individual subjects.

AVERAGING MAGNITUDE ESTIMATES

A large number of statistical measures of central tendency are available, and the choice of the most appropriate statistic depends to a major extent on the type of scaling procedure employed and on the form and variability of the resultant data. Let us consider magnitude estimation. The arithmetic mean is one of the simplest measures of central tendency, and it has the advantage of being the unbiased estimate of expected value. However, the arithmetic mean is quite precarious to use in experiments where no modulus is assigned, since the value of an arithmetic mean of a set of numbers will be determined to a disproportionate extent by very large numbers if they occur. For example, the estimates given by a single subject may dominate the arithmetic mean if his estimates are proportionally the largest of those in the group. The arithmetic mean is used relatively infrequently now in the reduction of scaling (magnitude estimation) data; when it is used (e.g., by Bartoshuk, 1968), the experimental procedure typically includes a modulus, the same for each subject. Use of a modulus helps to ensure that no subject will use a range of numbers very much different from those used by other subjects.

Without doubt the most commonly used measure of central tendency with scaling data, especially magnitude-estimation data, is the geometric mean. The geometric mean is particularly suitable because the distribution of magnitude estimates given to a single simulus level often approximates log normality, i. e., approximately normality after logarithmic transformation (J. C. Stevens, 1957). The geometric mean gives an unbiased estimate of the expected value of the logarithms of the magnitude estimates. The approximate log-normal distribution of magnitude estimates is another way of saying that there is a long tail at the high end of the distribution. Even when a modulus is employed, occasional very large estitates may be expected. These especially large estimates are precisely the ones that make the use of the arithmetic mean potentially hazardous. From a practical point of view, the main advantage of the geometric mean is that it prevents an aberrant judgment from overly swaying the result. An attractive feature of the geometric mean is that although it preserves the (geometric) average ratio of the estimates given to every pair of stimuli, it is not overly influenced by any single extraordinarily large ratio. If four subjects assign the number 1 to stimulus A and 10 to stimulus B, and a fifth subject assigns the numbers 1 and 100, the geometric means are 1 and 16, but the arithmetic means are 1 and 28. The single 100 : 1 ratio therefore influences the geometric mean to a smaller extent.

Another statistic that avoids undue influence of extreme judgments is the median. Unfortunately, the median avoids the influence of much of the data; since the median is the point that divides the distribution of responses into equal numbers greater and smaller, but takes no consideration of the form of those semi-distributions, the median is wasteful of information. Even so, the median is frequently called upon to get an experimenter out of a hole. That happens, for example, when some subjects cannot detect all of the stimuli, and give estimates of zero. The geometric mean of a set of numbers any one (or more) of which is (are) zero is also zero. It is unrealistic to take as zero the "average" of a large number of positive values plus one or two zeros. In such a situation, the median becomes the statistic of choice.

In the warmth experiment of Stevens and Marks, 7 of the 60 stimuli produced occasional zero judgments; because they constituted only a small fraction of the total number of stimuli, the experimenters decided to drop those seven stimuli from the analysis, rather than resort to taking medians. In that way, the investigators were able to employ the geometric mean as the measure of central tendency. It should be made clear that there are no hard and fast rules to follow in this sort of situation. In a very similar experiment in which zero judgments were made to a more sizeable fraction of the stimuli, the median was used (Marks & J. C. Stevens, 1973). A general guideline might be: Use the geometric mean whenever possible, but in a pinch, use the median.

Describing the Data

Once some sort of average, such as the geometric mean, is decided upon, the data can be plotted. Figure 2.2 shows geometric means of the

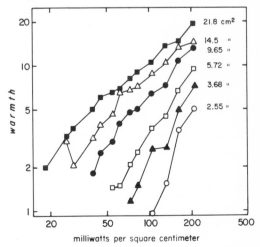

Figure 2.2. Average magnitude estimates of warmth produced by thermal irradiation of the forehead, as functions of stimulus irradiance in milliwatts per square centimeter. Both coordinates are logarithmic. Line segments connect data points obtained at constant areal extents of stimulation. Area varies from 2.55 to 21.8 cm². [Data from J. C. Stevens and Marks (1971).]

magnitude estimates of warmth plotted against corresponding stimulus irradiances. A different symbol is used to represent each different stimulus
size. Note that both of the coordinates of Figure 2.2 are logarithmic.
Magnitude estimates often plot as linear functions of stimulus intensity on
double-logarithmic coordinates. Such a linear plot would indicate a simple power function between stimulus and sensation. It is clear from Figure 2.2, however, that warmth did not follow a simple power function of
irradiance at any of the stimulus areas.

In many experiments, it is sufficient to indicate the general trend of the
data by plotting a free-hand curve or by connecting the data points by
straight-line segments. Straight-line segments are shown in Figure 2.2. On
the other hand, there are several things to be said in favor of attempting
to summarize the results in terms of some sort of quantitative statement.
By this is meant the description of the data by some mathematical expression, preferably a simple one. Although again there are no hard and
fast rules for deciding when any particular mathematical function provides a satisfactory fit to the data, nevertheless there are some criteria.
One is that the deviations of data from function should be random, not
systematic. Another is that the function should be simple, if possible.
This assumes that the function is purely empirical and is not based on
any theoretical considerations. It is often useful to have an empirical,
quantitative statement, if for no other reason than to provide a convenient summary of the results. There are many examples of relatively simple equations that often describe relations between magnitude estimates
and stimulus intensity. These include linear, power, logarithmic, and quadratic functions.

DETERMINING PSYCHOPHYSICAL FUNCTIONS

The results of Figure 2.2 cannot be described adequately by any equation that contains only two estimable parameters (e.g., by simple power
or logarithmic equations). For any given stimulus size, the curvature of
the data suggests that the estimates of warmth are approaching zero at
some positive levels of irradiance. Remember also that data for some additional stimulus levels were discarded because of zero judgments. The
curvature displayed is not an uncommon finding in stimulus–response
functions of psychophysics. Had the data suggested straight lines, they
would have conformed to the power equation

$$\psi = k\phi^{\beta} \tag{2.1}$$

where ψ is estimated sensory magnitude and ϕ is stimulus intensity. The
curvature suggests that an additional parameter needs to be added to Eq.

(2.1). There are several different hypotheses concerning the mathematical nature of the inclusion of another parameter. Some of these alternatives were reviewed by Marks and J. C. Stevens (1968a). Most of the hypotheses rest on the assumption that the curvature is related to the absolute threshold of sensation. That is, psychophysical functions may steepen because as stimulus intensity approaches threshold, sensation approaches zero. At that point the functions might become vertical lines. However, other interpretations, such as the operation of some sort of response bias (Irwin & Corballis, 1968), have also been put forth. Under that interpretation, the curvature is assumed to be the result of the subject adding or subtracting a constant value from his numerical judgments.

One of the hypotheses states that there is a threshold parameter (ϕ_0) that should modify the power equation by subtracting its magnitude from the value of each stimulus (Ekman, 1956; Luce, 1959; Stevens, 1959b). This view of the threshold implies that the power relation holds (the power transformation takes place) only after the absolute threshold is surpassed. Such a formulation reads

$$\psi = k(\phi - \phi_0)^\beta \qquad (2.2)$$

An alternative is that there is an addition that occurs on the scale of the response (sensory) variable (ψ_0) (Ekman, 1961; Fagot, 1963). Then the modified power equation reads

$$\psi + \psi_0 = k\phi^\beta$$

or

$$\psi = k(\phi^\beta - \phi_0^\beta) \qquad (2.3)$$

where $k\phi_0^\beta = \psi_0$. Other hypotheses include a proportionality between ψ and ϕ at low stimulus intensities (Ekman & Gustaffson, 1968; Zwislocki, 1965).

One unfortunate aspect of the proliferation of hypotheses is that they often cannot be distinguished from one another on empirical grounds. The more nearly linear the psychophysical function (the closer β is to 1.0), the more nearly identical are the various formulations. When the psychophysical relation is highly nonlinear, however, it may be possible to distinguish empirically among them (e.g., Marks, 1971a). Even so, distinction in a few instances would not resolve the issue, since different equations might apply to different continua (or to different stimulus conditions with the same continuum). Different equations might be appropriate, for instance, when masking stimuli are present and when they are not.

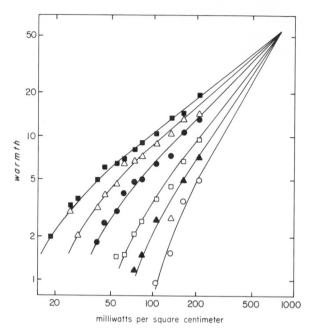

Figure 2.3. Average magnitude estimates of warmth produced by thermal irradiation of the forehead, as functions of stimulus irradiance in milliwatts per square centimeter, as in Figure 2.2. The fitted functions are threshold-corrected power functions that converge at a common point: warmth = 55, irradiance = 800 mW/cm². [After J. C. Stevens and Marks (1971). Courtesy of *Perception & Psychophysics*.]

The same data from Figure 2.2 are repeated in Figure 2.3. Here they are fitted by modified power equations. Because all of the exponents hover around 1.0, it turns out that either Eq. (2.2) or (2.3) can describe the results quite adequately; Marks and J. C. Stevens (1968a) reviewed some of the theoretical underpinnings of the different modifications. It would appear that the theoretical basis for selecting any one equation is not strong. Equation (2.2), (2.3), or any of a large number of others can easily be justified on some more or less reasonable grounds, and the selection of an equation is often mainly a matter of predilection and convenience.

The curves in Figure 2.3 were drawn in accordance with Eq. (2.2). However, the selection of values for ϕ_o was not based solely on goodness of fit. First of all, the sizes of ϕ_o were made to conform with the knowledge of how spatial summation operates at threshold. The sizes of ϕ_o are inversely proportional to the areal extents of stimulation. That inverse proportionality was a fixed feature of the data analysis. Sometimes it is possible to estimate best values for parameters like ϕ_o in Eq. (2.2) or (2.3) by maximum likelihood estimation or by the iteration of solutions according to the method of least squares (Fagot, 1966; Marks, 1966). However, such mathematical techniques leave estimated parameters at the mercy of variability in the data, especially to variability at low stimulus levels, so care should be taken to ensure that estimated threshold parameters are not unrealistic.

There was a second restriction placed on the computation of the curves in Figure 2.3. Preliminary analysis, consisting of the fitting of modified power functions to the data, suggested that the functions for the six different areas all tended to converge, upon extrapolation, on a single point. That point corresponded to a value of $\psi = 55$ and a value of $\phi \approx 800$ mW/cm². At levels of irradiance below 800 mW/cm², it is clear that warmth both increases with increasing area given a constant irradiance, and it increases with increasing irradiance given a constant area. At convergence, however, there would no longer be any variation of warmth with area. For quasi-theoretical reasons, this convergence is not surprising. An irradiance of 800 mW/cm² lies near the threshold for pricking pain (for a review of measurements of pain thresholds, see Adair, J. C. Stevens, & Marks, 1968), and pain seems to display little or no dependence on areal extent of stimulation (Greene & Hardy, 1958; Murgatroyd, 1964).

For these reasons, the curves of Figure 2.3 were calculated to satisfy the restriction that they all pass through the point $\psi = 55$, $\phi = 800$. Not all sets of data will suggest restrictions of this sort, of course, nor was it absolutely necessary in the present example to fit functions under such a restriction. As a matter of fact, there are often several alternative ways to present the data, in addition to those already mentioned. Besides drawing free-hand curves, connecting points by line segments, or fitting lines based on some simple empirical equation, it is sometimes possible to derive a set of curves from some theoretical considerations.

FITTING THEORETICAL CURVES TO PSYCHOPHYSICAL DATA

Careful inspection of the data in Figure 2.2, especially of the data points for the three largest areas, reveals some irregularities in the relations between warmth and irradiance. With several of the stimulus areas, the warmth functions seem to display inflections at a moderate level (about $\psi = 5$–7), suggesting a possible discontinuity in the functions. Similar, but more prominent, discontinuities appear in data obtained in the same experiment, but for the back rather than the forehead (see Chapter 5, Figure 5.6), and in data collected in other experiments (Marks & J. C. Stevens, 1973) (see Chapter 4, Figure 4.8). These discontinuities are somewhat reminiscent of discontinuities seen in vision, where they are often related to the transition between rod and cone func-

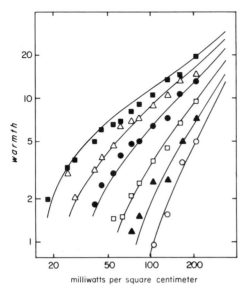

Figure 2.4. Average magnitude estimates of warmth produced by thermal irradiation of the forehead, as functions of stimulus irradiance in milliwatts per square centimeter, as in Figure 2.2. The fitted functions are calculated from a theoretical equation that hypothesizes two mechanisms whose outputs sum to yield total magnitude of warmth sensation. [Data from J. C. Stevens and Marks (1971).]

tioning. It is tempting to consider the possibility that warmth is mediated by two mechanisms and that the discontinuities in the warmth functions are not fortuitous, but reflect a true transition from one underlying mechanism to another.

The possibility that two mechanisms underlie warmth is enhanced by the fact that, as the data show, spatial summation is not constant throughout the dynamic range of the warmth sense. Since warmth grows at different rates for different stimulus sizes and the functions tend to converge at a very high irradiance, areal extent plays a smaller and smaller relative role at higher and higher irradiances. That is, spatial summation decreases as intensity increases. One possible way to account for this variation in summation is to assume that two mechanisms operate to mediate warmth: the one that operates at low levels displays rich spatial summation, whereas the one that dominates at high levels displays little or no summation.

Figure 2.4 repeats the data points of Figures 2.2 and 2.3, but shows a series of curves based on a model for warmth that supposes two mechanisms, one displaying complete spatial summation, the other no spatial summation (Marks, 1971b). The equation used to construct the curves is

$$\psi = 30 \left(\frac{.39A\phi}{1 + .39A\phi} + 2\phi^2 \right) - 2.5 \qquad (2.4)$$

where A is areal extent of stimulation and ϕ is in W/cm^2.

Each of the terms in the argument of Eq. (2.4) represents the contribution of one of the assumed underlying mechanisms. [Generation of Eq. (2.4) is discussed in greater detail in Chapter 5.] Although the fit to the data is far from perfect, the overall approximation is good, especially in light of the fact that all of the curves are derived from a single equation. Fitting a theoretical family of curves such as those of Figure 2.4 can often be revealing and valuable. On the other hand, the model that is implied by the curves and by Eq. (2.4) should be based on more than just one set of data. There is no doubt that a large number of other equations, and other theoretical models, could also describe the data well. However, it is the case both that the discontinuities in warmth functions appear in other experimental data (J. C. Stevens & Marks, 1971; Marks & J. C. Stevens, in press) and that others have found reason for suggesting the notion that warmth may be mediated by more than one mechanism (Hensel & Kenshalo, 1969; Herget et al., 1941).

Thus, we see that there may be several ways to deal with magnitude estimates as they relate to the primary stimulus variable; the number of alternatives available for analyzing data depends primarily, of course, on how the data turn out, and each alternative has its own advantages and drawbacks. Line segments between points remain most true to the data, but serve only to display basic trends. The fitting of free-hand curves can often serve to display trends somewhat more clearly, but at concomitant sacrifice of precise description. The major advantage to fitting some simple equation to the results stems from the capacity of a small number of parameters of the equation to characterize and summarize the entire psychophysical function. Often implicit in the use of an equation, even an empirical one, is the assumption that the equation gives a better indication of the "true," underlying psychophysical relation than do the individual data points. When simple equations are fitted under auxiliary constraining restrictions, the summary properties become even more potent. A theoretically based set of curves provides the greatest surfeit of information to the results, but with the ever-present, usually strong, and almost always correct possibility that the theory or model upon which the curves are based is false.

The analysis of results does not by any means end with the description of the primary psychophysical relation, e.g., of warmth to irradiance. It is also possible to repeat the entire analysis and description in terms of the second stimulus variable—in the present example, warmth re areal extent. Figure 2.5 shows the average estimates of warmth plotted as functions of stimulus size. The investigators found that it was not feasible to make the original ensemble of stimuli contain both a limited number of areas, each at a large number of intensities, and also a limited number of

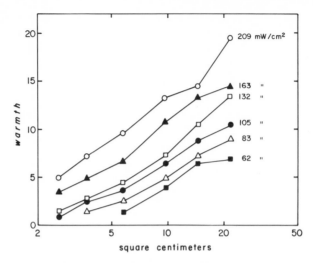

Figure 2.5. Average magnitude estimates of warmth produced by thermal irradiation of the forehead, as functions of stimulus size in square centimeters. The abscissa is logarithmic. Line segments connect data points obtained at constant stimulus irradiances. Irradiance varies from 62 to 209 mW/cm². [Data from J. C. Stevens and Marks (1971).]

intensities, each at a large number of areas. We have, therefore, only a few examples where the same irradiance was presented over several different areal extents. Even so, the general trends of the results are clear. In semi-logarithmic coordinates warmth is nearly linearly related to area. Thus, the relation between warmth and area is approximately logarithmic. These logarithmic functions are shown in Figure 2.6. Of course, the same options for description of the data are available for the second set of psychophysical relations (warmth versus area) as were available for the first set (warmth versus irradiance).

It is often true (and these data present no exception) that it is simpler to quantify the relation of estimated magnitude to one variable (e.g., irradiance) than to the other (e.g., area). In the present case, it appears that the logarithmic relations are fair approximations only at moderate levels of irradiance. One possibility that suggests itself is to plot for this second set of functions the values predicted from the first set. That is, we can read off from Figure 2.3 the values of warmth predicted by the modified power functions, and we can then make the curves relating warmth to area pass through those points. To the extent that we believe the modified power equations to be a better indicant of the underlying, primary psychophysical relations that are the data points themselves, then so too should the predicted values be better indicants of the secondary psycho-

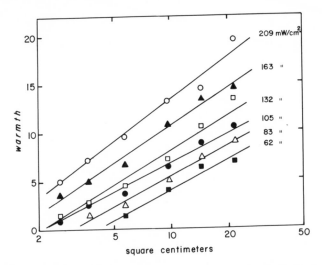

Figure 2.6. Average magnitude estimates of warmth produced by thermal irradiation of the forehead, as functions of stimulus size, as in Figure 2.5. The fitted functions are logarithmic. [Data from J. C. Stevens and Marks (1971).]

physical relations. Consequently, the points in Figure 2.7 were read off from the modified power functions in Figure 2.3. Of course, a theoretical equation like that used to produce the warmth–irradiance curves of Figure 2.4 will also produce theoretical curves relating warmth to area.

SENSORY-PHYSICAL ANALYSIS OF SCALING DATA

There remains yet another type of analysis of scaling data that pertains to multivariate stimuli. This final analysis can be very important; it consists basically of eliminating the sensory (response) variable and relating values of the primary to values of the secondary stimulus variable. The result is a series of sensory-physical relations, such as those produced by the more traditional psychophysical procedures, e.g., by methods such as intrasensory matching. In the present example of warmth, we eliminate the magnitude of warmth sensation as a variable for consideration and instead direct our attention to the relations between irradiance and areal extent, given constant levels of warmth sensation. Viewed in this way, sensory magnitude becomes a parameter rather than a variable.

It would take an enormous stroke of fortune for it to happen that scaling data (at least, magnitude-estimation data) would yield sensory-physical functions without requiring some interpolation between data points. In order to find irradiances that produce constant levels of warmth for different areas, it is necessary to make interpolations from the psycho-

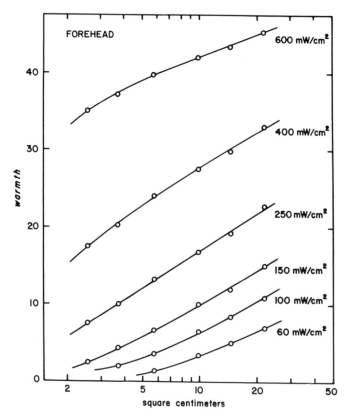

Figure 2.7. Warmth as a function of stimulus size. The values were read from the threshold-corrected power functions in Figure 2.3. [From J. C. Stevens and Marks (1971). Courtesy of *Perception & Psychophysics*.]

physical data. Thus, the precise outcome of sensory-physical analysis will depend upon whence the interpolation derives. The most conservative basis for making those interpolations is the use of line segments to connect data points (such as those of Figure 2.2). We can read off, for example, all of the irradiances and areas that produce estimations equal to 5. These irradiances would be 40 mW/cm² at 21.8 cm²; 57 at 14.5; 83 at 9.65; 140 at 5.72; 160 at 3.68; and 210 at 2.55. This series of points yields a single constant-warmth contour. If we change from one level of warmth to another, we obtain a new curve of constant warmth. A set of these curves, based on the line segments of Figure 2.2, is shown in Figure 2.8. To be consistent with the use of line segments in Figure 2.2, the points plotted in Figure 2.8 should also be connected by line segments.

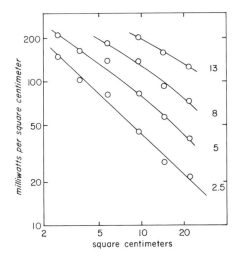

Figure 2.8. Contours of equal warmth, showing how irradiance and areal extent must covary to maintain the magnitude of warmth sensation constant. The greater the area, the lower need be the irradiance. The values were read from the line segments in Figure 2.2.

Free-hand curves were drawn, however, to facilitate comparison to other sets of constant-warmth contours. For comparison, Figures 2.9 and 2.10 show curves derived, respectively, from the modified power functions of Figure 2.3 and from the theoretical curves of Figure 2.4. In the present

Figure 2.9. Contours of equal warmth, showing how irradiance and areal extent must covary to maintain the magnitude of warmth sensation constant. The values were read from the threshold-corrected power functions in Figure 2.3. [From J. C. Stevens and Marks (1971). Courtesy of *Perception & Psychophysics*.]

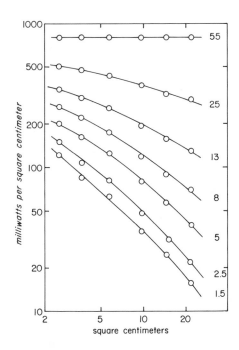

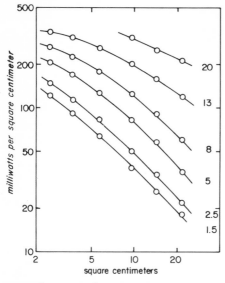

Figure 2.10. Contours of equal warmth, showing how irradiance and areal extent must covary to maintain the magnitude of warmth sensation constant. The values were read from the theoretical functions in Figure 2.4.

example, it can be seen that the similarities among these three sets of derived curves are much more prominent than are their differences.

In cases where the sensory-physical relations are of special importance, and where results based on interpolations may be worrisome or questionable, there is then good reason to prefer the use of magnitude production, rather than magnitude estimation, as a scaling procedure. One major advantage that production has is that the sensory-physical functions can be determined directly from the data, with no interpolation needed. Had it been feasible to use magnitude production for warmth, for example, the subjects would have adjusted irradiances at each of the various areas in order to produce particular levels of warmth sensation. Viewed in this respect, magnitude production is very much like a complex intrasensory matching procedure, a matching procedure that contains numerical judgments as intermediaries.

All three sets of constant-warmth curves lead to the same conclusions regarding the sensory physics of spatial summation in the warmth sense (Figures 2.8–2.10). At low levels of warmth (near the absolute threshold), the relation between area and irradiance is one that approaches complete reciprocity. Spatial summation is maximal at low irradiances. At higher and higher levels the curves tend to flatten out more and more, reflecting the fact that spatial summation is diminishing. The diminution in summation is progressive, until the pain threshold is reached, at which point summation has disappeared (Figure 2.9). It is also noteworthy that the summation functions are curvilinear in log–log coordinates. The nature of the curvature and its importance cannot be considered here, but it is discussed in Chapter 5.

It is important not to downgrade the analysis of the sensory-physical results from scaling experiments. In a certain sense, the sensory physics may generally be the most dependable and reliable aspect of the results. This is true because the sensory-physical relations are to a large extent independent of the ways the subjects go about assigning numbers to stand for their sensations (or some aspect of their sensations). The primary assumption for sensory-physical analysis is that the subject always assigns the same number (on the average) to stand for the same sensation, regardless of how that sensation was produced. Thus, the subject is assumed to assign the same number both to a large, weak thermal stimulus and to a small, intense one, as long as the two stimuli produce identical warmth. This outcome can obtain no matter how idiosyncratic a subject's use of numbers may be.

Recall that earlier in this chapter we noted how the steepness of psychophysical functions (and, therefore, the range of numerical responses) depends somewhat on procedural variables, such as stimulus range, presence and location of standard stimulus, numerical value assigned to standard, and so on. Presumably, these variables have a much smaller, if not negligible, effect on sensory-physical relationships. As previously noted, Stevens (1961b) has termed the tendency of subjects to constrict the range of whatever variable is under their control, regression. It is the range of numbers that is constricted when the procedure is magnitude estimation, but the range of stimuli that is constricted when the procedure is magnitude production. Given a single sensory modality and the same subjects, therefore, magnitude production typically yields larger exponents of power functions than does magnitude estimation (Stevens & Greenbaum, 1966). Regression appears to operate in all types of matching experiments, including intrasensory matching of loudness to loudness and of brightness to brightness.

To the extent that regression or range constriction takes place in scaling experiments, the psychophysical functions that are obtained will deviate to some extent from the "true" underlying functions. It is a difficult task to attempt to discover what the "correct" underlying psychophysical functions are, if in fact correct functions exist. There are some reasons, both empirical and theoretical, to suggest that they do (Marks, 1972). However, scaling experiments in which the stimuli vary over several stimulus dimensions are not always the most appropriate places to look for these functions. In particular, it seems to be the case that when the subject's task gets more and more difficult, the degree of range compression or regression increases, too. Whereas the basic exponent for the brightness of a not too short or small flash presented to the dark-adapted eye seems to lie near 1/3, experiments investigating the effects of variables such as locus of stimulation on the retina report much smaller exponents (Marks, 1971a). A probable reason for the smaller exponent is the difficulty of the task. It appears that judging a peripherally viewed stimulus is a difficult task, and the range of numerical responses suffers. That the effect is not a sensory one is shown by the fact that foveally viewed stimuli also have their response range constricted when they are interspersed with peripherally viewed stimuli. But it is important to note that the sensory physics of brightness vision in the peripheral and foveal retina appears to be fundamentally unaffected by range constriction (Marks, 1966).

It can even happen that the results of a scaling experiment provide a better measure of sensory-physical relations than do results of direct matching. When matches are not obtained in a balanced experiment, and when degree of regression is large, matching data may give a misleading picture. Take as an example the brightness matching of foveally and peripherally viewed stimuli (Marks, 1966). Average matching luminances, plotted in double-logarithmic coordinates, follow a straight line. If the line has unit slope, then the rates of growth (power-function exponents) for foveal and peripheral brightness are equal. It turned out, however, that the slope depended strongly on whether the subject adjusted the foveal or the peripheral luminance: the subjects strongly constricted the range of whichever stimulus was under their control. Although the average result suggested matching functions of unit slope, results of either matching foveal to peripheral brightness or vice versa suggested sizeable (about 20% on the average) differences between exponents. In matching, regression influences the picture of sensory physics, whereas in scaling, regression influences the psychophysics, but leaves the sensory physics relatively untouched.

Chapter 3

SENSORY MAGNITUDE AND
THE SENSITIVITY OF SENSORY SYSTEMS

The concept of sensitivity is one of the cornerstones of sensory psychology. A very high frequency sound put out by a whistle (20,000 Hz) may be heard by a dog, but not by a human being. When sound frequency is very high, the dog has greater auditory sensitivity. It is not that a person cannot hear a 20,000-Hz tone, but that he needs much more energy in the tone to hear it than the dog needs. Simply defined, sensitivity is the inverse of the intensity of the stimulus—be it energy of light or energy of sound, concentration of molecules in water or concentration in air—that is needed to produce some criterial response by the sensory system.

The Concept of Relative Sensitivity

Sensitivity connotes more, however, than just the inverse of the stimulus intensity under some specified set of conditions. Sometimes the term *relative sensitivity* is used to describe how the intensity of the stimulus (or its inverse) depends on some other stimulus properties. For instance, the person who requires a great intensity at 20,000 Hz in order to hear the tone will require much less energy (perhaps one-tenthousandth as much) at 2000 Hz. It follows that his relative sensitivity is 10,000 times as great at 2000 Hz as at 20,000 Hz. The present chapter together with Chapters 4 and 5 all deal almost entirely with relative sensitivity. In Chapters 4 and 5 we will take up, respectively, the parameters of temporal and spatial distribution of the stimulus, and the effects of these variables on the sensitivity of sensory systems. In the present chapter, we look at some of the other important variables that determine sensitivity, in particular what might loosely be called "compositional" variables; these are the variables that also determine to a major extent perceived differences in qualities of our sensory perception. Compositional variables include wavelength and frequency of light and sound, and chemical structure of olfactory and gustatory stimuli.

Differences in sensitivity manifest themselves in all of the sense departments. The eye, for instance, is not equally sensitive to all wavelengths. Blue (short wavelength) and red (long wavelength) lights do not appear to be as bright as a green light of the same intensity. The fact that the visual system is very sensitive to electromagnetic radiation over a small range of wavelengths, but is very insensitive to wavelengths outside that range (e.g., insensitive to ultraviolet and infrared radiation), makes the eye vulnerable to damage, such as manifests in the danger entailed in attempting to view a solar eclipse through sunglasses or poor filters. In a similar fashion, the ear's sensitivity to different frequencies varies greatly: relatively low and high frequencies may be inaudible, even though intermediate frequencies at the same intensity are clearly heard. The other senses—taste, smell, touch—also show these sorts of differences. Most of us can taste quinine salts at molar concentrations 100,000 times lower than those needed to detect ordinary table salt (sodium chloride), and most of us can smell compounds like ethyl mercaptan (a major constituent of skunk odor) and musk at concentrations millions of times lower than those needed to detect odorants such as methanol and ethanol.

Recall that sensitivity was defined as the inverse of stimulus intensity required for a given criterial response. Since one of the traditional measures of sensory function is the "absolute threshold," it is not surprising that a traditional measure of sensitivity is the reciprocal of threshold intensity. Indeed, some investigators used to take such a measure of sensitivity as the equivalent to a measure of the intensity of sensation. (This view of sensitivity as a measure of sensation traces back to Körber, 1746.) For example, a particular substance might be detected (have its absolute threshold) at a concentration one-tenth that of another substance. Then, (a) not only would sensitivity to the first substance be 10 times that of the second, but (b) the sensation produced by any concentration of the first substance would be assumed to be 10 times that produced by the same concentration of the second. This extension to sensation magnitude of the concept of sensitivity, when it is based only on threshold measurements, is wholly unwarranted and usually incorrect. It can be correct if and only if sensory magnitude is wholly proportional to stimulus intensity for any given substance.

But the concept of sensitivity need not be limited to sensitivity at absolute threshold. As a matter of fact, recall that the original definition referred only to some criterial response by the sensory system. Thus, the response need not be limited to one of stimulus detection; in fact, it need not be even directly related to sensory magnitude. This freedom of criterion is especially important for studies of visual processes, where sensitivity has been measured as a function of the wavelength of stimulating light for a wide range of criterial responses, including absolute threshold, equal brightness, minimal border between two lights, minimal flicker for intermittent light, and constant levels of visual acuity. Since we limit our concern here to the examination of sensory magnitude, it is within the domain of suprathreshold sensations that we shall look at relative sensitivity.

LEVEL-INDEPENDENT AND LEVEL-DEPENDENT RELATIVE SENSITIVITY

The extension of the concept of sensitivity to suprathreshold sensory relations is of fundamental import. That extension permits us to understand how stimulus variables interact in determining sensations not only at threshold, but over the entire ranges of stimulation to which sensory systems respond. Thus, we should properly speak not of the relative sensitivity of a sense department, but of its relative sensitivities, i.e., of the entire family of sensitivity functions, infinite in number, whose lower limit is threshold sensitivity. In some instances to be sure, the family consists of a set of parallel curves. In such examples, relative sensitivity may be said to be *level independent,* since the same sensitivity function obtains both at threshold and at every suprathreshold level of subjective magnitude. But often the curves of relative sensitivity are not the same at threshold as above. In such instances, relative sensitivity may be said to be *level dependent.* The type of family of sensitivity functions that obtains—level independent or level dependent—is intimately connected to the corresponding family of psychophysical functions.

Take as an example auditory sensitivity with respect to frequency. Curves of equal loudness are flatter at high loudnesses than they are at low or threshold loudness. The systematic variation in curves of auditory sensitivity (sensory physics of loudness) corresponds to variation in the form of the loudness function (psychophysics of loudness). Loudness grows more rapidly with sound pressure at some frequencies than at others. Since the question of level dependence versus level independence is psychophysical as well as sensory-physical, it can be approached from the point of view of psychophysical functions. Given any particular sensory dimension, the set of psychophysical functions (applicable regardless of stimulus composition) may be expressed

$$\psi = \Gamma(\phi) \tag{3.1}$$

In general, relative sensitivity is independent of level if there exist some transformations of ψ and/or of ϕ that make all of the psychophysical functions Γ plot as parallel lines. Given Γ as a simple power function, $\psi = k\phi^\beta$, then logarithmic transformations of ψ and of ϕ make Γ plot as a straight line. If, in addition, the function's exponent β is constant, all of the straight lines will be parallel and sensitivity will be level independent. [Level-dependent variation is possible, however, even given constant exponent, if the power function is modified to include an additive (e.g., threshold) constant that varies with composition.] Similarly, given Γ as logarithmic, $\psi = \beta \log \phi + k$, logarithmic transform of ϕ makes Γ plot as a straight line. Again, if the slope β is constant, then all of the straight lines will be parallel, and sensitivity will be level independent.

Thus the properties of the function Γ determine whether relative sensitivity is level dependent or level independent. Consider the example of auditory sensitivity. Were all Γ for different sound frequencies proportional to one another, then auditory sensitivity would be level independent. Given Γ as a simple power function, constancy of the exponent is a sufficient condition to ensure that relative sensitivity is level independent. On the other hand, as is the actual situation, exponents for loudness vary with sound frequency, and, concomitantly, relative sensitivity is level dependent.

In the case of scotopic vision there is no dependence of spectral sensitivity on level of brightness, and, correspondingly, brightness increases with light energy at the same rate at all wavelengths. The dichotomy between level independent and level dependent relations is not limited to those variations in sensitivity that are due to the composition of the stimulus. The same question—Does sensitivity depend on level of sensory magnitude?—may be asked when sensitivity varies with other stimulus dimensions, such as duration, area, or site of stimulation on the receptor surface. So the question will arise in later chapters as well as the present one.

Spectral Sensitivity in Vision

Scotopic and Photopic Sensitivity

The visual system is strikingly nonuniform in its sensitivity to light of different wavelengths. If we measure the minimal energy needed to detect a small patch of light directly viewed by an eye that has been adapted to darkness, we find differences in energy as great as on the order of 40,000-to-1 among wavelengths in the range 365–750 nanometers (nm). The open circles in Figure 3.1 show some such threshold measurements made by Wald (1954). The stimulus was a 1° field centered on the fovea, which is a small pit near the center of the retina. The maximum in sensitivity (the smallest amount of energy) occurred at a wavelength about 560 nm. At that wavelength, stimuli of higher intensities typically look greenish-yellow. The fall-off of sensitivity is great at very short and at very long wavelengths. As a matter of fact, the values of 400 and 700 nm are very often given as the spectral limits of vision, mainly because, as the steep declines show, relative sensitivity has already become extremely poor at these wavelengths. But Figure 3.1 also shows that sensitivity by no means ends at those limits. Other experiments (Griffin, Hubbard, & Wald, 1947; Pinegrin, 1944, 1945) have extended measurements farther into the ultraviolet and the infrared. The decline in sensitivity at very short and long wavelengths is rapid, but continuous.

The filled circles in Figure 3.1 show additional threshold measurements by Wald, made not at the foveal center, but at a point 8° eccentric from the fovea. There are two interesting features of the results. First, at all but the very longest wavelengths, it took less energy to see this peripheral light than to see the foveal light. Second, the shape of the sensitivity curve for the periphery is different from that for the fovea: the maximum in sensitivity appears here at a wavelength about 500 nm. This curve of sensitivity for the periphery of the eye is an example of what is called *scotopic spectral sensitivity*. The curve for the fovea is an example of *photopic spectral sensitivity*. Thus, the eye contains two systems, each of which displays a different relative sensitivity to lights of various wavelengths.

Figure 3.1. Relative visual sensitivity (reciprocal of absolute threshold) to different wavelengths. Open circles: sensitivity of the fovea (cones); filled circles: sensitivity of the dark-adapted peripheral retina (rods). [After G. Wald, Human vision and the spectrum. *Science*, 1945, **101**, 653–658. Reproduced by permission of the author and the American Association for the Advancement of Science.]

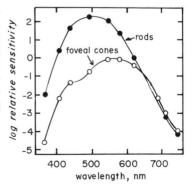

The difference between scotopic and photopic sensitivities is usually ascribed to differences between two types of receptors in the retina: rods and cones. The periphery of the eye contains both types of receptor. However, when the eye is adapted to darkness and test fields are not too small, the rods are capable of responding at much lower energies than are the cones; thus, absolute thresholds in the dark-adapted peripheral retina are mediated by rods. (An exception occurs at very long wavelengths, where the cones may operate at lower energy levels than do rods.) When the scotopic sensitivity curve is corrected for selective losses in the eye (through absorption of light by the cornea, humours, etc.), the resulting curve matches almost perfectly the absorption spectrum of rhodopsin, which is the photoreceptive substance within rods.

The fovea of the eye contains only cones, no rods, so the photopic sensitivity curve reflects cone sensitivity. However, there is evidence that at least three different types of cones are present in the retina of the human eye, and each of these cones has a different photoreceptive substance, i.e., a substance with a different absorption spectrum (for a review, see Rushton, 1972). Thus, photopic sensitivity curves probably reflect the sum or the profile of three receptor substances, each with a different spectral sensitivity.

SENSORY PHYSICS AND PSYCHOPHYSICS OF
VISUAL SPECTRAL SENSITIVITY

One important question is whether these two sensitivity functions, one for cones, the other for rods, remain invariant when the sensory criterion is changed. For example, what if, instead of absolute threshold, the criterion is some suprathreshold level of brightness? One reason this question is important is the following: If the spectral sensitivity curves are invariant with brightness level, then lights of different wavelengths and energies, but equal brightnesses, will remain equal in brightness when their energies are augmented or diminished in constant proportion. For instance, a red light 10 times as intense as a green one might appear just as bright. If we double (or halve) the energies of both lights, they would remain equally bright. Neglecting mutual interactions of brightness contrast, when a cloud passes overhead or when additional light is added to a room, the *relative brightnesses* of spectrally different objects will not change.

Invariance of spectral sensitivity was assumed by the Commission Internationale de l'Eclairage (CIE), which has defined two curves to be used for evaluating the effectiveness of light of different wavelengths.

These *luminosity functions,* one scotopic, the other photopic, were taken from averages of a number of experiments. The scotopic function displays maximal sensitivity at 505 nm, the photopic function, a maximum at 555 nm. The two functions are similar in overall shape, but the scotopic function displays relatively greater sensitivity to short wavelengths, the photopic function relatively greater sensitivity to long wavelengths. Not all of the measurements from which the functions were derived involved a brightness criterion, however. Some, for example, used a criterion of minimal flicker. Nevertheless, a basic assumption of the CIE system is the invariance of the photopic and scotopic functions when the level of output, e.g., brightness, changes.

The invariance with level of the sensitivity of the scotopic system is well known (e.g., Weaver, 1949). Ekman and Künnapas (1962) scaled the brightness of monochromatic lights under scotopic conditions, and they found no systematic variation of the form of the brightness–energy relation with wavelength. There is little reason to expect any variation at scotopic levels, however, since only one type of receptor, rods, mediates scotopic responses. The specific shape of the scotopic spectral sensitivity curve results from the spectral absorption of rhodopsin plus the spectral transmission of light by the eye. It is as if the eye filters out some wavelengths more strongly than others.

Photopic sensitivity is another matter. At least three types of cones mediate photopic responses, so it is important to know whether the sensitivity they produce varies with brightness. Studies aimed at answering this question have used two types of procedure: heterochromatic brightness matching and numerical estimation of brightness. Sloan (1928) performed a brightness-matching experiment in which monochromatic lights and white lights were flashed for 2-sec exposures, and the subject adjusted the energy of each monochromatic light so that its brightness equaled that of a white light serving as a standard. Sloan found that the photopic sensitivity of a 57' (foveal) field did change with brightness level. At higher and higher brightnesses, the spectral sensitivity curve narrowed; the relative sensitivity decreased markedly at long wavelengths.

Sloan's results suggest differences in operating characteristics among three cone receptors (or interactions among them). The results also suggest that the psychophysical brightness function (the function relating brightness to stimulus energy) varies with wavelength. For white light, evidence from direct scaling shows that brightness (B) is related to intensity (ϕ) by the power relation

$$B = k\phi^\beta \qquad (3.2)$$

where $\beta = 1/3$ when the eye is dark-adapted and stimulus duration is 1–2 sec. If photopic spectral sensitivity does not change with brightness level, then Eq. (3.2) would apply equally well to all monochromatic lights. Only the constant k would vary, and its value would reflect the (invariant) spectral sensitivity. A variation in sensitivity with brightness, however, would imply more complex relations between brightness and energy. Possibilities would include changes in the value of the exponent β in Eq. (3.2) as wavelength varies, or departures from power-function form. Sloan's results in particular imply, as one possibility, that the exponent for long wavelengths (i.e., red light) is smaller than the exponent for intermediate wavelengths (e.g., green).

Half a dozen studies have determined brightness-intensity functions for stimuli of different spectral compositions. Onley (1960) compared brightness functions obtained for white and for broad bands of blue, green, and red light. All stimuli were foveal (1° visual angle). Onley reported that the exponent did not seem to change with wavelength; but the data do suggest a slightly smaller exponent for blue and red light than for green. However, that must remain an uncertain point for those data. Ekman, Eisler, and Künnapas (1960b) obtained magnitude estimates of the photopic brightness of monochromatic lights in the range 459–672 mm. Brightness grew as a power function of energy at each wavelength, and all of the exponents were in the vicinity of one-third. There were some differences in exponent from one wavelength to another, and *in general,* the exponent decreased as wavelength increased. That is, the smallest exponents occurred for the longest wavelengths. Unfortunately, the experiment examined only a small range of energies (for most wavelengths, a range of a little more than 10:1), much smaller than the range of photopic vision. Furthermore, later experiments, which used both the methods of magnitude estimation (Ekman & Künnapas, 1962) and magnitude production (Ekman, Eisler, & Künnapas, 1960a) failed to show similar variation in exponents.

The most extensive examination of brightness functions for monochromatic lights was made by Wilson (1964), who obtained magnitude estimates for wavelength of 430–700 nm. Figure 3.2 shows representative results. All of her data showed brightness to grow as a negatively accelerated function of energy when the data were plotted in double-logarithmic coordinates. Thus, the functions deviate from power functions. The reason was probably that Wilson used a simultaneous comparison field that bordered on the test field. The comparison field was light of 550 nm at an intermediate level of energy; such a comparison field will typically exert an inhibitory effect on other stimuli, especially ones that would normally be less bright (see Chapter 5).

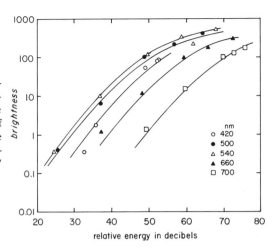

Figure 3.2. Magnitude estimates of the brightness of monochromatic light presented to the dark-adapted eye, as functions of stimulus energy in decibels. The stimulus subtended a visual angle of 2°. [Data of subject JW from Wilson (1964, Table 4a).]

Wilson reported that the form of the brightness function was independent of the wavelength of the test field. All of the curves in Figure 3.2 are drawn in accordance with that conclusion. A corollary is that spectral sensitivity was independent of brightness level. That outcome and those of Ekman and his colleagues appear to be in conflict with that of Sloan (1928), who found clear evidence of change of spectral sensitivity with change in brightness level. Recall that Sloan found that sensitivity to long wavelengths decreased when the criterial level of brightness increased. On the other hand, Bedford and Wyszecki (1958) found no variation at long wavelengths, but clear variation at short: as brightness increased, relative sensitivity to short wavelengths increased. Interestingly, Thomson (1947) obtained the effects reported both by Sloan and by Bedford and Wyszecki: as brightness increased, sensitivity to long wavelengths decreased and sensitivity to short wavelengths increased.

Variation in spectral sensitivity similar to that found by Sloan was also reported by Bornstein and Marks (1972). In that experiment, however, the criterion was threshold flicker. It is perfectly possible for spectral sensitivity to vary with level for one sort of criterion, but not for another. Guth, Donley, and Marrocco (1969) have argued that brightness (and spectral sensitivity determined thereby) is mediated by systems that contain opponent processes as well as nonopponent processes, whereas flicker is not usually mediated by opponent processes. It could therefore be that flicker spectral sensitivity would vary with level, but brightness spectral sensitivity would not (or that both would vary, but in different ways). The evidence at hand is ambiguous.

An extensive study of spectral sensitivity was conducted by Decker (1967), who obtained brightness matches (46′ field) with stimuli that

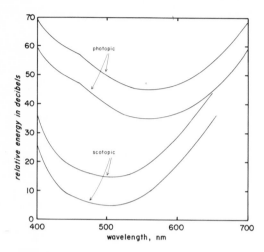

Figure 3.3. Contours of equal brightness. The two lower curves show how energy must covary with wavelength to maintain constant brightness at scotopic levels. The two upper curves show how energy must covary with wavelength to maintain constant brightness at photopic levels.

varied in intensity over a range of 50 dB. Decker's data showed spectral sensitivity to be independent of brightness level.

The bulk of the evidence suggests that photopic spectral sensitivity, like scotopic sensitivity, is independent of brightness level. Assuming that to be the case, Figure 3.3 shows a few equal-brightness curves at scotopic and at photopic levels. The curves are based on the CIE luminosity functions for a 2° field, except for short-wavelength photopic sensitivity, which is based on Judd's (1951) recommendation. (At short wavelengths, the CIE curve markedly underestimates spectral sensitivity.) It is worthwhile keeping in mind that these curves are at best averages and approximations. Individual subjects differ in their spectral sensitivities, at least in part because of individual differences in ocular pigmentation.

One condition under which spectral sensitivity is certainly level dependent is the viewing of stimuli that range from scotopic through photopic levels. When the stimulus is extrafoveal (larger than the fovea) or peripherally viewed, low-intensity stimulation will excite only rods, high-intensity stimulation both rods and cones. Thus, spectral sensitivity can be primarily scotopic or photopic, depending on brightness level. Furthermore, the transition from scotopic to photopic sensitivity is relatively gradual: even at rather high intensities the scotopic contribution to brightness is not negligible (Walters & Wright, 1943).

One implication of this outcome is that psychophysical brightness functions obtained with extrafoveal or peripheral stimuli that vary over a wide range of intensities must differ in form from one wavelength to another. Whether the differences entail a change in exponent of a power function, or a departure from power-function form, is not known.

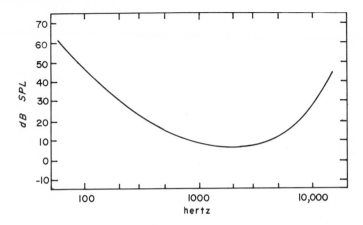

Figure 3.4. Relative auditory sensitivity (absolute threshold) to different sound frequencies. [After L. J. Sivian and S. D. White, On minimum audible sound fields. *Journal of the Acoustical Society of America*, 1933, 4, 288–321. Courtesy of *The Journal of the Acoustical Society of America*.]

Spectral Sensitivity in Audition

The Audibility Function

Perhaps even more extensively studied than the eye's sensitivity to different wavelengths is the ear's sensitivity to different frequencies. Like the visual system, the auditory system is not equally sensitive to all wavelengths or frequencies. The sensitivity of the ear is greatest to frequencies in the region 1000–5000 Hz, and the fall-off in sensitivity at higher and lower frequencies is quite rapid. Figure 3.4 shows measurements of auditory thresholds made by Sivian and White (1933). At 60 Hz, the threshold was 50 dB greater than it was at 2000 Hz (where sensitivity was maximal). That is to say, the energy that was needed to make a sound just detectable at 60 Hz was 100,000 times as great as the energy needed at 2000 Hz.

For both the auditory and visual systems, a large, if not the major, portion of the variation in spectral sensitivity is a product of nonuniform transmission of sound and light energy to the receptors. In vision, scotopic sensitivity is determined by the absorption spectrum of the visual pigment, rhodopsin, that resides in retinal rods, plus the nonuniform transmission of light of different wavelengths through the ocular media; the shape of the photopic spectral sensitivity curve is determined to a large extent by absorption spectra of pigments in retinal cones plus the (same) nonuniform transmission characteristics of the ocular media. Similarly, the audibility curve of the ear has a form whose shape depends largely on

how sound is collected and reflected by the pinna (outer ear) and
transmitted to the cochlea through the external auditory meatus and b
the bones of the middle ear. There is a great deal of nonuniformity o
transmission: sounds of different frequencies are attenuated or augment
ed to different degrees. Zwislocki (1965) has argued that when the
threshold audibility curve is corrected for losses due to the attenuation
characteristic of the middle ear, the remaining curve is quite different; it
shows a continuous increase in sensitivity (for frequencies greater than
about 200 Hz) that amounts to 3 dB (a doubling of energy) for every
doubling of frequency. That remaining inconstancy in sensitivity Zwislocki attributes to temporal properties of auditory processing. His theory
of temporal summation can account for the 3 dB/octave change in sensitivity. For frequencies below 200 Hz, the change in sensitivity is even
more than 3 dB/octave, and that greater variation in sensitivity may reflect decreased innervation (density of primary neurons) of the basilar
membrane in the region corresponding to low frequencies.

SENSORY PHYSICS AND PSYCHOPHYSICS OF AUDITORY SPECTRAL SENSITIVITY

The major concern of this section is the sensitivity of the auditory system to suprathreshold stimuli. Fortunately, several experiments have attempted to answer the question how the ear's sensitivity depends on level
of loudness. Their results concur in the demonstration that the audibility
curve is not invariant, but changes with loudness. Most of these studies
have employed matches for loudness between sounds of different frequencies; results of such studies directly give curves of equal loudness.
An example of such a series of curves is given in Figure 3.5 (Fletcher &
Munson, 1933). Over the range of middle and high frequencies, the
equal-loudness curves show little variation with level and strongly resemble audibility at threshold. Over low frequencies, however, the curves
tend to flatten out at higher and higher loudnesses. That is, there is less
difference in sensitivity from one frequency to another when the criterion
is a high level of loudness than when the criterion is low or threshold
loudness.

The flattening of the equal-loudness curves at low frequencies has
been noted in a large number of experiments (Fletcher & Munson, 1933;
Kingsbury, 1927; Robinson & Dadson, 1956; Ross, 1967). Stevens (1966e,
1972) reviewed a number of the relevant experiments and concluded
that the flattening reflects a change in the exponent of the power function for loudness: below about 400 Hz the exponent increases as frequency decreases. The increase in exponent is substantial: the exponent
at a frequency of 20 Hz is twice as large as that at frequencies beyond

400 Hz. Since the exponent for middle and high frequencies equals about one-third (Stevens, 1972) when the stimulus is reckoned in terms of sound energy, the exponent at 20 Hz would equal about two-thirds.

The dependence of exponent on frequency was determined directly in two experiments that utilized direct scaling procedures. Schneider, Wright, Edelheit, Hock, and Humphrey (1972) found that the exponent determined by magnitude estimation was larger at 100 Hz than at frequencies of 200 Hz and higher. There was also a suggestion that the exponent increased again for frequencies beyond 2500 Hz. Hellman and Zwislocki (1968) employed the method of *numerical loudness balance* (a combination of results obtained by magnitude estimation and magnitude production) to determine the relation between loudness and sound pressure at three frequencies—100, 250, and 1000 Hz. Figure 3.6 shows their results: the loudness function is steepest in double-logarithmic coordinates (i.e., has the largest exponent) at 100 Hz and is least steep at 1000 Hz. Note also that once sound pressure levels of about 60–80 dB are reached, the functions become less steep and more nearly parallel.

Stevens (1966e) also described that transition from more to less rapid

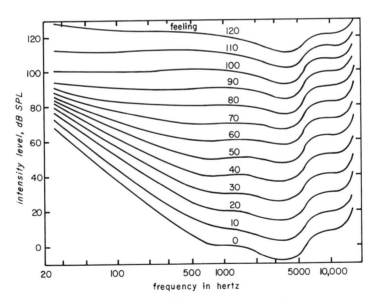

Figure 3.5. Contours of equal loudness. The curves show how sound energy and frequency must covary to maintain constant loudness. The parameter is the loudness level of each contour. [From H. Fletcher and W. A. Munson, Loudness, Its definition, measurement and calculation. *Journal of the Acoustical Society of America*, 1933, **5**, 82–108. Courtesy of *The Journal of the Acoustical Society of America*.]

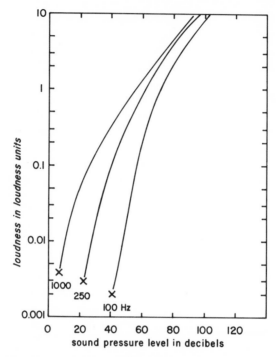

Figure 3.6. Psychophysical functions showing how the loudness of pure tones of 100, 250, and 1000 Hz increases with sound pressure level in decibels. Over low sound pressures, the lower the frequency, the steeper the log-log slope. [From R. P. Hellman and J. J. Zwislocki, Loudness determination at low sound frequencies. *Journal of the Acoustical Society of America,* 1968, **43**, 60–64. Courtesy of the authors and *The Journal of the Acoustical Society of America.*]

growth of loudness. It appears that once high enough intensities are reached, the loudness functions for all frequencies may have the same exponent. However, there is a suggestion (from results of Robinson & Dadson, 1956; see Stevens, 1972) that even at very high intensities the loudness functions for very low frequencies (less than about 100 Hz) may still have larger exponents than higher frequencies have.

The increase in the exponent of the loudness function at low frequencies, and the concomitant flattening at high loudnesses of the equal-loudness contours, is of some special interest in light of Zwislocki's (1965) account of the audibility curve that was described above. It was only at very low frequencies that the shape of the threshold audibility curve deviated from the prediction made on the basis of middle ear transmission plus temporal summation. Sensitivity to very low frequencies was lower than predicted. The change in exponent for low frequencies implies, however, that at high levels of loudness relative sensitivity to low frequencies is improved. Thus, at high loudness the deviation from prediction may be diminished or even disappear.

The difference between the growth of loudness at low and at higher frequencies may relate to fundamentally different underlying processes. One possibility is that the shift, which takes place near 400 Hz, reflects a transition from one mode of activity of the basilar membrane to another. A "place" principle might operate at the higher frequencies (each frequency producing a maximal displacement at a particular place on the basilar membrane), but a different (nonplace) principle at lower frequencies (the entire basilar membrane vibrating more or less in unison).

As an alternative, a place principle might operate over all frequencies. The extremely poor threshold sensitivity displayed at low sound frequencies may then be ascribed to decreased neural innervation of the basilar membrane (Zwislocki, 1965). One conclusion derived from the flattening of the equal-loudness contours is that decreased innervation results in diminished relative sensitivity only when the stimuli are weak. When the stimuli are strong, differences in degree of innervation seem to have less effect on sensory magnitude and, concomitantly, on sensitivity. We shall encounter other examples where low neural innervation corresponds to low sensitivity near threshold, but practically "normal" sensitivity at sensation levels well above threshold. Békésy (1955, 1958) has pointed out the possibility that this is a general principle of sensory functioning.

The relatively rapid growth of loudness with sound pressure at low frequencies has been termed *low-frequency recruitment*. Recruitment per se refers to a phenomenon encountered in certain types of deafness, especially nerve deafness. A recruiting ear will have an abnormally high threshold (i.e., a large hearing loss at low sound intensities), but a much smaller loss at high intensities. Binaural loudness matches between a normal and a recruiting ear suggest that the recruiting ear has a loudness function characterized by a larger exponent at low sound pressures and a normal, i.e., lower, exponent at higher sound pressures. Stevens (1966e) attempted to account for both deafness and low-frequency recruitment in terms of a transformation of the power function. Lochner and Burger (1962), on the other hand, attempted to account for low-frequency recruitment by means of a subtraction of constant loudness. Such a subtraction was proposed earlier by Steinberg and Gardner (1937) to account for recruitment in nerve deafness.

We may write the psychophysical loudness equation

$$L = kP^{2\theta} \tag{3.3}$$

where L is loudness, P is sound pressure, and θ is the exponent. (Since under most experimental conditions sound energy is proportional to the square of sound pressure, the exponent θ is equivalent to the exponent of the loudness function when the stimulus is measured in terms of sound energy.) The major source of variation in the loudness equation from one frequency to another is the value of k. This variation serves, in effect, to determine the shape of the audibility and equal-loudness curves at all but the very low frequencies. Over middle and high frequencies, the value of θ is one-third; over low frequencies (f) and not too high sound pressure levels, θ varies approximately according to the equation

$$\theta = .33 + .0009\,(400 - f) \tag{3.4}$$

where $f < 400$ Hz.

Olfactory Sensitivity

Sensitivity to Different Odorants

We turn now to the question of sensitivity in the chemical senses. It is a sometimes surprising fact that human olfactory sensitivity can be quite keen. For example, Moncrieff (1944) estimated that ethyl mercaptan (the odor of skunk) is detectable when there is only about 1 molecule of mercaptan per $5 \cdot 10^{10}$ molecules of air; furthermore, de Vries and Stuiver (1961) estimated that for the threshold of mercaptan to be reached, at most 8, and probably only 1, molecules need stimulate any single olfactory cell, and that probably 40 cells in all must be stimulated.

On the other hand, we are not nearly that sensitive to many other olfactory stimuli. Just as in vision and audition, we find that olfactory sensitivity varies as a function of some fundamental properties of the stimuli. But here lies a great difference: in vision and hearing, we know what the fundamental property is—it is the wavelength, or more properly the frequency of the light and sound. And though all details are not clear, yet even so, we have a fair understanding why the frequency domain is important and exerts the effect that it does. But the situation is quite different for the chemical senses. In spite of the fact that sensitivity (absolute threshold) can vary over a range of 10 billion or more to 1 from substance to substance, yet the bases for these variations remain elusive. [Davies and Taylor (1959) give the threshold concentration of skatol as $1.8 \cdot 10^9$, that of ethane as $1.3 \cdot 10^{19}$ molecules/cm^3 of air.] To sum it up succinctly, we still do not know what is the stimulus for olfaction.

What makes some molecules odorous, some less odorous, others not so at all? What makes some odorous molecules smell fruity, others putrid, others pungent? There are many theories, some of which go back to ancient Greece—e.g., Democritus's supposition that atomic shape is the variable crucial to taste and odor quality. Recent years have seen some suggestion of movement toward answers: Some of these suggestions arose, at least in part, from experiments involving the scaling of odor intensity. The earliest studies of olfactory intensity (e.g., Jones, 1958a & b; Reese & Stevens, 1960) showed that odor intensity grew as about the square root of odorant concentration. Jones (1958b) obtained magnitude estimates of the subjective intensity of benzene, heptane, and octane; he found that the data for all three substances could be accounted for by a single power equation with an exponent equal to .55. This outcome—constancy of exponent—implies that the relative sensitivity to these three substances is the same throughout their subjective ranges.

Jones (1958a) also measured responses to two other sets of three substances: n-butanol, pyridine, and cyclohexene, and iso-butanol, sec-butanol, and ethyl acetate. The latter three compounds gave almost identical psychophysical scales, with exponents of about .55; thus, they too gave nearly constant relative sensitivity. However, clear differences were manifest among the former three substances. The exponent for pyridine was .58, but for n-butanol and cyclohexene it was slightly over .4. Unfortunately, Jones reported his stimuli only in terms of the ratios of stimulus vapor pressure to vapor pressure at threshold; therefore, it is not possible to compare sensitivities to different stimuli in terms of molecular concentration. Nevertheless, it is clear that sensitivity to pyridine, as compared to sensitivity to n-butanol and cyclohexene, is relatively greater when the measurement is made at a high level of odor intensity than it is when the measurement is made at a low or threshold level. Changes in sensitivity are also implicit in results obtained by Drake, Johannson, von Sydow, and Døving (1969), who found a 2-to-1 range of exponents among 13 different odorants, and by Berglund, Berglund, Ekman, and Engen (1971), who found a 6-to-1 range for 28 odorants.

SENSORY PHYSICS AND PSYCHOPHYSICS FOR ALIPHATIC ALCOHOLS

Engen (1965) derived magnitude-estimation scales for propanol and octanol. These substances are two aliphatic alcohols, which contain three and eight carbon atoms, respectively, in their carbon chains. The two odorants produced quite dissimilar psychophysical functions re concentration: propanol gave an exponent of .42, octanol an exponent of .14. Thus, propanol grows three times as rapidly with increases in concentration. At 100% concentrations, propanol is subjectively stronger, but as the two odorants are diluted in equal proportion, the subjective intensity of propanol diminishes relatively rapidly, whereas that of octanol changes very slowly. On the basis of these results among others, Engen suggested that the length of the carbon chain of homologous alcohols is an important variable in determining sensitivity.

Engen's suggestion was taken up by Cain (1969), who obtained magnitude estimates of odor intensity for four alcohols—propanol, butanol, hexanol, and octanol—each in a separate experiment. These substances have 3, 4, 6, and 8 carbon atoms in their chains. Cain found that as carbon-chain length increased, the exponent of the power function for odor intensity decreased. Propanol yielded a high exponent of .38, octanol a low of .15. Cain also obtained matches of subjective intensity between pairs of odorants, and the matching results substantiated these differences in rates of growth. Henion (1971c) extended the study to alcohols with as few as 2 (ethanol) and as many as 10 (decanol) carbon atoms, and he found a similar trend: Longer chains of carbon atoms yielded lower exponents.

The olfactory intensity of undiluted alcohols was scaled by Henion (1970a). As the number of carbon atoms increased, odor intensity first increased, then decreased. The maximum in subjective intensity occurred for butanol (four carbon atoms). This result is in general agreement with that of Kruger, Feldzamen, and Miles (1955), who obtained matches for olfactory intensity among different alcohols. Engen, Cain, and Rovee

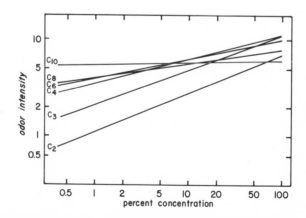

Figure 3.7. Psychophysical functions showing how the odor intensity of seven aliphatic alcohols increases with concentration of the odorant. The longer the chain of carbon atoms, the lower the exponent of the power function.

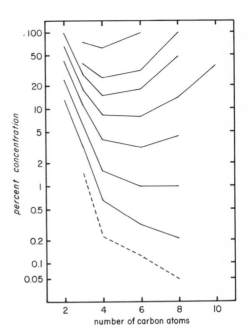

Figure 3.8. Contours of equal odor intensity. The curves show how odorant concentration and carbon-chain length of aliphatic alcohols must co-vary to maintain constant odor intensity. The curves were derived from the psychophysical functions in Figure 3.7. The dashed line gives threshold sensitivities measured by Cain (1969). As criterial level of odor intensity increases, maximal sensitivity shifts from alcohols with long to shorter chains of carbon atoms.

(1968) reported that odor intensity decreased monotonically as chain length increased. On the basis of the results obtained by Engen, Engen *et al.*, Cain, and Henion, we can construct two sets of functions. The first (Figure 3.7) shows how odor intensity grows with concentration for aliphatic alcohols of different carbon-chain lengths. Relative sensitivities to the odorants are given in the second set of functions (Figure 3.8). At low concentrations, sensitivity to the longer chains is relatively high—the dashed line shows relative threshold measurements made by Cain (1969). At threshold, concentration is inversely related to chain length. At higher and higher levels of subjective intensity, the relation of sensitivity to chain length undergoes systematic variation, variation that reflects the differences in the exponents of the psychophysical functions. At intermediate levels, there are relatively small differences in sensitivity among the alcohols, and sensitivities to particular pairs of alcohols become identical. Finally, at high concentrations, sensitivity is no longer monotonically related to chain length; sensitivity is greatest to alcohols with intermediate numbers of carbon atoms.

What remains important to discover is how these changes in sensitivity might relate to actual stimulation of olfactory receptors. There is no guarantee that any given increase in odorant concentration leads to a proportional increase in stimulation of olfactory receptors. Recall that in vision and hearing, the transmission of light and sound energies to receptors—through the ocular media and through the middle ear—are fundamentally linear processes: the proportion of light quanta that reaches photoreceptor cells is the same regardless of the intensity of the external light. But linearity of transmission in olfaction cannot be taken for granted. Differences in sensitivity, and even differences in the rate of growth of odor intensity with concentration, might be the outcome of nonlinear transmission of odorous molecules to the olfactory receptors. Even if such nonlinearities are not responsible for the differences, should they exist they should require modification of the psychophysical relations if the stimulus were measured in terms of number of molecules absorbed by primary receptors.

Cain (1969) enumerated a number of physicochemical variables that correlate with carbon-chain length and that might be responsible for differences in sensitivity and for changes in the differences. These variables include molecular size, boiling point, vapor pressure, and water solubility. Cain selected the last, water solubility, as the most likely candidate for relevance, in part because differences in exponent, albeit small, that Jones (1958a) reported (as described above) also correlated well with water solubility. An attempt has been made by Laffort (1970) and by Dravnieks and Laffort (1970) to relate the relative sensitivity to many odorants to their physicochemical properties. Water solubility can be only one of a number of relevant physical variables, as Cain (1971) pointed out. Berglund et al. (1971) obtained psychophysical functions for 28 odorants by the method of magnitude estimation, and they found a 6-to-1 range in the size of the average exponent. However, water solubility could account for only a small portion of the variation (correlation coefficient between exponent and water solubility equaled .27). Clearly, other physicochemical properties are important.

A word should be said about the actual sizes of exponents obtained in olfactory scaling experiments. In a basic respect, the absolute sizes are at present of much less interest than relative sizes. This is so because the absolute size depends strongly on the method of stimulus presentation. Cain (1969) showed that sniffing odorants from cotton swabs that were diluted in liquid (a common diluent is the odorless diethyl phthalate) produced exponents that are only about one-half as large as those produced by sniffing odorants diluted in air (olfactometric presentation). Although one reason for the differences might be variation in degree of regression or response construction (see Chapter 1), we cannot eliminate the possibility that the differences relate to different efficiencies in getting odorous molecules to receptors. Nevertheless, the *relative* sizes of exponents do not seem to depend on method of stimulus presentation. This implies that relative sensitivities to different odorants are independent of the method of stimulus presentation.

Taste Sensitivity

Many of the generalizations concerning olfaction pertain also to taste. For taste, too, there is a paucity of knowledge concerning the nature of the stimulus. There are two major questions we may ask. First, what are the requisites for a stimulus to be sapid? Second, what are the requisites for a stimulus to have a particular taste quality? Obviously, these two questions and their answers may not be independent. The most definite relation known at present is that between the presence of hydrogen ions and the sour taste; even so, as we shall see, hydrogen ion concentration, per se, fails to predict quantitatively the magnitude of the sour taste. Other correlates exist. Many, but by no means all, organic compounds are bitter; several anions and cations, such as sodium (Na^+) and chloride (Cl^-), seem related to the salty taste; most sugars are, to a greater or lesser degree, sweet. But also certain salts, e.g., some lead salts, are sweet, and certain sugars have bitter perceptual components. Understanding the stimulus for taste sensation is a difficult task; to some extent, at least, a knowledge of gustatory sensitivity, both at threshold and suprathreshold levels, should help eventually to clarify our understanding.

PSYCHOPHYSICS FOR BASIC TASTE SUBSTANCES

Some of the early attempts to measure how the taste intensity of various substances depends on concentration used the method of fractionation. Lewis (1948) examined the tastes of sodium chloride, sucrose, quinine sulfate, and tartaric acid. Over most of their ranges of concentration, these compounds produce pure salty, sweet, bitter, and sour tastes, respectively. Lewis asked his subjects to find concentrations of all four substances that were subjectively half as strong as standard solutions. For all four substances, the reductions in concentration amounted to decreases of about one-third of a logarithmic unit, i.e., to about 50% of the standard concentrations. Thus, Lewis's data suggested that taste intensity for these compounds are nearly linear functions of concentration. Beebe-Center and Waddell (1948) had subjects match the taste intensity of quinine sulfate, tartaric acid, and sodium chloride to that of .56% sucrose. They combined their data with those of Lewis to produce a general scale of taste intensity—the *gust scale*. One gust was defined as the subjective intensity of 1 gm sucrose/100 cc water (equivalently, to a .03M solution of sucrose). Their functions have been calculated in terms of molar concentrations and are shown in Figure 3.9. (For the present, ignore the plotted data points.) The functions for quinine sulfate and tartaric acid are nearly straight in double-logarithmic coordinates, thereby suggesting power functions; the functions for sodium chloride and for sucrose are slightly curvilinear. More important, the figure shows how relative sensitivity to these substances depends on subjective level. Sensitivity is greatest throughout the range to quinine, and it is poorest to sucrose.

Shown also in Figure 3.9 for comparison are results (data points) obtained by Meiselman (1968a), using the method of magnitude estimation. He scaled the taste intensity of sodium chloride, hydrochloric acid, sucrose, and quinine sulfate relative to a standard solution of .36 M sodium chloride. Thus, three of the substances are the same ones used by Lewis and by Beebe-Center and Waddell. The results show a general similarity to the scales of Beebe-Center and Waddell, particularly with respect to sensitivity to the different compounds. The psychophysical functions differ somewhat, however, in that Meiselman's function for sodium chloride is steeper, that for sucrose shallower, than those of Beebe-Center and Waddell. These results are also consistent with matches obtained by Bujas (1937) for the subjective intensity of sodium chloride and sucrose. There is clear evidence in Meiselman's data for differences

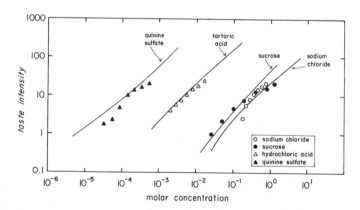

Figure 3.9. Psychophysical functions for the taste intensities of quinine, acid, sucrose, and sodium chloride. The lines show how taste magnitude increases with molar concentration of each compound. [After Beebe-Center and Waddell (1948). By permission of *The Journal Press*. The data points are average magnitude estimates of taste intensity obtained by Meiselman (1968a).]

in rates of growth of different substances; subjective intensity of sodium chloride grew more rapidly than did that of hydrochloric acid and sucrose, and slightly more rapidly than that of quinine. For example, at high taste intensities, sucrose and sodium chloride tasted equally strong when the molar concentration of sucrose was twice that of sodium chloride; but at low taste intensities, they tasted equally strong when the concentration of sodium chloride was three times as great as that of sucrose. Near .4 M both solutions tasted about equally strong.

PSYCHOPHYSICS OF SWEET SUBSTANCES

The fractionation method was also used by MacLeod (1952) to compare sweetness of glucose and sucrose. The results suggested some difference in the way the subjective intensities of these sugars increase with concentration: glucose appeared to grow more rapidly. The most extensive comparisons of sensitivity to sweet substances are those of Moskowitz, who compared psychophysical functions for 16 (Moskowitz, 1970a) and for 43 (Moskowitz, 1971c) sugars. In the two series of studies (10 and 8 experiments), subjects gave magnitude estimates of the sweetness of various subsets of the 16 and 43 sugars. Moskowitz concluded that the rate that sweetness grows with concentration is about the same for all sugars—a power function with an exponent about 1.3. (This is the same value as that reported by Stevens, 1969b, for glucose and sucrose and is a little smaller than the 1.5 reported by Ekman and Åkesson, 1965, for sucrose.) If this constancy of exponent holds up, then *relative* sensitivity to all sugars will be the same regardless of the criterion level of subjective sweetness. If we write

$$S = kC^{1.3} \tag{3.5}$$

where S is sweetness and C is sugar concentration, then the variations in sensitivity among sugars are subsumed under variations in the size of the constant k.

Of all the sugars Moskowitz (1970a) examined, sucrose turned out to be the sweetest (i.e., gave the greatest sensitivity) and fructose the next sweetest. However, the order of sensitivity of the remaining 14 sugars depended on the measure of concentration. When the measure was molarity, raffinose, maltose, and lactose followed in sweetness, and xylose was least sweet. When the measure was percentage concentration (grams solute per unit volume of solution), however, raffinose, maltose, and lactose were among the least sweet. (The change is due to the differences in molecular weight of different sugars. The heavier the sugar, the smaller the number of molecules—consequently the lower the molar concentration —given a constant mass, and therefore a constant percentage concentration.)

There does not seem to be any simple relation between stimulus attributes of various sugars and their gustatory sensitivity. Moskowitz (1971c) pointed to eight prominent features of molecular structure of sugars that appear important. One complication of any scheme is that it is not completely clear that the exponent of the sweetness function is exactly the same for all sugars. Mannose, for example, gave psychophysical functions that could not be described by power functions (and in fact, they were not wholly monotonically increasing). Raffinose, as another example, had its sweetness scaled in three experiments (Moskowitz, 1970a) and gave exponents less than 1.3 all three times; the average was .63. Given that sweetness grows as a power function of sugar concentration, then relative sensitivity to different sugars is level independent only if the exponent is constant. As was noted earlier, however, MacLeod (1952) found the sweetness of glucose to grow more rapidly with concentration than did the sweetness of sucrose. Taste matches obtained by Cameron (1947) suggest the same conclusion. Cameron found evidence that the sweetness of fructose, lactose, and galactose also grow more rapidly than does sweetness of sucrose. These results imply that relative sensitivity to sugars may be somewhat level dependent.

Although most of the exponents obtained by Moskowitz (other than those for mannose and raffinose) fell close to 1.3, it remains possible, if not likely, that small differences do exist, thereby implying that relative sensitivity among sugars does depend on the criterial level of sweetness. A possible reason for some of the differences in exponent comes about from the complex taste of many sugars. As concentration increases, other taste qualities in addition to sweet (such as bitter) also increase in some sugars. The bitterness could suppress (mask) sweetness and might act thereby to depress the size of the exponent for sweetness.

Sweetness functions for substances other than sugars definitely display different psychophysical functions (Moskowitz, 1970b). Cyclamate salts grow less rapidly in subjective strength than does sucrose, and saccharin grows less rapidly than cyclamate. Approximate exponents found were 1.5 for sucrose, .9 for cyclamate, and .45 for saccharin. At low levels of sweetness, sensitivity to saccharin is extremely high, and a $.00001\ M$ concentration produces the same sweetness as $.06\ M$ sucrose. But $.005\ M$ saccharin and $.6\ M$ sucrose also produce the same (but higher) level of sweetness. A possible reason for the differences is that cyclamate and especially saccharin have strong bitter components to their tastes at high concentrations. Interactions between sweetness and bitterness, if nonlinear, could serve to change the overall relation between taste intensity and concentration.

PSYCHOPHYSICS OF SOUR SUBSTANCES

Turning now to sourness, Moskowitz (1971b), in four experiments, scaled the sourness of 24 acids by the method of magnitude estimation. Sourness grew as a power function of concentration with an average exponent equal to about .85, when measured both in terms of molar and percentage concentration. As was true for sugar sweetness, different substances gave somewhat different exponents, and it is difficult to know for certain whether observed differences are reliable. When the measure of log concentration was pH (reflecting number of hydrogen ions), the data could also be described in terms of power functions, but with exponents equal to −1.70, on the average. With pH as the measure, acids of low molecular weight tended to be the more sour. But no relation to molecular weight occurred when either molar or percentage concentration was used. When molarity (but not percentage concentration) was used, addition of extra chemical groups to a simple acid increased its sourness. With pH, the same addition decreased sourness. The position of COOH group could make as much as a sixfold change in sourness, except when pH was used.

For taste as for smell, the method of stimulus presentation can affect such aspects of results as the exponent of the psychophysical function. All of the results described thus far employed what has been called the "sip and spit" method. The expression is self-explanatory. Some investigators have used a gustometric method that flows solutions over the tongue. These two procedures bear some superficial parallel to liquid-dilution and air-dilution olfactometry. In the case of taste, the flow technique appears to give a much lower exponent. McBurney (1966) reported an exponent of about .4 for sodium chloride, in comparison to values of 1.6 reported by Ekman and Åkesson (1965), 1.0 by Jones and Marcus (1961), and 1.4 by Stevens (1969b), all using "sip and spit." Feallock (1965) obtained gustometric exponents around .3 for quinine hydrochloride, whereas Stevens (1969b) obtained 1.0. Recent comparisons of the two procedures by Meiselman (1971), who examined exponents of psychophysical functions obtained for sodium chloride, quinine, sucrose, saccharin, citric acid, and tartaric acid, showed clear-cut effects of procedure. Different degrees of regression or response constriction may account for much of the variation (Meiselman, Bose, & Nykvist, 1972).

Tactile Sensitivity

Last for consideration is relative sensitivity of the sense of touch. The responsiveness of the tactile sense to vibratory stimulation is far from uniform, but varies markedly with the frequency of stimulation. The dependence of absolute sensitivity on frequency was demonstrated by several investigators (Goff, 1967; Sherrick, 1953, 1960; Verrillo, 1962, 1963), who showed the typical result, namely that sensitivity is maximal in the frequency region around 250 Hz, with lower absolute sensitivity at higher and at lower frequencies. However, variation in sensitivity depends strongly on the areal extent of the stimulation. When a small enough vibrator is used, the vibratory threshold does *not* vary with frequency (Verrillo, 1963, 1966a & b).

Sensitivity curves for different areas are shown in Figure 3.10 (Verrillo, 1963). Below 250 Hz, sensitivity decreases with frequency for large stimuli; the increase in threshold amounts to 12 dB for every doubling of frequency. Above 250 Hz, sensitivity decreases at a rate of 9 dB per doubling. For small enough stimuli (on the order of 1 mm²), the curve is flat: sensitivity hardly varies with stimulus frequency over the range 25–700 Hz.

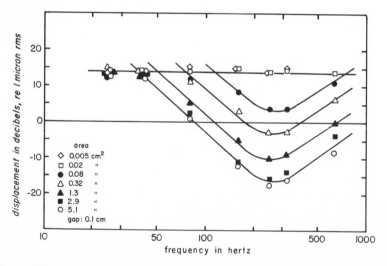

Figure 3.10. Relative tactile sensitivity (absolute threshold) to different vibration frequencies. When the stimulator has an area 1 mm² or greater, sensitivity is greatest near 250 Hz. When the stimulator is smaller, sensitivity hardly varies with frequency. [From R. T. Verrillo, Effect of contactor area on the vibrotactile threshold. *Journal of the Acoustical Society of America*, 1963, 35, 1962–1966. Courtesy of the author and *The Journal of the Acoustical Society of America*.]

What about suprathreshold sensitivity? Goff (1967) obtained two contours of constant apparent vibration magnitude on the finger, covering a frequency range 25–800 Hz. One contour was measured at 20 dB, the other at 35 dB above the measured vibratory thresholds. Goff found the sensitivities above threshold similar to measured threshold sensitivity: maximal sensitivity occurred around 200 Hz both at and above threshold. But the results also suggested some systematic variation in the shape of the sensitivity curve with increasing sensation level. Hahn (1960) had previously noted that Goff's equal-sensation functions suggested some change in sensitivity, and, concomitantly, different rates of growth relating apparent vibratory intensity to stimulus amplitude at different frequencies.

Three studies have examined how the relation between vibration magnitude and stimulus amplitude depends on frequency; their results also show how vibration sensitivity varies with sensation level (Franzén, 1969a; Stevens, 1968; Verrillo, Fraoli, & Smith, 1969). Franzén obtained magnitude estimates for apparent vibration magnitude of 50 mm² stimuli applied to the finger and to the toe. Frequencies were 50–250 Hz. Franzén found a regular change in the exponent of the power function for vibration: at 50 Hz the exponent was near unity, but it declined to about .6 at 250 Hz. At low frequencies, the apparent magnitude of vibration is approximately proportional to the amplitude of the vibratory stimulus, but at higher frequencies the relation is curvilinear. Thus, between the two frequencies 50 and 250 Hz, the sensitivity function flattened more and more as sensation level increased. This flattening, and the concomitant variation in exponent, is reminiscent of the way auditory sensitivity varies with frequency.

Franzén's data are in good agreement with direct matches obtained by S. Ross and reported by Stevens (1968). Ross used a 28 mm² stimulator applied to the finger tip, and he studied frequencies in the range 20–320 Hz. The curves of constant sensation in Figure 3.11 show how sensitivity varied with frequency at eight levels of apparent vibration magnitude. The flattening of the contours as level increases is a reflection of the change in the exponent of the psychophysical function for vibration.

A larger range of frequencies (25–700 Hz) was explored by Verrillo et al. (1969). They used the method of numerical magnitude balance—a combination of magnitude estimation and magnitude production that was developed by Hellman and Zwislocki (1963)—to scale perceived vibration magnitude of a 290 mm² vibrator presented to the thenar eminence of the hand. The results differed from those of Goff, Franzén, and Ross, in that the psychophysical functions at different frequencies had almost identical exponents: the exponent equaled .9 over the range 25–350 Hz,

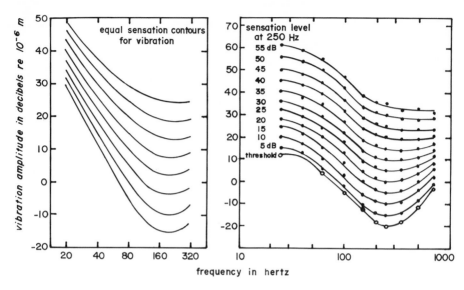

Figure 3.11. Contours of equal vibratory sensation, showing how stimulus energy and frequency must covary to maintain constant sensory magnitude. Left: contours derived from direct matches, using a 28 mm² stimulator applied to the finger. [Data of S. Ross, reported by Stevens (1968). Courtesy of the author and *Perception & Psychophysics*.] Right: contours derived from numerical magnitude balance, using a 290 mm² stimulator applied to the thenar eminence of the hand. [From Verrillo, Fraoli, and Smith (1969). Courtesy of the authors and *Perception & Psychophysics*.]

then rose to .95 at 500 Hz, and 1.2 at 700 Hz. Because of the relative independence between the exponent and vibratory frequency, the results of Verrillo *et al.* imply that vibration sensitivity changes little when sensation level changes. Their contours of constant sensation, shown in Figure 3.11 for comparison to those of Ross, show that invariance of sensitivity with sensation level. The shapes of these contours were verified by results of direct matches of apparent vibration.

What are the reasons for the differences among results? Verrillo *et al.* suggest that the differences reflect important variations in procedure. For example, one variation was the locus of stimulation: thenar eminence versus fingertip or toe. Perhaps more important was the degree of spatial spread of the induced vibratory patterns. Verrillo *et al.* used a rigid surround to prevent the spread of traveling waves. Goff (1967) used a 6.5 mm diameter stimulator centered in a 19 mm diameter hole, Franzén (1969a) used a 8 mm stimulator through a 12 mm hole, and Ross apparently used no surround. Verrillo and Chamberlain (1972) found that introduction of a surround increases both threshold and the rate of growth of vibration (slope of the psychophysical function). Both effects appear to be due to the reduced number of receptors stimulated when a surround limits the spread of waves in the skin. Although it is not difficult to hypothesize why degree of spread should affect the way sensory magnitude grows with amplitude (e.g., by affecting degree of lateral summation and/or inhibition), it remains mysterious why the magnitude of the effect should vary with vibration frequency.

Summary

Relative sensitivity (sensory physics) is intimately related to sensory dynamics (psychophysics). Give a general psychophysical equation $\psi = \Gamma\ (\phi)$, the question whether relative sensitivity is level dependent is equivalent to the question how the function Γ depends on stimulus composition (wavelength, frequency, chemical composition, etc.). If Γ is a simple power function, sensitivity is level dependent only if the function's exponent varies with composition.

In vision, relative spectral sensitivity appears to be independent of level both when stimuli are viewed scotopically (low brightness, peripheral vision) and when they are viewed photopically (foveal vision). The relation of brightness to energy is a power function whose exponent is independent of wavelength. However, when the entire dynamic range of extrafoveal and peripheral vision is considered, sensitivity is level dependent: scotopic at low brightness (maximal sensitivity at 505 nm), photopic at high (maximum at 555 nm).

Spectral sensitivity in audition is markedly level dependent. Relative sensitivity to low sound frequencies (lower than 400 Hz) improves as criterial level of loudness increases. This variation in sensitivity reflects the fact that the loudness of low frequency sounds grows more rapidly with increasing sound intensity than does the loudness of high frequency sounds.

The chemical senses display strong level-dependent variations in relative sensitivity. In olfaction, the aliphatic alcohols, for example, yield a systematic change in sensitivity: when odor intensity increases, relative sensitivity to alcohols with long carbon chains decreases. This change in sensitivity correlates with the systematic decrease in the exponent of the psychophysical function with increase in chain length. In taste, relative sensitivity to sucrose, sodium chloride, tartaric acid, and quinine sulfate (substances that produce the four basic taste qualities) depends on level. For instance, relative sensitivity (inverse of molar concentration) to quinine is much greater than sensitivity to sodium chloride, but the difference is not so large at high, as at low, levels of taste intensity. Within a class of compounds that produce primarily one quality (e.g., sugars that produce sweet tastes), level-dependent variations seem less prominent.

Sensitivity of the somesthetic sense to vibratory stimulation depends on vibratory frequency. When the stimulator is surrounded by a rigid annulus (to prevent spread of waves in the skin), relative sensitivity is independent of level. When no surround is present, however, sensitivity does vary with level of sensation: relative sensitivity to low frequencies increases as sensation level increases. Under this condition of no rigid surround, the rules governing vibratory sensation (sensory physics and psychophysics) bear marked similarities to rules governing auditory sensation.

Sensitivity in Relation to Responses to Stimulus Mixtures

Related to the question of how sensitivity varies in each of the sensory systems is the question of how perceived intensity depends on the components of stimulus mixtures. What, for example, is the brightness of a light compounded of different monochromatic bands? What is the loudness of a mixture of pure tones? What are the perceived intensities of mixtures of gustatory and of olfactory stimuli?

It has been traditional to attempt to divide sensory modalities into the classes of *synthetic* and *analytic*. These categories purport to describe how the senses deal with mixtures of stimuli. Vision is often given as a prototypical example of a synthetic sense: combinations of wavelengths, e.g., red and yellow, produce a resultant orange, which cannot be decomposed subjectively into its constituents. Although the orange appears more "similar" to red and to yellow than to green or to blue, nevertheless red and yellow are not separable components of the perceived orange. Audition, on the other hand, is typically considered to be analytic; we have a rather good capacity to "hear out" the individual pure tones in a mixture (although there are some severe limitations to this capacity). As we shall see, however, this binary system of classification has not always easily been applied to the other senses, e.g., olfaction. Even the auditory system behaves in some circumstances like a synthetic sense. It may be preferable, if the terminology is to be retained, to refer not to the senses as being analytic or synthetic, but to particular modes of response under particular conditions for each sense being more or less analytic or synthetic.

A related issue concerns how the perceived magnitude of a compound stimulus relates to its components. One possibility is that the subjective magnitudes of the components add. Consider a mixture that contains two components, one of which, when presented alone, yields sensory magnitude ψ, the other 2ψ. From the assumption that subjective magnitudes add, we predict the mixture to yield 3ψ. Another possibility is that sensitivities to the components add. For example, the physical intensities of the components might be ϕ and 4ϕ (if ψ equals the square root of ϕ). From the assumption that sensitivities add, we predict the mixture to yield 2.24ψ (i.e., $\sqrt{5}\psi$).

A simple hypothetical rule might be that subjective magnitudes add when a sensory system behaves analytically, whereas sensitivities add when it behaves synthetically. However, these two alternative types of addition are not exhaustive and, as shall be seen, cannot describe responses to stimulus mixtures on all sensory modalities. Even where they do apply, the simple rule just described does not apply universally.

Vision

The brightness responses to complex lights, to lights that contain substantial contributions of energies at more than one wavelength, depend fundamentally on visual spectral sensitivity. If we mix a monochromatic red light, say, of 640 nm with a monochromatic yellow (580 nm) of about the same brightness, the combined light will appear an orange that is brighter than either of the constituents. The original components (red and yellow) are transformed in the mixture: hence, the reference to vision as a synthetic sense. Colors blend, and only a total brightness is seen, not individual brightnesses of the components. (The other side of this coin is the fact that a given percept, for example an orange of particular saturation and brightness, may be the outcome of any of a number of different combinations of spectral lights; different light combinations that produce identical percepts are called *metamers*.)

When lights are added, how do their brightnesses combine? To a good first approximation, it is sensitivities to the components that add. In the case of the mixture of yellow and red that were equally bright, the synthesized orange will have a brightness approximately equal to the brightness of either original component at twice the original intensity. We may write the psychophysical brightness equation for monochromatic light of wavelength λ

$$B_\lambda = (k_\lambda E_\lambda)^{1/3} \qquad (3.6)$$

where E_λ is the energy at λ and B_λ is its brightness. Then the total brightness B_t of light composed of any combination of energies across the spectrum is

$$B_t = [\Sigma(k_\lambda E_\lambda)]^{1/3} \qquad (3.7)$$

The coefficients k_λ define the *luminosity* of the visual system, i.e., the spectral sensitivities.

Equation (3.7) gives brightness as a function of the summation of energies weighted by their coefficients, and it thereby reflects Abney's law. Abney's law was formulated originally in the nineteenth century, and it states that perfect additivity obtains for visual responses to energies that are weighted in accordance with luminosity coefficients. The quantity $k_\lambda E_\lambda$ defines the *luminance* of the light. If spectral sensitivity at wavelength a is twice that at b $(k_a = 2k_b)$, then the luminance of lights of those wavelengths will be the same when $E_b = 2E_a$ and $k_a E_a = k_b E_b$ Adding the two lights $(k_a E_a + k_b E_b)$ produces, according to Eq. (3.7), a light of luminance twice either $k_a E_a$ or $k_b E_b$.

Is Abney's law true? In the case of scotopic vision, which is mediated by a single type of receptor (rods), there is theoretical reason to believe that Abney's law holds perfectly. In the case of photopic vision, however, which is mediated by at least three types of cones, there is less reason to expect the perfect additivity of Abney's law. If we consider brightness to be the criterial response, it appears that Abney's law does not quite hold for photopic vision. With some combinations of lights, regular but usually small deviations from perfect additivity have been shown (Guth *et al.* 1969; MacAdam, 1950; Piéron, 1939; Sperling, 1958). The largest deviation seems to occur when the components of the mixture are complements (lights that when added in some ratio of energies will produce a hueless gray). It even can happen that addition of a second light can make invisible a light that previously was visible, i.e., can produce inhibition: the threshold for lights 475–550 nm was increased when a subthreshold 685 nm light was added (Guth *et al.*, 1969). Failure of Abney's law seems most marked when red and green lights are mixed.

The failure of additivity can be seen in results of Padgham (1971), who scaled the brightness of colored lights (test field 5°) that varied in their degree of spectral purity. Padgham was able to measure the extent to which relatively pure spectral lights appear brighter than lights composed of mixtures of wavelengths (more nearly colorless), when all lights had the same luminance. (Luminance was defined by a system that assumed perfect additivity.) In comparison to a "white" light, the highly colored lights were judged to be about 40% brighter. The lower brightness of desaturated (e.g., "white") lights is known as the Helmholtz-Kohlrausch effect. That result reflects the breakdown of Abney's law: Combinations of wavelengths (white) yield brightnesses not quite as great as the perfect additivity of Abney's law would predict.

It is interesting to note that additivity does seem to hold when criteria other than brightness are used. When the criterion is minimal flicker (Piéron, 1939; Sperling, 1958) or contrast detection (Boynton & Kaiser, 1968) Abney's law appears to be valid. Guth *et al.* (1969) argued that brightness perception probably involves visual mechanisms that include wavelength-related opponent processes, and that these opponent processes can produce cancellation, a reduction in the effectiveness of combined spectral lights. In the case of temporally modulated light, the opponent processes are less able than nonopponent processes to follow rapid changes in luminance. Thus only nonopponent processes normally mediate flicker perception, no cancellation takes place, and Abney's law holds.

Of related interest is the fact that the luminosity function defined by the CIE was based to a large extent on results obtained by flicker photometry; the methods for calculating luminance are based on the assumption that Abney's law is valid. For the evaluation of brightness, however, even in the simplest of stimulating conditions (e.g., no surround to produce brightness contrast) the CIE formulation can provide at best only an approximation.

AUDITION

Let us turn now to the question of what happens when two or more tones are presented to the ear. The situation is even more complex than that of presenting light of two or more wavelengths to the eye. The visual stimuli blend (synthesize), but the auditory stimuli often remain phenomenally separable. Under such circumstances audition tends to be *analytic;* the sounds of the individual components can be heard out.

The auditory analysis of sounds into components that relate to individual sound frequencies exemplifies Ohm's acoustical law. This law states that the auditory system performs something like a Fourier analysis (Fourier's theorem states that any complex, periodic waveform can be analyzed into a series of sinusoidal components of frequencies f, $2f$, $3f$, . . .). The analysis is not perfect, however. If two tones are very close together in frequency, the individual tones are not heard. Instead, we hear a single sound that waxes and wanes in loudness. Tones of 1000 and 1003 Hz produce a sound that increases and decreases in loudness three times per second. (The number of cycles is proportional to the difference in frequency.) These increases and decreases are called beats. If the tones are farther apart in frequency, but one is much more intense than the other, the less intense sound may not be heard at all, even though it is perfectly audible when presented alone. This decrease in loudness is termed masking. Both masking and beats provide examples of situations where the auditory system is not analytic. (Another example is noise, which produces a unitary, synthetic perception in which the multiplicity of components cannot be distinguished.)

For all but the lowest of frequencies, the effect of a pure tone is to produce a wave that travels along the basilar membrane, and the traveling wave has a maximal displacement on a particular portion of the membrane; the locus of maximal stimulation depends on the stimulus frequency (Békésy, 1949a & b). In other words, there is a mapping of the frequency domain in the sound stimulus onto a spatial dimension on the basilar membrane. When two or more tones are played simultaneously, and their frequencies are not too far apart, the overall sensitivity to sound energy equals the sum of the individual sensitivities. Put another way, the energies sum (Zwicker, Flottorp, & Stevens, 1957; Zwislocki, 1965). There are frequency limits to this summation, however: perfect summation of energy occurs only for a limited range of frequency. Centered around 1000 Hz, the range covers about 160 Hz. The range of frequencies around any point and over which energy sums is called a *critical band* (Zwicker *et al.*, 1957). It is curious that when tones fall within a critical band (are relatively close in frequency), the components may still be "analyzed," yet the energies sum, as they do for the "synthetic" sense of vision. When tones are well separated with respect to frequency, it is not energy but loudness that sums (Fletcher, 1940; Howes, 1950). Thus, for sound energies close together in frequency total loudness L_t is given by

$$L_t = k(\Sigma P^2{}_f)^\theta \tag{3.8}$$

where P_f is sound pressure at frequency f. But for sound energies far apart in frequency

$$L_t = \Sigma L_f \tag{3.9}$$

In both cases, the auditory system behaves analytically (components may be perceptually separable), yet two very different types of summation occur.

It is possible to look at energy summation within a critical band in terms of masking. That is to say, any component will appear less loud in a mixture than it would appear if it were sounded alone. This follows from the fact that the size of θ in Eq. (3.8) is much less than 1. Given two components of identical sound pressure, and equal loudness when sounded alone, the loudness of their mixture will be less than the sum of the isolated loudnesses. The questions of energy summation, loudness summation, and masking, and their relations to critical bands are taken up in greater detail in Chapter 5, which deals with spatial parameters that determine sensory intensity.

The treatment of loudness of complex sounds solely in terms of summation of energy and summation of loudness is an oversimplification. Noises, for example, fail to follow the psychophysical power equation exactly; like pure tones around 1000 Hz, noises give curvilinear functions between loudness and sound pressure in log–log coordinates (Stevens, 1972). In this regard, noises appear to show even greater deviations from adherence to a power law than do tones. The change in the form of the psychophysical function—from power to nonpower form—cannot be explained simply in terms of summation of energy or of loudness.

Even more striking are loudness functions obtained when the sound stimuli are very complex, but meaningful. Speech sounds, for instance, obey the psychophysical power equation, but with exponent θ equal about to .6, when stimulus magnitude is measured in terms of energy flow (Ladefoged & McKinney, 1963; Mendel, Sussman, Merson, Naeser, & Minifie, 1969). That value is nearly twice the size of the exponent for pure tones. This difference in exponent may be related to the higher exponent obtained for judgments of self-produced (autophonic) loudness (Lane, Catania, & Stevens, 1961). When subjects produce sounds vocally and then estimate their loudness, the size of θ also approximates .6.

When more than one tone is sounded, the resulting auditory perception can be rich and complex in comparison to that for single tones. Most of the sounds that we encounter daily are of such a complex sort: noises, speech sounds, music, etc. It is only complex sounds that enjoy the musical attributes of timbre and consonance. Plomp and Levelt (1965) obtained some fascinating results with ratings of the tonal consonance of pairs of tones. They found that consonance is maximal under two conditions—when the frequencies are the same (i.e., a single frequency is sounded) and when they are very far apart. In fact, the degree of consonance depended on how the difference in frequency between two tones related to the critical band: when both tones were separated by less than a critical band, the tone pair appeared relatively dissonant; when separated by more than a critical band, they appeared relatively consonant. Maximal dissonance appeared at frequency separations approximately one-fourth the width of a critical band. The prototypical relation between consonance and frequency appears in Figure 3.12. The ratings of consonance even permitted Plomp and Levelt to make rough estimates of how the size of the critical band depends on frequency.

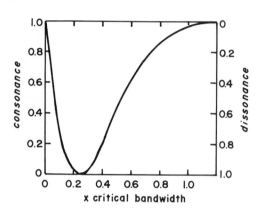

Figure 3.12. Prototypical relation between tonal consonance and the frequency separation between members of a tonal pair. Consonance is maximal when the frequencies are either identical or far apart. Consonance is minimal (dissonance is maximal) when the tones are separated by one-fourth of a critical bandwidth. [From R. Plomp and W. J. M. Levelt, Tonal consonance and critical bandwidth. *Journal of the Acoustical Society of America,* 1965, **38,** 548–560. Courtesy of the authors and *The Journal of the Acoustical Society of America.*]

This research on consonance was extended by Kameoka and Kuriyagawa (1969a & b), who introduced the term dissonance intensity to represent an assumed physical dimension to underlie the perception. An additional assumption was that the perceived degree of dissonance grows as some power of dissonance intensity. They incorporated these assumptions into a model that they applied to ratings of consonance (actually, to ratings of differences in consonance between pairs of sounds). Their results suggest that consonance and dissonance may not be so simply related to critical bandwidth as Plomp and Levelt reported.

OLFACTION

A number of studies have examined how the magnitude of an odorous substance presented together with a second substance varies with concentration. Jones and Woskow (1964) obtained magnitude estimates of the perceived intensity of six pairs of odorants (all possible pairs from benzene, zylene, and toluene, and from cyclohexene, *sec*-butanol, and amyl acetate). Power functions were fitted to results both for pairs of odorants and for single odorants. The subjective intensities of mixtures were generally smaller than the sum of the individual intensities, but greater than their average. Jones and Woskow concluded, apparently on the basis of subjects' comments, that olfaction is an analytic sense, that subjects are able to separate out the smells of individual components. They pointed out that in an extreme case, such as the mixture of ethyl acetate and cyclohexene, subjects found it difficult to judge the total sensory magnitude, so salient were the components. Jones and Woskow's conclusion concurs with Pfaffmann's (1951) statement that olfaction is analytic.

The generality of the conclusion that olfaction is analytic must be questioned. Mitchell and McBride (1971) asked subjects to estimate the subjective magnitude of eugenol in the presence of propanol. The estimates of odor intensity were greater for the mixture than they were when eugenol was presented alone. It appears that subjects were unable, for these odorants, to separate out the components. They perceived the stimulus as unitary and estimated its subjective magnitude. Cain (1969) scaled the subjective intensities of propanol and heptanol, both separately and together. He found that the exponent of the power function for the mixture fell intermediate in size between the exponents for the individual odorants. Other aspects of the mixture interaction, such as degree of odor summation, cannot be determined, however, because each condition appeared in a separate session.

Finally, Berglund, Berglund, and Lindvall (1971) measured interactions between dimethyl monosulfide and dimethyl disulfide. Subjects matched apparent finger span to odor intensity of single odorants and of mixtures. Later, the same subjects gave magnitude estimates of finger span, so the experimenters were able to use those estimates to calculate perceived odor intensities. Power functions fitted the relations between odor intensity and concentration, and all of the exponents were of about the same size (.5). The intensity of a mixture was, as in the experiment of Jones and Woskow, not equal to the sum of the individual intensities. It was smaller by about 40%. Thus, the subjective intensities did not add. [Similar results obtained with mixtures of dimethyl disulfide, hydrogen sulfide, and pyridine (Berglund, Berglund, Lindvall, & Svensson, 1972).] The experimenters also attempted to examine whether sensitivities added: they calculated perceived intensity on the assumption that the stimulus concentrations added. Such an additivity is equivalent to additivity of sensitivities. The data agreed only fairly well with the prediction. In fact, it can be shown that sensitivities did not add. First, all of the exponents were about .5. Then, for the case where the two components are equally strong subjectively, additivity predicts that the mixture should appear $2^{.5}$, or 1.4 times as strong as either of the components presented alone. However, mixtures of equal concentration appeared only 1.2 times as strong. Thus, the results indicate less sensitivity than is predicted by additivity of component sensitivities.

Berglund et al. (1972) proposed a vector model of odor interaction. Components of mixtures are considered as vectors, and the angle between them is presumably related to differences in odor quality. Intensity of a mixture equals the resultant of the vectors. For components equal in subjective strength, the model predicts, and the data support, linear addition but not proportionality. That is, component intensities add linearly, but, as discussed above, not completely. It is worthwhile to keep in mind, however, that tests of summation of intensities by direct scaling will be sensitive to any nonlinearity in subjects' use of numbers. Change in power-function exponent, for example, will affect the degree of intensity summation. On the other hand, tests of summation of sensitivities are not influenced by such nonlinearities.

TASTE

When sapid stimuli of different taste qualities are combined, the resulting taste usually maintains the qualities of the original components, unless the initial magnitudes of some of them were subjectively very weak. Equally (subjectively) strong salty and sweet substances produce a salty–sweet mixture. However, the sensory magnitudes of the analyzed components generally are suppressed (Beebe-Center, Rogers, Atkinson, & O'Connell, 1959; Kamen, Pilgrim, Gutman, & Kroll, 1961; Kiesow, 1896; see especially Fabian & Blum, 1943, and Pangborn, 1960). Perceived taste intensities of different qualities do not always add linearly. Sometimes, however, there are reports of enhancement of one substance by another (e.g., Kamen et al., 1961). Weak salt, for instance, enhances the sweetness of sucrose. That is not surprising, since weak sodium chloride itself tastes sweet. One of the difficulties encountered in evaluating results of experiments on taste mixtures is that selective adaptation may have influenced the outcome. Distilled water can take on a taste, depending on previous stimulation: for example, water tastes sour–bitter after sodium chloride and sweet after quinine (Bartoshuk, 1968). Unless care is taken to rinse the tongue sufficiently between stimuli, these "water tastes" may appear to be components of the subsequent stimuli (McBurney & Bartoshuk, 1972).

Moskowitz (1971a) examined how the taste magnitude of simple stimuli–sodium chloride, quinine sulfate, tartaric acid, and sucrose–grows with concentration when each substance appears alone and also in combination with every other (masking) substance. (In the latter case, the subjects were asked to ignore the taste of the masking stimuli.) Moskowitz concluded that power functions related taste intensity to concentration under all conditions, and that the exponent of the power function did not seem to depend on whether masking substances were present. There is a suggestion, however, that both for quinine and tartaric acid the exponents may be slightly greater for single substances than for combinations. Any conclusion is weakened, however, by the fact that each experimental condition appeared in a separate session. There is some possible danger in attempting to compare results for different conditions: it is not possible to decide to what extent the simultaneous presence of a second masking taste substance produced a general suppression of the subjective intensity of the test substance.

Touch

We end with a very brief consideration of stimulus mixture in touch. The vibratory sense gives evidence of being synthetic when more than one stimulus frequency is applied. Sherrick (1960) asked subjects to estimate the apparent magnitude of a 50-Hz vibration presented to the finger tip together with a 20–100 Hz band of vibratory "noise." The simultaneous presentation of the noise yielded an *increase* in sensory magnitude, i.e., the two stimuli added their effects, at least to some extent. In contrast, when the noise was presented to a different but nearby site on the skin (the palm), the subjective intensity of the 50-Hz stimulus was not only less than it was for superimposed noise, but actually reduced in comparison to its magnitude in the absence of noise. This decrease in magnitude was especially noticeable at low amplitudes of stimulation. Sherrick pointed out the similarity of the latter result to the auditory masking effect of a noise on a tone.

When signal and noise are superimposed at the same site in the skin, subjects appear incapable of separating them out, and their effects add to form a unitary sensory experience. When signal and noise are spatially separate, they are perceptually distinguishable, even though their effects interact. Under the former condition, the vibration sense appears synthetic, under the latter, analytic. Stimulus mixtures in the vibration sense appear, therefore, to be able to result in at least two very different types of sensory response.

Summary

Sensory systems are frequently classified as analytic or synthetic, depending whether or not the components of stimulus mixtures are perceptually separable. It is usually more appropriate, though, to refer to sensory systems as analytic or synthetic under specific conditions of stimulation.

Sensitivity to a compound stimulus is sometimes directly relatable to the sensitivities to the stimulus components, other times to the subjective intensities of the components. The rule—sensitivity summation or intensity summation—that applies given any particular set of conditions is not always simply related, however, to whether the system acts synthetically or analytically.

Vision is synthetic: there exist metamers (different mixtures of wavelengths), which produce identical sensations. Brightness is approximately the resultant of the addition of luminances, i.e., sensitivity to a mixture of lights approximates the sum of the sensitivities to the components.

Under many circumstances, audition is analytic: individual component tones in a complex may be perceptually distinguishable. The loudness of sounds composed of frequencies separated by less than a critical bandwidth depends on the sum of energies (sensitivities add), but the loudness of sounds composed of frequencies separated by much more than a critical bandwidth depends on the sum of the component loudnesses.

Olfaction can be either synthetic or analytic, depending on the odorants that are mixed. It does not appear that the perceived intensity of a mixture depends either on the simple sum of the sensitivities to the components or on the simple sum of the perceived intensities of the components.

Taste is often analytic, but strong interactions take place between effects of different substances. Usually, strong tastes of one quality are seen to suppress weaker tastes of another.

Addition of vibratory noise to a vibratory signal can produce either an increase in subjective strength of the signal (signal and noise presented to the same body locus) or a decrease in subjective strength (noise presented to a different, but nearby, locus).

Chapter 4

TEMPORAL FACTORS IN SENSATION

When a stimulus of constant intensity is turned on, the sensation it produces usually goes through two stages, which are at least partially separable. The first stage consists of the build-up of sensory magnitude. This build-up may be quite rapid—a matter of a fraction of a second—or more prolonged—a matter of seconds. In the second stage the sensation diminishes in magnitude over time, either reaching a final, asymptotic value, or, in some instances, disappearing entirely. Between these two stages there may be a period of time during which the sensory magnitude remains relatively constant. Figure 4.1 is a schematic diagram of the temporal course of sensation magnitude. The figure is intended solely to indicate the early rise and later fall in sensory intensity. The exact shape of the curve, i.e., the way sensation grows (temporal summation) and then decays (adaptation), are among the matters of experimental inquiry to which we now turn.

Temporal Summation

It is instructive to listen, let us say, to a series of bursts of noise as the duration of each burst progressively decreases. As long as the duration remains greater than about .1 sec, the loudness of the noise will appear to remain fairly constant. But when the duration goes below about .1 sec, loudness will decrease; with shorter and shorter durations, the bursts will appear softer and softer until eventually, at short enough durations, they will not be heard at all. The dependence of sensory magnitude on stimu-

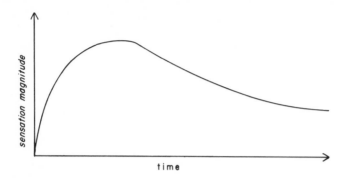

Figure 4.1. Schematic diagram of the time course of sensory magnitude. Given a stimulus intensity that remains constant over time, sensation first increases in magnitude (temporal summation), then declines (adaptation).

lus duration, more precisely, the build-up of sensation over short durations, is often termed *temporal integration,* or *temporal summation.*

Temporal summation is almost certainly a universal sensory phenomenon. The fact that temporal summation occurs in most if not all sensory systems is itself of striking interest. But of even greater interest is the question of whether temporal summation obeys the same general rules in one sensory system as in another. In other words: Is the way the magnitude of a sensation depends on how long the stimulus lasts the same in vision, for example, as in hearing?

VISION

A large number of studies, conducted in various sensory modalities, have looked at the nature of temporal summation at the absolute threshold. These studies ask, "How does the intensity of a stimulus that is re-

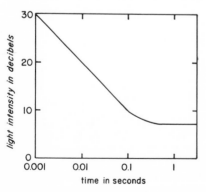

Figure 4.2. Temporal summation at the absolute threshold for vision. Over very short stimulus durations (up to .1 sec), the longer the flash of light, the lower need be the light intensity to arouse a sensation. Over long durations, the threshold intensity is independent of flash duration.

quired to produce a just-detectable sensation depend on how long the stimulus lasts?" One example appears in Figure 4.2, which shows how the luminance and duration of a light interact to maintain a minimal sensation of brightness. When the duration of a flash of light is very short (1 msec), a high level of luminance (30 dB) is needed in order to detect the stimulus, but, as the duration of the flash increases, the necessary level of luminance decreases. At relatively long durations—greater than .5 sec—the luminance required for a threshold response becomes independent of duration.

Both coordinates of Figures 4.2 are logarithmic. At short durations the function is a straight line whose slope is equal to −1. This outcome implies that perfect reciprocity exists between luminance and duration—the absolute threshold depends on the product of luminance (L) and duration (t), that is, on the luminous flux integrated over time. This relation is simply

$$Lt = \text{constant} \tag{4.1}$$

Equation (4.1) is known as Bloch's law. Bloch's law has been verified under several different conditions of stimulation for threshold vision (Barlow, 1958; Graham & Margaria, 1935; Karn, 1936).

Figure 4.2 shows that Bloch's law holds at the absolute threshold of vision for durations up to about .1 sec. The temporal limit of complete summation is often referred to as the *critical duration* (although that term is sometimes applied more generally, e.g., to the point in time at which all summation ends). At slightly longer durations, temporal summation continues to occur, but to a lesser degree than perfect reciprocity. Finally, at long enough durations (here, approximately .5 sec and longer), the luminance needed for threshold becomes independent of stimulus duration. The exact form of the luminance-duration curve depends on other stimulus parameters, such as stimulus size.

Although it is interesting and important to understand the sensory-physical relation between intensity and time at threshold, it is at least equally important to know the psychophysical relation between *sensory magnitude* and stimulus duration. Perhaps the first attempt to use direct scaling to measure temporal summation was a study conducted by Raab (1962), who investigated temporal summation of brightness by means of the method of magnitude estimation. A similar experiment was carried out by J. C. Stevens and Hall (1966). The results of both studies were very much the same. We shall examine here the data of Raab's study. These data appear in a reanalyzed form in a paper by Stevens (1966b). It is worthwhile to mention that the same conclusions can be reached from the data of J. C. Stevens and Hall.

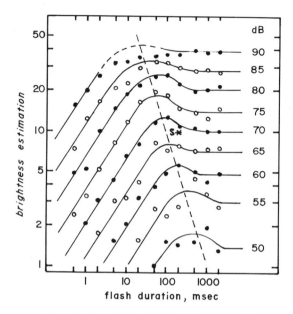

Figure 4.3. Magnitude estimates of the brightness of flashes of light, as functions of flash duration. The parameter is the luminance of the white light in decibels *re* 10^{-6} cd/m². Both coordinates are logarithmic. The star indicates the standard stimulus, whose brightness was assigned the value 10. The dashed line shows how the temporal locus of the Broca-Sulzer enhancement decreases as luminance increases. [Data of Raab (1962), as presented by Stevens (1966b). Courtesy of Stevens and *Perception & Psychophysics*.]

Figure 4.3 shows how the brightness of a flash of light grows with duration for each of several levels of luminance. The horizontal portions of the curves (i.e., at the right-hand side of the figure) imply brightness is independent of duration. The rising portions show that at short durations brightness increases with increasing duration. Both coordinates of Figure 4.3 are logarithmic; thus, the straight lines fitted to the rising portions of the data imply that brightness (B) grows as a power function of duration

$$B = a_1 t^\alpha \qquad (4.2)$$

Were we to plot brightness as a function of luminance for some short, constant duration, we would find that brightness grows also as a power function of luminance. Futhermore, the size of the exponent is the same as that describing the growth of brightness with duration:

$$B = a_2 L^\alpha \qquad (4.3)$$

In other words, for brief flash durations, brightness grows as the same power of both luminance and duration. For example, the brightness of a brief flash is approximately doubled when either the luminance or the duration is multiplied by a factor of 4.

The relations given by Eqs. (4.2) and (4.3) can be combined into a single statement that includes the effects of both luminance and duration, namely,

$$B = a(Lt)^{\alpha} \qquad (4.4)$$

From Eq. (4.4) another simple relation can be seen. That is, in order to produce any constant level of brightness, a perfect reciprocity exists between luminance and duration:

$$Lt = (B/a)^{1/\alpha} = \text{constant} \qquad (4.5)$$

The relation expressed by Eq. (4.5) is, of course, Bloch's law, the same relation found to hold at the absolute threshold and described by Eq. (4.1).

An important feature of Figure 4.3 is the variation in the limit over which temporal summation occurs. It is readily apparent that at the lowest levels of luminance, summation takes place (brightness grows), with increases in duration up to about .5 sec, but, as luminance increases, the limit on temporal summation decreases. Thus, at the highest level of luminance the limit of summation was only about .01 sec. A similar variation can be shown to exist in the temporal limits for Bloch's law (the critical duration). At low levels of suprathreshold brightness, as at the absolute threshold, *complete reciprocity* obtains between luminance and duration up to about .1 sec, but at higher levels of brightness, the critical duration is reduced to about .01 sec. It is of interest that a similar phenomenon occurs for incremental (differential) thresholds: Baumgardt and Hillmann (1961), Biersdorf (1955), Graham and Kemp (1938), Herrick (1956), and Keller (1941) have shown that Bloch's law holds for threshold increments superposed on luminous backgrounds, but that as adapting luminance increases, the temporal limit on summation decreases.

The fact that the limits on temporal summation depend on luminance has important consequences for the psychophysical (brightness) function. Specifically, the change in limits of summation implies that brightness grows more rapidly with increasing luminance when duration is short than when duration is long (Anglin & Mansfield, 1968; J. C. Stevens & Hall, 1966; Stevens, 1966b). If the exponent of the brightness function equals one-third at relatively long durations (greater than about 1 sec), then the exponent will equal approximately one-half at very short durations (less than about 10 msec). At intermediate durations, the brightness function will be bisegmented, with exponent one-half at low luminances, and one-third at high.

We have described how brightness grows with increasing exposure duration up to a critical duration, and we have mentioned that at longer durations, brightness is relatively independent of duration. But a curious phenomenon occurs in the region of the critical duration, as can be seen in Figure 4.3. This phenomenon consists of a hump in the function, sometimes described as an "overshoot" of brightness, and is typically referred to as the Broca-Sulzer effect. [Broca and Sulzer (1902a & b) were among the first to describe the phenomenon.] In the region of the Broca-Sulzer effect, brightness is greater than at shorter and at longer durations. The existence of the Broca-Sulzer overshoot is apparent in the data of J. C. Stevens and Hall, as well as in the data of Raab, and was also evidenced in several brightness-matching experiments (Aiba & Stevens, 1964; Katz, 1964; Nachmias & Steinman, 1965; Raab & Osman, 1962; Stecher, Sandberg, & Minsky, 1970).

At approximately the same time that Stevens published his reanalysis of Raab's data and J. C. Stevens and Hall published their data, Ekman (1966) reported another reanalysis of Raab's data. Ekman's analysis, shown in Figure 4.4, plots brightness on a linear scale as a function of duration on a logarithmic scale. Comparison of Figure 4.4 to Figure 4.3 makes clear the difficulty in deciding on the basis of goodness of fit whether brightness grows as a logarithmic function or as a power function of duration. (The primary reason for the difficulty is the relatively short range over which temporal summation usually can be measured. If temporal summation proceeded over several usable decades of duration, no doubt the two alternatives could be discriminated on the basis of goodness of fit.)

It would appear to be necessary to use some criterion other than goodness of fit in order to decide between the two alternative descriptions of temporal summation. Although not absolutely conclusive in nature, there does seem to be strong reason for preferring the power relation. Simply stated, if brightness is related to luminance by a power function (for brief but constant durations), then Bloch's law [Eq. (4.5)] can hold only if brightness grows also as a power function of duration. Alternatively stated, if brightness grows as a power function of luminance, but as a logarithmic function of duration, then Bloch's law cannot hold. Yet direct brightness matches also appear to confirm Bloch's law (Aiba & Stevens, 1964; Brindley, 1960).

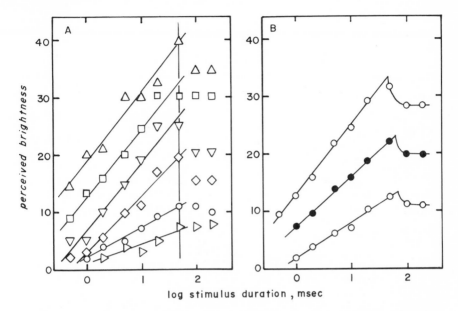

Figure 4.4. Magnitude estimates of the brightness of flashes of light, as functions of flash duration. The abscissa only is logarithmic. [Data of Raab (1962), as presented by Ekman (1966). Courtesy of Microforms International Marketing Corporation.]

A criticism of Ekman's analysis was made by Baumgardt (1967). Ekman extrapolated to absolute threshold, that is, to zero brightness, his logarithmic functions relating brightness to duration. These extrapolations suggested that Bloch's law holds at the absolute threshold for durations up to only about 1 msec. But we have already seen that measurements of absolute threshold show complete temporal summation to take place for durations up to about .1 sec (100 msec). Baumgardt argued that the failure of Ekman's extrapolated thresholds to predict Bloch's law correctly provides prima facie evidence against the validity of direct scaling procedures. However, Stevens's analysis of the same data does not make this erroneous prediction. The failure to predict correctly the nature of the relation between luminance and duration at the absolute threshold would seem to be the result of the particular (logarithmic) function that Ekman used, rather than any intrinsic lack of validity of the method of magnitude estimation.

The temporal variations in brightness can be accounted for quite well in terms of a simple model (Marks, 1972). The essence of the model is the concept of feedback: the basic assumption is that brightness is processed in a series of stages, where each stage acts like a low-pass filter.

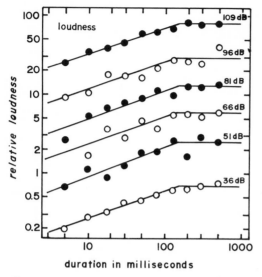

duration in milliseconds

Figure 4.5. Magnitude estimates of the loudness of bursts of noise, as functions of stimulus duration. The parameter is the sound pressure level of the noise in decibels. [From J. C. Stevens and Hall (1966). Courtesy of the author and *Perception & Psychophysics*.]

The sensitivities and time constants of the filters are modified by the magnitude of the final output. A two-stage feedback-filter can predict: (*a*) a cube-root relation between brightness and intensity for relatively long stimulus durations; (*b*) a square-root relation for brief durations; (*c*) Bloch's law over short durations; (*d*) inverse proportionality between the critical duration of summation and final brightness; and (*e*) the Broca-Sulzer overshoot of brightness. (The model will be considered with greater detail in Chapter 5, where it is expanded in order to account as well for effects of variations in spatial parameters of stimulation.)

AUDITION

Direct scaling procedures have been used to investigate temporal summation in several other sensory modalities. Notable among these have been the numerous studies of temporal summation of loudness (Berglund & Berglund, 1967; Ekman, Berglund, & Berglund, 1966; J. C. Stevens & Hall, 1966; Wright, 1965). The data that have been obtained for loudness show some striking similarities to data for brightness. Let us take as an example magnitude estimates of the loudness of white noise that were obtained by J. C. Stevens and Hall (1966). These magnitude estimates are shown in Figure 4.5 as functions of duration (in double-logarithmic coordinates). Although the data are rather variable, they demonstrate fairly clearly the general nature of temporal integration of loudness: loudness first increases with duration, then becomes relatively independent of duration.

Stevens and Hall fitted the data for each (constant) sound pressure level with two straight line segments. Thus, at short durations loudness (L) grows as a power function of duration

$$L = b_1 t^\gamma \qquad (4.6)$$

Equation (4.6) is the auditory analogue to Eq. (4.2) for vision. Unlike the situation for vision, however, the critical duration for audition, i.e., the duration over which temporal integration takes place, does not seem to change appreciably from one intensity level to another.

The horizontal line segments imply that at durations greater than the critical duration, loudness is independent of duration. It is not clear, however, that the data deserve the sharp discontinuity implied by the use of two line segments. Such a use would be justified if it were based on some theoretical consideration. But, in fact, at least one theory of temporal summation of loudness (Zwislocki, 1960, 1969) predicts a more gradual transition from temporal integration to no temporal integration. The abruptness of the transition between summation and no summation remains in doubt. Loudness matches obtained by Small, Brandt, and Cox (1962) do suggest that the transition may be rather sharp.

The size of the exponent γ in Eq. (4.6) was found by Stevens and Hall to equal about .34 at each level of intensity. When the same data are

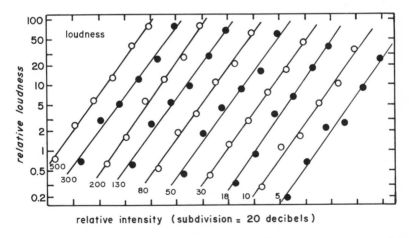

Figure 4.6. Magnitude estimates of the loudness of bursts of noise, as functions of stimulus energy. The parameter is the duration of the noise burst in milleseconds. The functions are displaced horizontally from one another in order to improve clarity. [From J. C. Stevens and Hall (1966). Courtesy of the author and *Perception & Psychophysics.*]

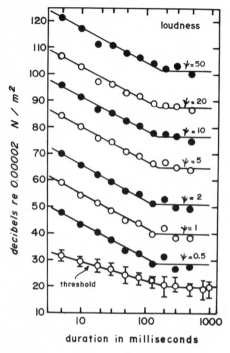

Figure 4.7. Contours of equal loudness, showing how sound pressure level and duration of noise must covary to maintain constant loudness. The lowest function gives temporal summation at absolute threshold; the other functions give temporal summation at suprathreshold levels of loudness. [From J. C. Stevens and Hall (1966). Courtesy of the author and *Perception & Psychophysics*.]

plotted in terms of stimulus energy (E), power functions are again able to describe the functions relation (see Figure 4.6)

$$L = b_2 E^{\theta} \tag{4.7}$$

As a function of sound energy, however, the value of θ was .27. Therefore, in contrast to the situation for brightness, loudness does not grow at the quite same rate with increases in sound duration as with increases in sound energy. This outcome is apparent in Figure 4.7, which shows how sound energy and duration can be traded one for the other in order to maintain a constant level of loudness. If loudness grew at the same rate with sound energy as with sound duration, the functions in Figure 4.7 would obey Bloch's law. (We have already seen that this was the case for brightness.) The line segments in Figure 4.7, however, do not have a slope of -1.0, but rather -1.25 (the ratio .34 : .27). An exception to this last statement occurs at the absolute threshold, where the slope is equal to $-.7$. The difference may be summarized as follows. In order to maintain constant suprathreshold loudness, a given change in sound energy can be offset by a smaller percentage change in sound duration (an example of "supersummation"), but in order to maintain a just-detectable sensation of loudness, a given change in sound energy must be offset by a larger percentage change in sound duration (partial summation). Furthermore, the critical duration for summation at threshold is a little longer (about 230 msec) than at suprathreshold loudness (about 150 msec).

The trading functions of Figure 4.7 agree with direct matches of the loudness of bursts of noise (Port, 1963; Small et al., 1962; Zwicker, 1966). They are also in close agreement with the prediction from Zwislocki's (1969) theory of temporal summation. Zwislocki hypothesized two underlying processes: one, a process of summation, assumed to occur peripherally in the auditory system; the other, a process of decay, assumed to occur more centrally.

Two studies of temporal summation of loudness have been conducted by Ekman and his associates (Berglund & Berglund, 1967; Ekman et al., 1966). In both of these studies the growth of loudness of a 1000-Hz tone with stimulus duration was described in terms of a logarithmic function. However, Zwislocki (1969) noted that the trading functions predicted by the data of Ekman et al. diverge systematically from trading functions obtained by direct matching of the loudness of tones (Port, 1963; Zwicker, 1966).

It is interesting, and somewhat curious, that temporal summation proceeds in quite a different manner when the stimulus consists of a pair of tone bursts, separated in time. Under these conditions, summation takes place over an interval of only about 40 msec (Irwin & Zwislocki, 1971). This is a much smaller interval than the 200 msec typically found for continuous tone or noise.

It is perhaps worth mentioning Berglund and Berglund's (1967) suggestion of still another possible equation to describe the growth of sensory magnitude (loudness) with stimulus duration. In a slightly different, but mathematically equivalent form, their equation reads

$$L = b(1 - e^{-t/\tau}) \tag{4.8}$$

Equation (4.8) is of particular interest, since it describes the behavior of a simple linear integrator, i.e., a low-pass filter with time constant τ. In this respect, it resembles the model (Marks, 1972) outlined above, which was used to account for temporal summation of brightness. However, Eq. (4.8) describes temporal summation in a linear system, whereas the model for brightness incorporates nonlinearity (feedback). Nothing further will be said about Berglund and Berglund's formulation here. We are certainly far from deciding with certainty what is the correct form of the function relating sensory magnitude to stimulus duration. It is possible, of course, that the function differs from one sensory modality to another. In any case, it will be necessary for the function to be consistent with other relevant empirical data and, most likely, with any pertinent theoretical considerations.

TASTE

Bujas and Ostojčić (1939) measured temporal summation of taste intensity by a direct matching procedure. Taste intensity continues to increase over several seconds of stimulation, the exact duration of summation depending on the compound. Sodium chloride takes about 3 sec to produce maximal subjective strength, sucrose takes about 4 sec, and quinine hydrochloride about 6 sec. For both sodium chloride and quinine, the growth of sensation over time followed the same course at all stimulus concentrations; for sucrose, however, the time to reach maximum itself increased as concentration increased.

PROPRIOCEPTION

Let us now look at temporal summation in a sensory modality that has received relatively scant psychophysical attention. Brown (1966, 1968) employed the methods of magnitude estimation and cross-modality matching to investigate perceived magnitude of angular rotation. Subjects were rotated at constant angular acceleration, and, for each of several levels of acceleration, they judged perceived magnitude of rotation at each of several durations of acceleration. The results showed quite clearly the nature of temporal summation: perceived rotation increased linearly with increasing angular acceleration, but increased as the 1.3 power of duration. In other words, doubling the duration produced a greater increase in perceived rotation than did doubling the acceleration—an example, if you will, of temporal "supersummation," like that found for loudness of noise.

A notable feature of these data is that the limit of temporal summation appears to vary inversely with level of acceleration. This is qualitatively similar to the way the critical duration for brightness summation depends upon luminance. In the case of angular acceleration, however, the temporal limits are relatively large. For the smallest acceleration that Brown examined ($.3°/sec^2$) temporal summation proceeded for nearly 40 sec.

An interesting example of temporal summation, if in fact that term is proper here, occurs in muscular sensations. When subjects are asked to maintain a constant force over time, the sensation of effort builds up dramatically. Concomitantly, in order to maintain sensation constant over

time, the subject must decrease continuously the force that he exerts
(Cain & J. C. Stevens, 1971). J. C. Stevens and Cain (1970) obtained
magnitude estimates of the perceived effort involved in producing iso-
metric handgrip contractions that ranged in force from 22 to 220 N and
in duration from 4 to 60 sec. At all force levels, the perceived effort grew as
about the square root (.57 power) of duration, and at all durations as
about the 1.4 power of force. Thus, increasing the duration is not as potent as increasing the force in producing an increment in effort. From the
two power equations, the trading relation between force (F) and duration (t) was computed. Rather than a reciprocity law (Ft = constant),
the result is $Ft^{.41}$ = constant. Of special interest was the discovery that
the same relation described results relating force and the maximal duration that a contraction can be held (endurance time).

The build-up of muscular fatigue seems by its nature a process somewhat different from temporal summation in other sensory modalities.
From an operational and descriptive point of view, however, the psychophysics of muscular effort does display features in common with temporal summation in other senses. The apparent difference resides in the fact
that the subject must produce the stimuli for muscular sensations, and as
fatigue or effort builds up, he becomes less and less capable, in some
way, of producing the stimuli. There is perhaps some semantic confusion,
in that the term "fatigue" is sometimes used to describe decremental responses (adaptation) in other senses. The receptive organs for perceived
effort, whether in muscles, skin, or joints, may display summative, rather
than decremental, properties over time.

Somesthesis

Ekman and his associates examined temporal integration in some of
the skin senses, and they again applied logarithmic functions to describe
the relation between sensory magnitude and stimulus duration. Berglund,
Berglund, and Ekman (1967) determined how subjective magnitude of a
250-Hz vibratory stimulus applied to the finger varies with vibration amplitude and duration. Verrillo (1965) had shown that the critical duration for temporal summation at absolute threshold is about 1 sec. Berglund et al. found the critical duration for suprathreshold summation to
vary a good deal from one subject to another. For a few subjects, sensation reached maximum after .2-sec stimulation; for others, magnitude
increased up to 1.2 sec, the longest duration examined. The median subject showed a maximum at about .7 sec. This contrasts with Békésy's
(1959) report that suprathreshold vibratory sensation produced by a
100-Hz stimulus reaches maximum at 1.2 sec.

Of some interest was the way the perceived vibration magnitudes depended on vibration amplitude in Berglund *et al.*'s study. A power function could describe the results at every duration, but the size of the exponent was not constant. As duration increased from .03 to .4 sec, the exponent decreased from 1.4 to .8. With further increase in duration, the size of the exponent remained constant. This result is reminiscent of the change with duration in the exponent of the psychophysical function for brightness, and it suggests that, as in the case of brightness, the critical duration for temporal summation of vibration magnitude may change with the amplitude of the stimulus. In fact, 6 of the 10 subjects showed shorter critical durations at low, than at high, vibration amplitudes. This relation is opposite to the direction of change found for brightness, however, and it implies that the relative rate at which vibration magnitude increases with duration itself varies systematically with amplitude.

From the graphs presented by Berglund *et al.*, it is possible to determine the extent of temporal summation, that is, the percentage change in intensity needed to offset a given change in duration. In general, summation was less than complete: a threefold to sixfold increase in intensity was as effective as a tenfold increase in duration.

Two experiments examined how the subjective magnitude of electric shock varies with duration and magnitude of electric current (Ekman, Frankenhaeuser, Levander, & Mellis, 1966; Ekman, Fröberg, & Frankenhaeuser, 1968). In both studies the subjects were asked to estimate the *unpleasantness* of the sensations, rather than subjective intensities per se. However, it appears that unpleasantness and subjective intensity of shock are proportional to each other. Unpleasantness continued to increase with duration up to the longest duration examined—1.7 sec. In contrast with results for vibration, however, there was no evidence that the exponent of the power function relating unpleasantness to current varied with duration. All of the exponents equaled about 1.5.

Finally, let us look at temporal summation in the warmth sense. A strong radiant stimulus can produce a sensation of warmth whose intensity increases over time. If stimulation is prolonged, pain may eventually be felt. Marks and J. C. Stevens (in press) used the method of magnitude estimation to measure the increase in warmth over time when a portion of the forehead was irradiated. Warmth increased most rapidly when the irradiance of the stimulus was high, less rapidly when it was low. More interesting in many respects are the functions showing how warmth increases with irradiance at constant duration. Figure 4.8 displays these functions in logarithmic coordinates. Only when the duration was relatively long (3 or 6 sec) do the functions approximate linearity in

the log–log plot, and thereby conform to a power relation. When duration was short (.25–1 sec) the data show inflections, and the results can
be described by bisegmented functions.

Inflections in the psychophysical functions for warmth suggest, as one
possible explanation, that warmth sensations are mediated by two
mechanisms. In Chapter 2 similar, but smaller, inflections were seen in the
psychophysical functions obtained for 3-sec stimulation of the forehead.
The possibility that two mechanisms (perhaps two sets of receptors) underlie the perception of warmth will be taken up in the next chapter. Of
more immediate concern is how these functions relate to temporal summation. The lower segments of the functions all have about the same
slope. That means that the time–irradiance trading remains constant from
one low level of warmth to another. Furthermore, the degree of trading is
one of reciprocity, a parallel to Bloch's law. The upper segments clearly
differ in slope, however. Warmth increases more rapidly at short durations
than at long; at high levels of warmth, therefore, the trading functions are
not parallel: they get flatter as warmth increases. Thus, summation at high
warmth is only partial. Taking the entire dynamic range of warmth into
consideration, it appears that temporal summation is level dependent.

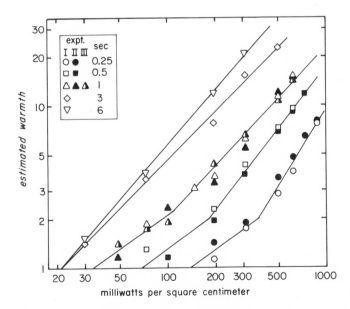

Figure 4.8. Magnitude estimates of the warmth produced by thermal irradiation
of the forehead, as functions of irradiance in milliwatts per square centimeter. The parameter is the duration of the stimulus. [From Marks and J. C. Stevens (in press).]

Degree of summation is greatest at low levels of warmth and is smaller at high levels.

From a psychophysical point of view, warmth displays temporal summation, in that the effect of a long-lasting radiant stimulus is greater than the effect of a short-lasting one of the same intensity. What is probably most important for production of warmth sensation is not level of radiation per se, however, but the change in tissue temperature at the site of warmth receptors. Since the locus (in fact, the existence) of warmth receptors remains obscure, there is no way to ascertain whether warmth receptors display temporal summation. But it is possible to calculate the changes in tissue temperature that are produced by irradiation of the skin (Stolwijk & Hardy, 1965). Calculations show that the changes in tissue temperature with time (at depths greater than .2 mm below the surface of the skin) are much more rapid than are the corresponding changes in warmth with time. Put another way, if warmth receptors are situated at depths greater than about .2 mm below the skin surface, then it is unlikely that the receptors display temporal summation. In fact, it is possible that they display adaptation even over the first few seconds of stimulation. Given a constant change in temperature, the longer the time needed to produce that change, the weaker the sensation.

SUMMARY

Not only does temporal summation appear to be a general sensory phenomenon, but, furthermore, the rules of summation appear similar from one sensory modality to another.

1. *Psychophysics.* Sensory magnitude increases as stimulus duration increases, but usually only up to a point. In vision, the temporal limit of brightness summation (peak of the Broca-Sulzer enhancement) depends on luminance: from a high value of about .5 sec at low luminance, it decreases to about 10 msec at high luminance. In audition, loudness increases with duration for about .15 sec, independent of stimulus intensity. Vibratory sensation reaches a maximum at about .7 sec, on the average, but the critical duration may be somewhat smaller when intensity is low than when it is high. Taste sensations summate over longer intervals—3–6 sec—depending on the compound. Sensations of movement produced by angular acceleration can summate for intervals up to 40 sec, when rate of acceleration is small; lower summation limits result when the rate is high. There seems to be no temporal limit on the growth of muscular effort except for the limit of endurance. Similarly, there may be no temporal limit on growth of warmth when the skin is heated with intense irradiation; there is a limit of about 3 sec when irradiance is low.

2. *Sensory physics.* The same subjective magnitude can be aroused by

intense, brief stimulation or by a weaker, longer stimulation. The relation between intensity (ϕ) and duration (t) required to produce constant sensation is often approximated by the hyperbolic equation ϕt^{μ} = constant. When μ = 1, summation is complete, as in the case in brightness vision (Bloch's law). For loudness of white noise, μ = 1.25; for perception of velocity under angular aceleration, μ = 1.3. Both of these values imply "supersummation," that is, increases in duration are more effective than proportional increases in intensity. Partial summation (increases in duration less effective than proportional increases in intensity) seems to be the rule for several other modalities: muscular effort yields μ = .4; vibration yields μ = .5–.8; and warmth yields μ = .5 – 1.0.

Adaptation

We have just seen how sensory magnitude grows with time of stimulus exposure over relatively short durations. But with prolonged stimulus exposure, sensations often decline in magnitude. It is a common experience, for example, to place a hand in warm water and feel, after an initial surge of warmth, the sensation begin to decline and perhaps eventually even disappear. Similarly, it has probably been everyone's experience to leave a movie theater and be dazzled by bright sunlight. Within a few moments, however, the dazzle disappears and the brightness diminishes.

VISION

Time-dependent changes in the sensitivity of a sensory system, such as those described, are often termed *sensory adaptation*. Strangely enough, little inquiry has been made in the time course of visual light adaptation. Two studies, one by Geldard (1928), the other by Wallace (1937), measured brightness adaptation by direct matching. That is, the subjects matched for brightness the luminance of a brief flash of light presented to one eye to a steady light presented to the other eye. We can transform the measured values of matching luminance into values of brightness by means of the psychophysical power law. The relation employed was $B = kL^{1/3}$. Results obtained from Geldard's data are shown in Figure 4.9. There are two important features of these functions. First, the degree to which brightness declines itself varies with stimulus intensity: the brightness of the most intense stimulus decreased by 50%, whereas the brightness of the least intense decreased only 20%. Second, the time needed for brightness to reach a steady level also depends on intensity. The brightness at the highest intensity became stable after 30 sec, but at the lowest only after about 2 min.

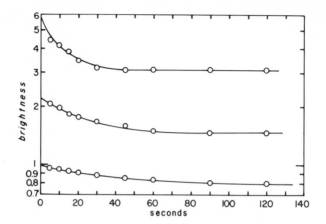

Figure 4.9. The course of brightness adaptation. A white light of constant luminance yields a brightness that declines rapidly at first, then more slowly until an asymptote is reached. The greater is the initial brightness, the faster and more extensive is its decline. [Derived from data of Geldard (1928).]

The curve drawn through each set of points follows the simple exponential equation

$$B = ce^{-t/\tau} + j \tag{4.9}$$

where τ is the time constant of adaptation. The values of τ in the three curves equal 40, 23, and 11.5 sec, from lowest to highest. This 4 : 1 ratio in time constant from largest to smallest is striking and of theoretical significance. Short (e.g., 1 sec) flashes at these intensities produce a similar, though somewhat larger (6 : 1) ratio of brightnesses. Although the correspondence is not perfect, there is the possibility that the rate of brightness adaptation is proportional to the initial (unadapted) brightness. In any case, the decrements in brightness that are displayed in Figure 4.9 are not very large. Complete adaptation, i.e., decrements in brightness to zero, can occur under a very special condition, namely when the image is stabilized on the retina (Riggs, Ratliff, Cornsweet, & Cornsweet, 1953); when an image is stabilized, eye movements cannot affect the location of the image on the retina. Thus, it appears likely that the lack of complete adaptation found under normal viewing conditions is the result of eye movements that move the border of the image over the retina.

Even under normal conditions of viewing, more extensive degrees of adaptation than those shown in Figure 4.9 occur when stimulus intensity is very high. For luminances greater than about 100 dB *re* 10^{-6} cd/m², the asymptotic brightness is practically constant regardless of stimulus intensity (Craik, 1940; J. C. Stevens & Stevens, 1963). Because of the smaller changes in brightness that ensue at low intensities, it is unlikely that the time course of visual adaptation could be studied effectively by means of direct scaling procedures. Brightness matching seems to be the most useful approach.

AUDITION

If it has proved difficult to study visual adaptation by means of scaling procedures, it appears impossible to study auditory adaptation that way! The reason, however, is somewhat different. If a subject is exposed to a sound at constant intensity, and he is directed to attend carefully to its loudness, he will typically report that the loudness undergoes no change. That is, loudness will not appear to diminish over time. In fact, if the sound pressure is changed slowly over time by the experimenter, and the subject is instructed to maintain loudness constant, he will offset the experimenter's changes and keep sound pressure constant (Harris & Pikler, 1960). Thus, even in a dynamic situation there is no evidence of diminution of loudness.

Nevertheless, some sort of auditory adaptation can be demonstrated by means of interaural loudness matching: present a constant-intensity tone to one ear, and, at various moments after its onset, adjust the sound pressure of a brief comparison in the other ear to effect a loudness match. One of the first extensive studies of this type was conducted by Hood (1950), who reported that the matching sound pressure showed a decrease as great as 35 dB at the end of 5 min of listening. (The greater the intensity of the adapting tone, the more extensive was the adaptation as measured in decibels.) Assuming a 2/3 power relation between loudness and sound pressure, a decline of 35 dB implies a decrement in loudness to one-fifteenth the initial value. In a similar manner, it is possible to estimate the time constant of decay (time for loudness to fall $1/e$ its total decline); the value turns out to equal approximately 60 sec. The extent to which loudness "adapts" is a function of the sound pressure of the stimulus, and not all levels produce so drastic a reduction. Even so, the decrements reported by Hood and others are generally much larger than the decrements in brightness depicted in Figure 4.9. Yet these changes in loudness are not readily perceptible to listeners except when comparison stimuli are presented to the other ear.

The nature of *perstimulatory* auditory adaptation, as it is called, is one of the intriguing problems in audition. Much of the relevant material has been reviewed by Small (1963). Perstimulatory adaptation is presently a battleground of controversy. One hypothesis states that loudness does not adapt, or adapts only very little, but the changes that are measured reflect adaptation in a system that mediates auditory localization. According to this first hypothesis, the subject does not match loudness, but sets the image of the sound heard binaurally to appear at the midline. A second hypothesis states that there is some other sort of binaural interaction between adapting tone and comparison, but not necessarily related to localization.

One way to circumvent possible confusion of localization and loudness is to use a comparison tone that differs in frequency from the test tone. Egan (1955) found that comparison tones of the same and of different frequencies yielded identical measures of adaptation. [Calculation of time constants from Egan's data yields a value of 60 sec, identical to that calculated from Hood (1950).] On the other hand, Fraser, Petty, and Elliott (1970) found little evidence for adaptation when matches were obtained for differing test and adapting sound frequencies.

Both hypotheses outlined above are directed toward the fact that the decrements in loudness seem to occur when simultaneous binaural matches are made. The question arises whether simultaneity is crucial for the appearance of decrements. An alternative procedure is to present the comparison stimulus after termination of the adapting stimulus. Under that condition, loudness matches can be obtained that presumably are uncontaminated by binaural interactions. Egan and Thwing (1955) found some evidence of perstimulatory adaptation using such a method of delayed matching. Similar experiments by Petty, Fraser, and Elliott (1970), Stokinger, Cooper, Meissner, and Jones (1972), Stokinger and Studebaker (1968), and Fraser, Petty, and Elliott (1970), however, found little evidence for any adaptation when delay was introduced. Furthermore, Stokinger, Cooper, and Meissner (1972) showed that the amount of adaptation measured with simultaneous matching depends on degree of temporal overlap between test and comparison stimuli.

In spite of the fact that a great deal of effort has gone into the study of perstimulatory auditory adaptation, firm conclusions remain elusive. It seems most probable that binaural factors are of great importance, and the question whether the sensory magnitude loudness actually undergoes a decrement over time is still not settled.

OLFACTION

At least two studies have attempted to assess directly how odor magnitudes vary as a function of time of adaptation. Ekman, Berglund, Berglund, and Lindvall (1967) examined the time course of the perceived intensity of odor. Their subjects were presented constant concentrations of hydrogen sulfide in air. The subjects were instructed to breathe at a normal rate throughout exposure to the stimulus (12 min in one experiment, 15 min in another). Odor magnitude was assessed by a cross-modality matching procedure: the subjects matched for subjective magnitude finger span to odor at various points in time during exposure. In a separate experiment, the subjective magnitude of finger span was assessed by the method of magnitude estimation, so the data for matching finger span could be converted to equivalent magnitude estimates. The final results are shown in Figure 4.10. Odor magnitude declines quite rapidly at first, then more gradually, and appears to approach a different asymptote for each level of concentration. In fact, for all but the highest concentrations, subjective odor magnitude reached a steady-state level within 4–10 min.

The curves that Ekman *et al.* fitted to the relation between subjective odor magnitude *(O)* and time of adaptation were all of a simple exponential form, namely

$$O = ce^{-vt} + j \qquad (4.10)$$

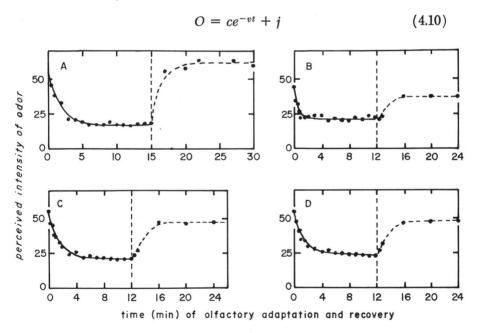

Figure 4.10. The course of odor adaptation and recovery. An odorant (hydrogen sulfide) of constant concentration yields an odor whose intensity declines rapidly at first, then more slowly until an asymptote is reached. When the odorant is removed, subsequent recovery of odor intensity appears to take about as long as did adaptation. The odorant concentrations were: A, .7 parts/million; B, .9 parts/million; C, 2.6 parts/million; D, 6.4 parts/million. [From Ekman, Berglund, Berglund, and Lindvall (1967). Courtesy of the authors and *Scandinavian Journal of Psychology*.]

Equation (4.10) is formally identical to Eq. (4.9) for brightness. The authors did not provide values of the constants used in Eq. (4.10), but these may be estimated at least roughly from Figure 4.10. Of particular interest is the size of v, since $\tau = 1/v$ may be taken as the value of the time constant for the olfactory system under the particular conditions of experimentation. The value of v appears about the same (approximately .01) for three of the levels of concentration (.7, 2.6, and 6.4 parts/million), but clearly larger (approximately .02) for the fourth level (.9 part/million). These values correspond to time constants of 100 and 50 sec, respectively. They are notably larger than values derived from Geldard's data on brightness adaptation. Although this irregularity (lack of complete constancy or regular variation with concentration) makes hazardous any conclusion concerning the nature of the time constant, its approximate constancy for three levels of concentration does suggest the possibility that the time constant is invariant with concentration.

A different conclusion is reached from a study of olfactory adaptation to methyl isobutyl ketone (Pryor, Steinmetz, & Stone, 1970). Pryor et al. traced the time course of odor intensity by magnitude estimation, and they found an approximate 4-to-1 ratio in the duration that odor sensation required to become constant. Adaptation was faster for low than for high stimulus concentrations. In fact, the odor of the lowest concentration (four times threshold) disappeared after about 300 sec exposure. Thus, the relation between rate of adaptation and intensity reported by Pryor et al. is opposite to that found for brightness. Approximate values of the time constant of odor adaptation (the time needed for odor magnitude to fall $1/e$ the total extent of decline) ranged from 100 to 200 sec. These values are roughly comparable in magnitude to values derived from data of Ekman et al.

TASTE

One interesting feature of adaptation in taste is that it often is complete (Abrahams, Krakauer, & Dallenbach, 1937; Krakauer & Dallenbach, 1937). That is, as long as solute concentration is not too great, taste sensation will not only decline over time, but eventually may disappear. This complete adaptation is almost unique to taste, although, as we just saw, very weak odors can disappear. So can very weak light sensations.

Adaptation curves have been measured for taste intensity. Meiselman (1968b) used the method of magnitude estimation to determine the time courses for sensations aroused by sucrose, sodium chloride, and quinine sulfate. Adaptation was not complete in this experiment, probably because of the procedure, which required the subject to sip the test solution and hold it in his mouth during the adapting interval. It is not certain that simple exponential functions are adequate to describe Meiselman's results. It is worthy of note, however, that the temporal decay of taste magnitude is negatively accelerated, and that taste intensity dropped $1/e$ the total extent of its decline in about 75 sec, similar to values obtained for olfactory adaptation.

The course of gustatory adaptation to sodium chloride was also examined by Bujas (1953), who flowed solutions on the tongue and employed a taste-matching procedure. It is difficult to calculate the time constant of decay (time for taste intensity to fall to $1/e$ the initial value), however, without knowing more precisely the nature of the psychophysical function. Exponents of power functions relating saltiness to concentration of sodium chloride range fom .4 when a flow procedure is used (McBurney, 1966) to 1.6 when "sip and spit" is used (Ekman & Åkesson, 1965). If we assume the exponent to equal .4, the time constant from Bujas's experiment averages about 100 sec; if we assume the exponent to equal 1.6, the time constant averages about 25 sec. Regardless of what the absolute values of the time constant may be, Bujas's data show clear dependence on level: the higher the adapting concentration, the slower the adaptation, i.e., the larger the time constant.

PROPRIOCEPTION

An interesting experiment was conducted by Clark and Stewart (1968), who obtained magnitude estimates of perceived velocity of rotation when subjects were rotated about a vertical axis. Three conditions were examined: constant acceleration, linearly increasing acceleration, and zero acceleration (constant velocity). When velocity was constant, subjects reported no perceived rotation. When velocity increased linearly (i.e., acceleration was constant), the perceived magnitude of rotation first increased and then declined. The nature of the increase has already been examined (in the section concerned with temporal summation) in our discussion of the data obtained by Brown (1966). With respect to the decrease in sensation magnitude, when level of acceleration was low and constant ($.5°/sec^2$), perceived rotation declined to zero after about 80 sec; when acceleration was higher, no rotation was perceived after about 120 sec. When acceleration itself increased linearly, the magnitude of perceived rotation increased slowly, then leveled off about 100 sec after the start of acceleration. There is a suggestion that near the end of the exposure duration (180 sec) perceived rotation magnitude began to decline.

It would appear, therefore, that the sensory response to angular acceleration displays both temporal summation and adaptation. Given a constant stimulus (constant angular acceleration), perceived rotation first increases and then declines. Given a linearly increasing stimulus, little or no decline in sensation magnitude is evidenced; presumably, this result is due to the counterbalancing effects of increasing stimulus intensity and adaptation. It does not appear feasible to attempt to fit precise mathematical functions to the data obtained by Clark and Stewart.

SOMESTHESIS

A study of the course of adaptation to vibrotactile stimulation was conducted by Berglund and Berglund (1970). Three stimulus levels (40, 47, and 54 dB above the absolute threshold) were presented for durations of 7 min; sensory magnitudes were scaled by matching to perceived finger span, and the matching data transformed in accordance with ob-

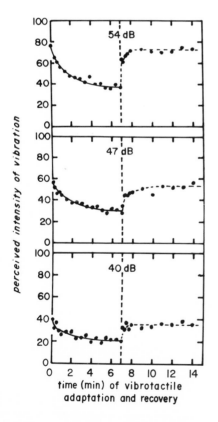

Figure 4.11. The course of vibrotactile adaptation and recovery. A vibratory stimulus (250 Hz applied to the finger) of constant displacement (40, 47, or 54 dB above threshold) yields a sensation whose intensity declines rapidly at first, then more slowly until an asymptote is reached. When the vibratory stimulus is removed, subsequent recovery of vibratory intensity appears to be more rapid than adaptation. [Reprinted with permission of author and publisher: Berglund, U., and Berglund, B. Adaptation and recovery in vibrotactile perception. *Perceptual and Motor Skills*, 1970, **30**, 843–853.]

tained magnitude estimates of finger span. The results are depicted in Figure 4.11. The curves follow simple exponential decays

$$V = ce^{-vt} + j \tag{4.11}$$

where V is sensory magnitude of vibrotactile stimulation. Equation (4.11) is formally identical to Eqs. (4.9) and (4.10) for adaptation to visual and to olfactory stimulation.

Berglund and Berglund noted that the rate constant of adaptation, v in the notation of Eq. (4.11), was approximately the same for all three levels of stimulation. The values of v for intensities 40, 47, and 54 dB re threshold were .012, .0088, and .01, respectively. They correspond to time constants $(\tau = 1/v)$ of 83, 114, and 100 sec. These results appear quite compatible with results obtained by direct matching (Hahn, 1966). It is paticularly intesting to note that these values are quite similar to the time constants estimated for olfactory and taste adaptation.

The last experiments to be considered concern the perception of warmth (Marks & J. C. Stevens, 1968b) and cold (Marks & J. C. Stevens, 1972). In the experiment on warmth, subjects were exposed to irradiant stimulation that heated nearly all of the front surface of the body. Several different levels of irradiant flux were used, and every level of flux was presented through each of several durations between 2 and 12 sec. The subjects judged the magnitude of warmth experienced at the end of each exposure.

The magnitude estimates of warmth as a function of duration are given in Figure 4.12. It is clear that for any constant level of irradiation, the perceived magnitude of warmth varied little between 2 and 12 sec after the onset of stimulation. For the highest levels of irradiation, warmth increased continuously (the extent of the increase was about 50% between 2 and 12 sec); for moderate levels of irradiation, warmth first increased a little, then remained relatively constant; for the lowest levels of irradiation, warmth first increased and then decreased.

These changes in warmth with duration are much smaller than changes described earlier in the chapter under the topic of temporal summation. The reason for the difference seems to be the levels of irradiance used. In the previously discussed study (Marks & J. C. Stevens, in press), only a small portion of the forehead was stimulated, so much more intense stimuli were required to elicit sensations of warmth. (The warmth sense displays a great deal of *spatial summation,* as discussed in Chapter 5.) In fact, the data for 3- and 6-sec exposures displayed in Figure 4.8 show that identical warmth was produced by an irradiance of about 50

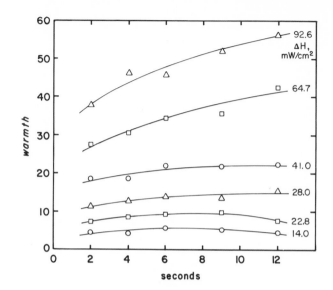

Figure 4.12. Magnitude estimates of warmth produced by thermal irradiation of the front surface of the body, as functions of the duration of the stimulus. The parameter is stimulus irradiance. [Data from Marks and J. C. Stevens (1968b). Courtesy of *Perception & Psychophysics*.]

mW/cm²; over comparable levels of irradiance, those data are in excellent agreement with the data in Figure 4.12.

At first glance, one might wonder what relation these results bear to adaptation. Only the lowest stimulus level at the longer durations shows the decline in sensory magnitude that is typically associated with adaptation. In fact, the initial increases in warmth would appear to be examples of temporal summation, not adaptation. How then are the data shown in Figure 4.12 related to the observations or processes typically termed "adaptation"?

The answer to this question resides in the nature of the stimulus for warmth. It is most certainly not proper to think of thermal radiation as a proximal stimulus for warmth. There does not seem to be any evidence for warmth receptors that are directly responsive to thermal irradiation. (Visual receptors, of course, are directly responsive to electromagnetic radiation.) Warmth can also be produced by touching to the skin an object whose temperature is higher than that of the skin or by exposing the skin to air that is at a relatively high temperature. This is all to say that we are almost certainly better off to look at a variable like skin temperature as the stimulus responsible for warmth sensations.

Figure 4.13. Magnitude estimates of warmth produced by thermal irradiation of the front surface of the body, as functions of the corresponding increases in skin temperature. Each dashed line shows how warmth increases with increasing irradiance, duration constant. Each solid line shows how warmth increases with increasing duration, irradiance constant. [From Marks and J. C. Stevens (1972). Courtesy of University of Illinois Press.]

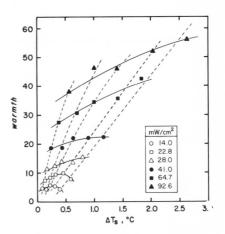

The same magnitude estimates from Figure 4.12 are plotted again in Figure 4.13, but here as a function of the increase in skin temperature produced by each stimulus exposure. It should be obvious from Figure 4.13 that any given increase in skin temperature can be produced in a number of ways, i.e., by heating the skin at a high level of irradiant flux over a short duration (rapid heating) or by heating at a lower level of flux over a longer duration (slow heating). It is apparent that rapid heating results in a larger magnitude of the warmth sensation than does slow heating, even though the increase in skin temperature is the same. One possible *hypothesis* to account for this outcome involves the notion of adaptation. That is to say, we may assume that slow heating provides the opportunity for adaptation to occur to a relatively greater extent than rapid heating provides.

The experiment just described is far from being an ideal paradigm for the study of thermal adaptation. We can only infer the possibility that adaptation (defined as a time-dependent variation in sensitivity) is an appropriate explanatory concept. Since skin temperature increases continuously over time as long as the radiant flux is applied, the "stimulus" (skin temperature) does not remain constant over time. This fact makes extremely difficult any attempt to extract theoretical values of the time constant for warmth adaptation.

Finally, mention shall be made of an experiment concerned with magnitude of cold sensation (Marks & J. C. Stevens, 1972). This experiment was conducted under conditions directly comparable to those used to study warmth. Again most of the front surface of the body was stimulated —in this case by rapidly reducing the level of irradiance that previously kept subjects thermally neutral in cold air.

Over the range of duration, 4–31 sec, sensations of cold seem to correlate much more closely with skin temperature than do sensations of warmth. Cold sensation increases markedly over time. When rate of cooling was low, no evidence of adaptation appeared (a given change in skin temperature, regardless of how rapidly it was brought about, produced constant cold). When rate of cooling was high, however, some evidence of adaptation appeared (rapidly produced changes in skin temperature produced greater cold); but adaptation seemed less extensive than that found for warmth.

SUMMARY

Like temporal summation, adaptation appears to be a nearly universal sensory phenomenon. Most modalities show decrements over time in sensory magnitude as stimulus duration extends over many seconds or minutes. An exception is sensation of muscular effort, which rises continuously over time when a constant force is maintained.

The time course of adaptation can often be approximated mathematically by an exponential equation of the form

$$\psi = ce^{-t/\tau} + j.$$

In the quantitative description of sensory adaptation, two important measures are the extent of adaptation (j) and the rate of adaptation $(1/\tau)$.

1. Only a few senses show complete adaptation, i.e., diminution of sensory magnitude to zero $(j = 0)$. Taste is an example of one that does. Lights (under normal conditions of viewing), sounds, odors, vibrations seem not to disappear, at least so long as the sensations were not extremely weak at the start.

2. The rate of adaptation (inverse of the time constant τ) appears remarkably constant across several sensory modalities. Time constants on the order of 100 sec are the rule for odor, taste, and vibratory adaptation. Auditory adaptation shows a slightly lower value—about 60 sec. Only in vision does the time constant of adaptation differ markedly from 100 sec: brightness adapts with a time constant whose size varies inversely with luminance. Results suggest a range in time constant at least 10–40 sec.

Prototypes for Models of Temporal Summation and Adaptation

Let us now take a rather general look at the nature of temporal summation and adaptation. Although in a basic respect the data are somewhat fragmentary, we are able to make some quantitative statements about the temporal course of sensation in a number of sensory modalities. It is true that only for vibrotactile sensation do we have direct scaling data for both the summative and adaptational periods. These data are given in Figure 4.14. (Note the different time scales; the gap corresponds to an interval over which data were not obtained in either experiment.) Figure 4.14 demonstrates quite clearly the large difference between the rates of temporal summation and adaptation. Summation takes place at most over durations of a couple of seconds; adaptation on the other hand continues for several minutes. The same generalization holds for brightness and loudness, and perhaps for other modalities.

The time course of sensations of visual brightness (ranging over both the summative and adaptational periods) can also be described, even though no direct scaling data for brightness adaptation have been obtained. The temporal course of brightness is complex, since the durations and extents of both the summative and adaptive stages depend on stimulus intensity. High luminances result in brief summation (a few tens of milliseconds) and rapid adaptation (a few tens of seconds), whereas low luminances result in more prolonged summation (a few hundreds of milliseconds) and slow adaptation (hundreds of seconds). Furthermore, high luminances produce less extensive summation and more extensive adaptation than do low luminances. The general nature of the initial rise and subsequent decay, however, resembles that found for vibrotactile sensation.

It would most certainly be flagrantly optimistic to expect all, or even most, sensory modalities to show quantitatively similar functions for the time course of sensory magnitude. We have already noted a few phenomena that appear to be somewhat idiosyncratic, or at least not universal. These phenomena include, in the realm of brightness, the Broca-Sulzer enhancement and the dependence of the temporal limit of summation on

Figure 4.14. The course of temporal summation and adaptation of vibrotactile sensation. Summation is much more rapid than is adaptation. (Note that the time scales in A and B differ.) [Reprinted with permission of author and publisher: Berglund, U., and Berglund, B. Adaptation and recovery in vibrotactile perception. *Perceptual and Motor Skills*, 1970, **30**, 843–853.]

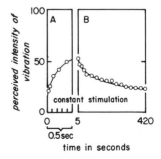

luminance level. Furthermore, it may well be that some sensory modalities will not even show evidence of adaptation; for example, there is evidence that thermally induced pain may not adapt (Greene & Hardy, 1962), and the basis of perstimulatory auditory adaptation is unclear. Even so, it may be thought likely that for several sensory modalities— e.g., olfaction—the time course of sensory magnitude will eventually be shown to follow qualitatively and even quantitatively similar functions to those seen for vibrotactile sensation and for brightness.

No attempt will be made here to generate quantitative models of temporal summation and adaptation for all sensory modalities. An attempt will be made, however, to show what the general nature of such models might be. It should be pointed out at the start that any model purporting to describe summation and adaptation will necessarily have to account for the results of numerous experimental studies in addition to those studies concerned with direct scaling.

The "proto-models" to be described here are conceptually founded and based upon some general principles elucidated in several models of temporal phenomena in vision (Fuortes & Hodgkin, 1964; Matin, 1968; Rushton, 1965; Sperling & Sondhi, 1968; Sperling, 1970). The basic features of the proto-models are depicted in Figure 4.15.

Two features are of major importance. The first is that the system contains feedback. Part of the output is assumed to feed back and thereby to modify the system's sensitivity. The feedback is inhibitory (−), and it serves to decrease sensitivity. The second important feature is that the input can be inhibitory (−) as well as excitatory (+). The single stage depicted in Figure 4.15 can be expanded if necessary, e.g., into a concatenation of stages, in order to account for the observed psychophysical behaviors.

Let us first consider the model as it may be applied to temporal summation. In a certain sense, the problem in understanding temporal summation may reside less in explaining why summation occurs than in explaining why it ends. The notions of inhibitory feedback and of inhibitory input provide two possible means by which summation could terminate. Marks's (1972) model incorporates feedback in order to account for temporal summation of brightness. That model employs two stages (like

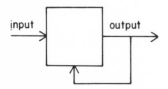

Figure 4.15. Prototype for feedback models of temporal summation and adaptation. Part of the output feeds back and acts as inhibitory input.

those of Figure 4.15) with the inhibitory feedback originating at the second stage only, but modifying the sensitivities of both stages. Zwislocki's (1969) model for temporal summation of loudness also fits the general scheme of the present proto-model. Zwislocki postulated two processes underlying auditory summation. One invokes complete summation over short durations, and that process can be described in terms of the integration of excitatory and inhibitory effects of input. Zwislocki assumed this summation takes place after the power transformation between sound pressure (P) and loudness. Prior to temporal summation, then, loudness equals $P^{2/3}$, and after temporal summation loudness equals $P^{2/3}t$. In order to account for the relation actually observed (loudness about proportional to $P^{2/3}t^{2/3}$) a second process was invoked. This process was purely inhibitory (adaptational) in nature, and was assumed to take place more centrally in the auditory system. We might ascribe the latter process to either inhibitory input or inhibitory feedback at a later stage.

Figure 4.15 is also useful as a model of adaptation. Such a model was proposed by Rushton (1965), which, in turn, was based upon a model set forth by Fuortes and Hodgkin (1964). The concept central to this model is that adaptation is the result of a mechanism that operates in the manner of an "automatic gain-control." In other words, the sensitivity of the sensory system is modulated by the momentary output that the system produces; by means of a feedback loop the system controls its own gain.

One of the suggestive features of visual light adaptation is its nonlinearity. Figure 4.9 showed that both the magnitude of decrement in brightness and the time constant of adaptation vary with stimulus intensity: the greater the intensity, the greater the decrement and the smaller the time constant. Whether the nonlinearity is the result of inhibitory processes at input (feed-in) or at output (feedback) is not as yet clear; the relatively small changes in brightness and in time constant, compared to the large range of stimulus intensities, suggest that it may be level of unadapted brightness that determines the course of brightness adaptation.

Although the proto-model depicted in Figure 4.15 appears quite simple, great complexity can arise from any attempt to apply it quantitatively to the results of psychophysical investigations. It is most certain, for example, that many stages of processing are involved. However, we may examine in a qualitative manner how the proto-model might be applied to the data obtained on warmth that were shown in Figures 4.12 and 4.13. Let us assume that the input to the proto-model is proportional to the increase in skin temperature (or, perhaps, to the increase in skin temperature raised to some power, if such a transformation precedes the stage of adaptation). Recall that a constant level of irradiant flux results in a continuous rise in skin temperature. Thus, the magnitude of the input increases constantly over time. Without any gain control we would expect the output of the system also to increase continuously over time. But gain control in the system means that the output is fed back continuously over time and modulates (decreases) the system's sensitivity. The result is that the output does not keep up with the input; the output may increase more slowly than the input, may level off, or may even begin to decline.

Note that Figure 4.12 shows a systematic variation in the time course of warmth sensation with change in level of irradiation: given high levels of flux, warmth increases continuously; given intermediate levels, warmth increases and then levels off; given low levels, warmth first increases, then decreases. Thus, adaptation of warmth appears to be level dependent. One possible means to account for this variation in time course is to postulate that the *rate of adaptation*, i.e., the time constant of the gain control, depends upon level of input. That is, as long as the increase in skin temperature is small, the gain control responds rapidly and leads to markedly decreased sensitivity; but as skin temperature continues to increase, the response of the gain-control mechanism itself slows. Alternative assumptions exist, however. One alternative is that the relative *degree of adaptation*, i.e., the extent to which the gain control can modulate sensitivity, depends upon level of input; as skin temperature increases, the relative decrease in sensitivity produced by the gain control itself decreases.

These two alternative assumptions would result in different predictions about how warmth should vary over time when the temperature of the skin is increased abruptly, then held constant. If we assume that the time constant of the gain control depends upon level of input, then we would predict adaptation to proceed more slowly the greater the increase of skin temperature. If we assume that the relative degree of decrease in sensitivity exerted by the gain control depends upon level of input, then we would predict adaptation to proceed at the same rate for any magnitude increase in skin temperature; however, larger increases in skin temperature should result in relatively greater asymptotic levels of output at long durations. (It is possible, of course, that changes occur both in rate and in extent of control of gain.)

Another alternative entails the postulation of two mechanisms, the outputs from which sum linearly. Both mechanisms would respond directly to the temperature of the skin. Only one mechanism would contain gain control, i.e., adapt, but the time constant would not change with level. The second mechanism would display little or no adaptation. If the proportion of outputs from the two mechanisms changed with increasing input irradiance, the total output would yield level-dependent variations in the time course of warmth. This alternative takes on special interest since a hypothesis of two mechanisms mediating warmth sensation is also suggested by the bisegmented warmth functions of Figure 4.8 and by data on spatial summation discussed in the next chapter.

Adaptation and Sensitivity to Subsequent Stimulation

The effect of adaptation is not limited to decreases in sensory magnitude with increasing duration of exposure. In addition, adaptation modifies the sensitivity of a sensory system to subsequent stimulation. In general, this modification consists of a decrease in sensitivity. although there are also certain conditions under which adaptation enhances subsequent sensitivity. The basic paradigm for the experiments to be considered here is as follows. First, the subject is exposed to an adapting stimulus (typically, the exposure duration is great enough to permit the magnitude of the sensation produced by that stimulus approximately to reach an asymptotic value). Second, either the adapting stimulus is removed and a (usually brief) test stimulus is presented, or the test stimulus is superimposed on the adapting stimulus. The magnitude of the sensation aroused by the test stimulus can be measured by one of the direct scaling procedures, and the effect of adaptation can be evaluated by comparison of these results to results obtained when test stimuli are presented without prior adaptation.

VISION

The effect of adaptation on sensitivity to subsequent stimulation has been examined in a number of sensory modalities. The first modality in which direct scaling procedures were used was vision, where J. C. Stevens (1957), Onley (1961), and J. C. Stevens and Stevens (1963) measured the effect of light adaptation on the brightness of short flashes of light. The results of these experiments show clearly that the effect of light adaptation on the brightness of subsequent light flashes depends critically on the luminance of the subsequent flash. The most prominent effect of adaptation is to raise substantially the threshold for subsequent flashes. The effect on the brightness of suprathreshold flashes diminishes, however, as the luminance of the test flash increases. That is to say, as the luminance of the test flash increases, the brightness aroused is more nearly the same as that aroused by a test flash presented to a non-light-adapted eye.

Onley's experiment demonstrated that these conclusions hold both for test flashes presented after the adapting field is removed and for test flashes superposed on the adapting field.

The effect of light adaptation on the relation between brightness and luminance can be seen most clearly in Figure 4.16. Each curve in Figure 4.16 shows how brightness grows with luminance just after the adapting field is extinguished. That is, the adapting field is turned off, a test stimulus is turned on (for a duration of 2 sec), then the adapting field reappears. The subject estimates numerically the brightness of the test flash. Each curve in Figure 4.16 gives a brightness function obtained under a specified condition of adaptation: from dark adaptation at the far left to high-luminance light adaptation (105 dB *re* 10^{-6} cd/m^2) at the far right. Each curve conforms to the psychophysical power equation

$$B = k(L - L_0)^\beta \tag{4.12}$$

L_0 is related to the absolute threshold, and its value increases dramatically with increasing level of adaptation. Adaptation also modifies the value

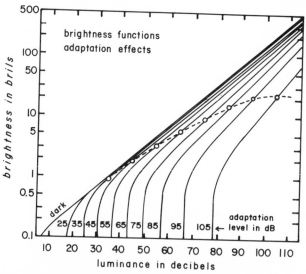

Figure 4.16. Psychophysical functions showing how brightness increases with luminance in decibels *re* 10^{-6} cd/m^2. The parameter is the level of light adaptation that preceded stimulation. The dashed line gives the locus of equilibrium brightness, i.e., the brightness produced by a stimulus equal in luminance to the level of adaptation. [From J. C. Stevens and S. S. Stevens, Brightness function: Effects of adaptation. *Journal of the Optical Society of America*, 1963, **53**, 375–385. Courtesy of the authors and *The Journal of the Optical Society of America*.]

of the exponent β. The value of β increases with increasing adaptation; thus, adaptation modifies the rate at which brightness grows as a function of luminance. High levels of adaptation decrease brightness of a subsequent flash to a greater extent than do low levels; the relative effect is greater for low- than for high-luminance test flashes. If we look at these results in terms of a gain-control model for adaptation, it can be said that the gain control primarily decreases sensitivity to small levels of visual input.

A family of functions similar to those seen in Figure 4.16 was reported by van den Brink (1962). His experiment utilized an interocular brightness-matching procedure in which one eye was light adapted, the other, dark adapted. Instead of parametrically varying the luminance of the adapting field, however, van den Brink kept adapting luminance constant (at 101.5 dB) and parametrically varied the time interval between extinction of the adapting stimulus and onset of the test stimulus. The shorter the time interval, the more the brightness function is influenced by adaptation (the larger the exponent and the higher the threshold). Over the range of intervals from about 700 to 1200 sec, the calculated brightness functions showed inflections. The inflections were interpreted to reflect transitions from rod to cone functioning.

OLFACTION

The results just described for the effects of visual adaptation have their counterpart in other sensory modalities. Similar experiments have been conducted in taste (Bartoshuk, 1968; McBurney, 1966; Meiselman, 1968a) and in olfaction (Cain, 1970; Cain & Engen, 1969). In their major respects, the results of these experiments are strikingly similar to the results obtained in vision. The effect of adaptation is primarily to diminish the magnitude of sensation aroused by weak levels of test stimuli; the effect on high levels is much smaller, and sometimes is negligible. However, some additional, rather interesting phenomena, emerge from these experiments.

Cain (1970) examined how adaptation to propanol (alcohol with three carbon atoms) and to pentanol (alcohol with five carbon atoms) affected odor intensity to subsequent stimulation with the same alcohol (self-adaptation) and with the other alcohol (cross-adaptation). Under conditions of self-adaptation, the two alcohols had equal adapting effects when the adapting concentrations produced equal subjective odor intensities. Thus, those results were consistent with the hypothesis that degree of adaptation is determined by the subjective strength of the adapting stimulus. Under conditions of cross-adaptation, however, results were more complex. Adaptation to pentanol produced a greater decrease in odor intensity of propanol than propanol produced in pentanol. Thus, cross-adaptational effects can be highly asymmetric.

Taste

Figure 4.17 shows the magnitudes of taste sensations produced by solutions of sodium chloride after adaptation to several different concentrations of sodium chloride. Test concentrations near the concentration of the adapting solution produce weak sensations, whereas high test concentrations produce sensations that are relatively independent of adapting concentration. But note that test concentrations below the adapting concentrations can also produce moderately strong taste sensations: Taste intensity increases as the test concentration decreases more and more below adapting concentration. It is important to point out that the taste quality depends upon whether test concentration falls above or below adapting concentration; test concentrations below adapting concentration produced a sour–bitter taste; concentrations above adapting concentrations produced a salty taste. The shift in taste quality has important implications for models of fundamental taste mechanisms.

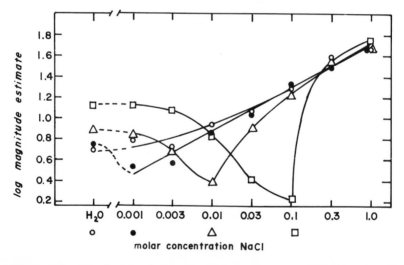

Figure 4.17. Magnitude estimates of the taste of sodium chloride, as functions of stimulus concentration. The parameter is the state of adaptation that preceded stimulation: open circles, adaptation to water; filled circles, adaptation to .001 *M* sodium chloride; triangles, adaptation to .01 *M* sodium chloride; squares, adaptation to .1 *M* sodium chloride. [From D. H. McBurney, Magnitude estimation of the taste of sodium chloride after adaptation to sodium chloride. *Journal of Experimental Psychology*, 1966, 77, 869–873. Copyright 1966 by the American Psychological Association, and reproduced by permission.]

As Figure 4.17 makes clear, the sour–bitter taste was strongest when the test solution was pure water. Bartoshuk (1968) showed that adaptation to sodium chloride and to sucrose produce bitter tastes when test concentrations fall below adapting concentration, and that hydrochloric acid and quinine hydrochloride produce sweet tastes when test concentrations fall below adapting concentrations. The subjective magnitudes of these tastes grew as concentration decreased, and they were strongest when the test solution was pure water.

McBurney and his colleagues have continued the study of the sensory magnitudes and qualities of taste after adaptation (McBurney, 1972; McBurney & Lucas, 1966; McBurney & Shick, 1971; McBurney, Smith, & Shick, 1972; Smith & McBurney, 1969). An important feature of their work is the elaboration of the method of magnitude estimation. In addition to estimating the total subjective magnitude of taste substances, their subjects are asked also to divide up the total estimates into submagnitudes appropriate to each component quality. This procedure produces a taste profile, which states the proportions of sweetness, saltiness, sourness, and bitterness of each substance under the experimental conditions. By comparing taste profiles obtained before and after adaptation, one can determine both the degree of change in taste magnitude and also which taste qualities show changes. Adaptation to initially sweet substances decreases the sweetness of subsequent stimuli, adaptation to bitter decreases bitterness, etc. The degree to which adaptation to one substance suppresses the taste of subsequent stimuli seems to depend on the extent that they have common taste qualities. An exception to this generalization must be made with respect to the bitter taste, which appears to be somewhat complex. Adaptation to quinine hydrochloride reduced the bitterness of quinine salts and caffeine, but not the bitterness of urea or magnesium sulfate; adaptation to urea, on the other hand, reduced the bitterness of quinine salts and magnesium sulfate, but not the bitterness of caffeine (McBurney et al., 1972). It may be that two separate mechanisms mediate the bitter taste.

In addition to suppression, adaptation also produces the "water tastes" measured by Bartoshuk (1968). It might be expected that substances that taste similar would, when used as adapting stimuli, induce similar tastes in water. Unfortunately, the outcome is not so simple. There does not appear to be any obvious, general relation between taste of adapting solution and subsequent water taste (McBurney & Shick, 1971).

Touch

Gescheider and Wright (1969) examined how adaptation to 60-Hz vibration affected the sensation magnitude of subsequent stimuli. They conducted the study in two stages. First, each subject adapted a finger for 10 min to a strong vibratory stimulus, then he adjusted the amplitude of vibration on the adapted finger in order to produce a sensory match to a stimulus applied to an unadapted finger. In the second stage, subjects gave magnitude estimates of perceived vibration to stimuli presented to the unadapted finger. On the basis of these estimates, the effect of adaptation (measured in the first experiment) could be assessed in psychological units. In line with results obtained on other sensory modalities, Gescheider and Wright found that adaptation had a greater relative effect on weak than on strong stimuli. That is, adaptation produced a steepening of the psychophysical function for vibration.

Recovery after Adaptation

Vision

Following prolonged constant stimulation, it is typically found that the sensitivity of a sensory system is depressed. (An example to the contrary is sensitivity to subadaptational concentrations of taste solutions discussed in the last section.) We have just seen one way in which the depression in sensitivity can be examined, namely by determining the nature of the psychophysical function after adaptation. One can also examine how sensitivity, decreased by prolonged stimulation, recovers over time after the adapting stimulus is removed. The increase in the eye's

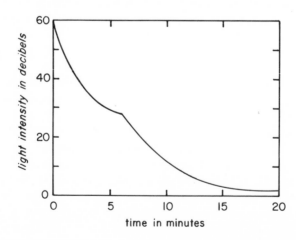

Figure 4.18. The course of visual dark adaptation (absolute threshold after adaptation to intense light). The initial rapid increase in sensitivity is due to recovery by cones, the later slow increase to recovery by rods.

sensitivity with time after entering a dark movie theater is an example of such recovery. In fact, this recovery, usually referred to as dark adaptation, is one of the most significant and exhaustively studied phenomena in the realm of vision. Figure 4.18 gives a typical example of the recovery of absolute sensitivity after adaptation to intense white light. Note that the increase in sensitivity is rapid at first, levels off, then proceeds more rapidly again before approaching its final asymptote. It has been known for a number of years that each of the two segments of the dark-adaptation curve corresponds to a separate receptor mechanism. The initial decline and plateau in threshold corresponds to recovery in sensitivity by the retinal cones; the later, slower, and more extensive decline in threshold corresponds to recovery by the retinal rods.

It is clear, therefore, that as time in the dark increases, the sensitivity of the eye, as measured by the absolute threshold, also increases. We might reasonably expect also that as time in the dark increases, the brightness aroused by a flash of light of constant intensity will increase. The way in which suprathreshold brightness increases as a function of recovery time has been scaled by Ekman, Hosman, and Berglund (1966). They preadapted each subject's eye to white light (luminance not specified), then measured how brightness of test flashes changed with increasing recovery time in the dark. Recovery followed very nearly the same course for the six levels of test luminance (32–39.5 dB) examined. The average recovery curve is depicted in Figure 4.19. The curve appears to consist of two negatively accelerated functions, and Ekman *et al.* presented evidence that two components, each a logarithmic function of time in the dark, can adequately account for the shape of the recovery curve. One component presumably operates over the entire duration of the recovery curve; the other makes its appearance only after about 7½ min of recovery.

Ekman *et al.* did not attempt to ascribe each of the two components of recovery to a separate receptor process. At first glance, we might interpret the function shown in Figure 4.19 to be a direct counterpart to recovery measured in terms of the absolute threshold. The next logical step would be to ascribe the initial component of recovery to the cones, the later component to the rods. It is puzzling, however, that the shape of the recovery curve was independent of test luminance. If we were to ascribe the second component simply to the rods, we might predict the inflection in the recovery curve to depend upon level of luminance. That is, with higher and higher levels of luminance the rod threshold should be surpassed at shorter and shorter recovery durations. It would appear, therefore, that the origin of the two components of recovery—the interaction of receptor mechanisms—may be quite a complex matter.

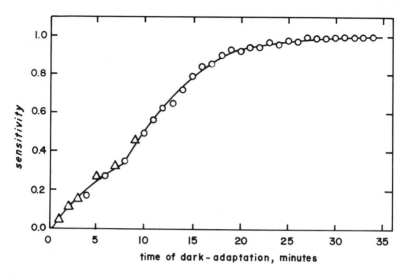

Figure 4.19. Course of visual dark adaptation after prior adaptation to intense light, as measured by the increase over time in brightness of a stimulus of constant luminance. Circles: averages of three levels of luminance. Triangles: average of one level of luminance. [Reprinted with permission of author and publisher: Ekman, G., Hosman, J., and Berglund, U. Perceived brightness as a function of dark-adaptation. *Perceptual and Motor Skills*, 1966, **23**, 931–943.]

The recovery of brightness was also measured by direct matching, in a study by van den Brink (1962). First, the eye was adapted to white light at 101.5 dB *re* 10^{-6} cd/m^2; then subjects adjusted the luminance of a light presented to the other (dark-adapted eye) to track the recovery of brightness of test lights presented to the light-adapted eye. (Test lights varied in luminance, 45–82.5 dB.) Van den Brink then converted values of matching luminance into units of brightness by means of a cube-root function. Brightness increased with recovery time according to a negatively accelerated function. Unlike the functions reported by Ekman *et al.*, however, van den Brink's functions showed no evidence for two components. Furthermore, the shapes of the recovery curves varied systematically with luminance in van den Brink's study: the higher the luminance, the more rapid the initial recovery. A possible reason for the differences is that van den Brink employed luminances to the adapted eye of 45 dB and greater, whereas Ekman *et al.* employed luminances less than 40 dB. Another difference is that van den Brink used a 6.5° field, Ekman *et al.*, 45° fields.

The data from both experiments imply that complete recovery—in terms of brightness—takes a relatively long time. In van den Brink's experiment, recovery was not always complete at the end of 8 min (the longest interval, examined; in Ekman *et al.*'s experiment, recovery was complete after about 20 min). This value contrasts with the 1 or 2 min (or less) required for brightness to adapt completely (Geldard, 1928). With brightness, then, recovery seems quite a bit slower than adaptation, at least when recovery is measured with stimuli less luminous than the adapting stimulus.

OLFACTION

It is likely that for some sensory modalities at least, the nature of recovery from adaptation, i.e., the form of the recovery curve, will turn out to be simpler than what appears to obtain for vision. Simpler recovery curves have in fact been obtained for odor and for vibrotactile sensation. Ekman *et al.* (1967) noted the general trend of olfactory recovery after stimulation with hydrogen sulfide (see Figure 4.10). Recovery, as measured by the increase in odor magnitude produced by a short presentation at the same concentration as used for adaptation, appeared to follow roughly the same time course as did adaptation. That is, to a first approximation the recovery curves look very much like the adaptation curves turned upside down.

TOUCH

The outcome for olfactory recovery is in marked contrast to results for recovery of vibrotactile sensation (Berglund & Berglund, 1970). Vibrotactile recovery seems extremely rapid in comparison to adaptation. (The reverse appeared to be the case for brightness.) The difference is apparent in Figure 4.20, which shows temporal summation, adaptation, and recovery for vibrotactile stimulation. Summation has a rather short time

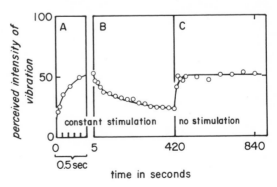

Figure 4.20. The course of temporal summation, adaptation, and recovery of vibrotactile sensation. Summation and recovery are much more rapid than is adaptation. [Reprinted with permission of author and publisher: Berglund, U., and Berglund, B. Adaptation and recovery in vibrotactile perception. *Perceptual and Motor Skills*, 1970, **30**, 843–853.]

span (only a few seconds at most); this is followed by a comparatively long-lasting period of declining sensation (adaptation). Seven minutes of constant stimulation fails to result in sensation magnitude reaching its asymptote. If stimulation is discontinued after 7 min, however, nearly complete recovery is evidenced after only another 1–2 min.

It is worthwhile to point out that, for the last two experiments described, recovery was measured in only one way, namely in terms of the increase in sensation magnitude of a stimulus whose intensity was the same as that of the adapting stimulus. We may recall that when we examined the effects of adaptation in terms of sensitivity to subsequent stimulation, we saw that (visual) light adaptation, for example, depressed sensitivity to weak stimuli to a greater extent than it depressed sensitivity to stronger stimuli. It might be expected then that a similar phenomenon occurs in recovery; and, in fact, there is evidence that this is precisely the case. An example is an experiment conducted by Hahn (1966), in which recovery from vibrotactile adaptation was seen to proceed more rapidly for suprathreshold vibrotactile stimulation (determined by direct matching) than for threshold.

More direct attempts to assess vibrotactile recovery in terms of sensory magnitude were made by Gescheider and Wright (1968, 1969). Recall that they used a two-step procedure: first, magnitude of vibrotactile sensation was scaled by the method of magnitude estimation; then, direct matches were obtained between the sensations aroused by stimulation of an unadapted finger and of a finger previously adapted to a high level of stimulation. The matches were obtained at several stimulus levels and at recovery intervals that varied from 5 sec to 6 min; the matching levels were transformed into values of sensation magnitude in accordance with the initial magnitude estimates. Since the test levels of stimulation were all lower than the adapting level, we might expect, and the results show, a longer time course for recovery than was found by Berglund and Berglund (1970). Given any constant time for recovery, the relation between sensation magnitude and vibration amplitude was a power function whose constants varied systematically with recovery time (Gescheider & Wright, 1969). It would appear from the data presented by Gescheider and Wright that even after 6 min, recovery was not complete. Vibrotactile recovery, then, is not always very rapid. It is tempting to hypothesize a simple relation between the effects of level of adaptation and of recovery time. Just as increasing the level of light adaptation systematically increases the threshold and the exponent of the psychophysical function for subsequent stimulation [see Eq. (4.12)], so decreasing recovery time after vibrotactile adaptation increased the threshold and exponent of the psychophysical function for perceived vibration. In a qualitative manner, the hypothesis may have some validity. There is some evidence, however, that light adaptation and recovery do not affect the form of the brightness functions in precisely the same way (van den Brink, 1962). For this reason, therefore, the hypothesis will not be further developed here.

Repetitive Stimulation (Flicker)

Some rather complex temporal interactions (interactions of summation, adaptation, and perhaps additional, emergent phenomena) occur with repetitive, intermittent stimulation. (Vibratory stimulation of the skin can be considered intermittent tactile stimulation; however, vibratory sensations are dealt with elsewhere in this chapter, and in other chapters, and therefore will not also be dealt with here.)

A light turned on and off rapidly enough appears steady in brightness. If the frequency of modulation is decreased, a point is reached at which some temporal variation of brightness can be detected, and as the frequency is lowered still further, the light appears to pulsate with increasing magnitude and at a slower rate. An extremely large number of experiments have been conducted whose purpose was to explore diligently all of the parameters that affect flicker thresholds (the transitions between perceptions of steady and nonsteady light).

A few investigators have been interested in the perception of suprathreshold flicker. J. C. Stevens and Shickman (1959) found perceived rate of flashes to be proportional to actual flash rate over the range 1–40 flashes/sec. (Similar proportionality was reported for perception of rate of intermittent auditory and tactile stimulation.)

In vision, there has been a good deal of interest in the phenomena termed *brightness enhancement.* We say "phenomena" because the term brightness enhancement has been applied to two somewhat different effects. When light is modulated very rapidly (above the flicker threshold), its brightness is equal to the brightness of a steady light whose luminance is the same as the time-average luminance of the intermittent light. In other words, if the intermittent light is on half the time, off half, its brightness would be the same as that of a steady light at half the peak luminance. At rates of modulation below the flicker threshold, there appears a primary, steady component with flicker superimposed. Studies by Bartley (Bartley, 1938; Schneider & Bartley, 1966) have shown that the brightness of this steady component can be greater than its brightness is at higher frequencies when no flicker is perceived. This increase Bartley termed brightness enhancement.

At much lower frequencies of modulation, the steady component disappears, and one sees a more or less discontinuous series of flashes. At some frequencies, the brightness of these flashes appears greater than the brightness that occurs when the light is not interrupted at all (Rabelo & Grüsser, 1961; Wasserman, 1966). This second phenomenon has also been called brightness enhancement. Several investigators (e.g., Wasserman, 1966) have suggested that this latter phenomenon is identical to the Broca-Sulzer phenomenon that was discussed earlier. If they are not identical, they are most certainly very closely related.

Most of the studies just mentioned examined the effects of intermittent light that was fully modulated. The perception of flicker is not limited to cases where a light is turned completely on and off. In one experiment, the amplitude as well as the frequency of modulation of white light was varied, and subjects were called upon to judge the "apparent depth of modulation," i.e., how *much* the brightness seemed to change from one moment to the next (Marks, 1970). (The subjects were instructed specifically to ignore changes in apparent rate of flicker.) Figure 4.21 gives the results for a typical subject at one of the luminance levels examined. For each curve, both average luminance and degree of modulation were held constant, while frequency was varied. (The modulation marked 44%, for example, indicates that the luminances of the peaks and troughs of the square wave were 44% greater and less, respectively, than the average luminance. For any given frequency, the greater the percent modulation, the greater the apparent depth of modulation.)

Large amplitudes of modulation (upper curves) are monotonic. They show that apparent depth of modulation decreases rapidly at high frequencies. Small amplitudes of modulation (lower curves) are nonmonotonic. They show a peak in apparent depth of modulation at frequencies around 8 Hz. That is to say, when the light is modulated to a small degree, apparent depth of very slow rates of modulation may be detected as very faint, if it is detected at all; the same is true for very rapid rates. At some intermediate rate, however, brightness changes are distinct and relatively large. The nonmonotonicities shown in Figure 4.21 are not unexpected, since measurements of flicker thresholds (for example, de Lange, 1961; Kelly, 1961a) show similar peaks. It is the monotonicity at high amplitudes of modulation that is rather curious. A strong possibility exists that high amplitudes of modulation also yield peaked functions, but have peaks at frequencies lower than those examined in the experiment.

Unfortunately, it is not possible to relate these results in any simple manner to the principles elucidated earlier concerning temporal summation and adaptation. Flicker, it seems, is quite a complex, complicated,

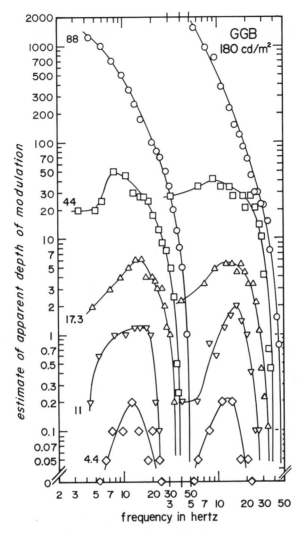

Figure 4.21. Magnitude estimates of "apparent depth" of temporal modulation of white light, as functions of modulation frequency. The parameter is the percentage that the square waveform was modulated. The left-hand section gives results for a series of decreasing frequencies; the right-hand section gives results for a series of increasing frequencies. Mean luminance was 82.5 dB *re* 10^{-6} cd/m². [From L. E. Marks, Apparent depth of modulation as a function of frequency and amplitude of temporal modulation of luminance. *Journal of the Optical Society of America*, 1970, **560**, 970–977. Courtesy of *The Journal of the Optical Society of America*.]

and sometimes bewildering phenomenon. A number of theorists have tried their hands at establishing quantitative models (see, for examples, de Lange, 1961; Kelly, 1961b; Matin, 1968; Sperling & Sondhi, 1968). These models often become quite complex, even though they often attempt only to account for threshold data. One possible application of a model is the extension of the feedback model of temporal summation. The two-stage feedback system outlined earlier, with the addition of low-pass filter stages, might suffice to account for the results. Such a model would be very similar to one proposed by Sperling and Sondhi (1968) to account for modulation sensitivity at threshold. The use of direct scaling procedures, such as magnitude estimation, would seem to open up a vast region of endeavor in the field of flicker that has been largely ignored. Whether data concerning suprathreshold flicker will simplify or complicate the job of theorists and model builders remains to be seen.

Summary

The general scheme outlined in Figure 4.1 applies to most sensory systems. After a stimulus is turned on, sensation grows rapidly in magnitude over a relatively short period of time (temporal summation), then declines more slowly over a longer period (adaptation). The actual extents of summation and adaptation vary from one sense to another, and, in some instances, vary within a single sense as a function of stimulus intensity. It is sometimes convenient to look at summation and, especially, adaptation in terms of gain-control mechanisms, which might operate by means of inhibitory feedback.

After a sensation has passed through the phases of summation and adaptation, removal of the stimulus leaves the sensory system in an "adapted" state. The magnitude of sensory adaptation may be assessed not only in terms of extent of adaptation itself, but also by the effect an adapting stimulus has on responses to subsequent stimulation.

Once an adapting stimulus is turned off, sensitivity begins to return over time. Recovery of subjective odor magnitude seems to take about as long as adaptation itself takes, at least when recovery is measured with a stimulus concentration identical to adapting concentration. Under similar conditions (equal test and adapting intensities), recovery of vibrotactile sensation seems to be more rapid than is adaptation. If vibrotactile recovery is measured with test stimuli of lower intensity than the adapting stimulus, complete recovery takes much longer than adaptation. Thus, speed of recovery depends (at least in some modalities) on the relative intensities of adapting and test stimuli. Not surprising, then, is the finding that recovery of brightness after light adaptation can take many minutes when test luminance is much lower than adapting luminance, although brightness adaptation itself takes no longer than a minute or two.

Adaptation acts to decrease (usually) sensitivity to subsequent stimulation: (a) the more intense is the adapting stimulus, the greater is the decrement in sensation; and (b) in general, the relative decrease in sensory magnitude is large when intensity of the test stimulus is much lower than the intensity of the adapting stimulus; as intensity of the test stimulus increases, the effect of adaptation diminishes. Both generalizations hold for visual, olfactory, and vibrotactile sensations. Apparent anomalies occur, however, in taste. For example, when concentrations of test solutions of sodium chloride are lowered more and more below concentration of the adapting solution, taste sensation *increases*. The quality of those subadaptational sensations (sour or bitter) differs from the quality (salty) of test concentrations above adapting concentration. Therefore, the generalization that adaptation serves to depress sensitivity to low stimulus intensities holds also for taste, if it is restricted to refer to sensations whose quality is the same as that of the adapting stimulus.

Large gaps remain in our knowledge of temporal summation, adaptation, and recovery. Even with the gaps that exist, however, it is not unreasonable to assert that marked similarities exist across sensory modalities with regard to these three processes. When these gaps are filled, we shall not only have a more firm understanding of the way each individual sensory system operates, but we shall most certainly also gain insight into the nature of properties common to all sensory systems.

Chapter 5

SPATIAL FACTORS IN SENSATION

We have seen how the duration of stimulation can have a profound influence on the perceived magnitudes of sensory stimuli. We shall now see how variations in spatial distribution of stimuli can also markedly influence the input–output relations of sensory systems. The effects of spatial variations are exceptionally important for two reasons. The first is that they play an invaluable role in the perceptions of the conscious, behaving, thinking person as he functions in his environment. The second is that elucidating these effects can lead to a deeper understanding of the mechanisms that underlie sensation. Take brightness contrast as an example. Because contrast serves to enhance brightness differences among spatially contiguous objects, it also serves to increase the salience of contours and boundaries. Contrast helps to make objects appear as objects. In addition, of course, we wish to understand the nature of the visual mechanisms that produce contrast.

It is worthwhile to keep in mind that functional significance of spatial variations may vary dramatically from one sensory system to another, and, indeed for certain systems such variations may be of negligible import. Let us be concrete. The visual system, to give one example, is capable of making exquisitely fine spatial discriminations. This is perfectly obvious from our abilities to perceive and discriminate objects in the visual world (consider the visual discrimination required in order to read this book). By comparison, systems such as the senses of warmth and cold can make only much cruder spatial discriminations. More important for the thermal senses is the fact that they provide information about thermal changes that take place over large portions of the body's surface. That is

146

to say, the thermal senses tend to signal the degree to which the body as a whole is gaining or losing heat with the end result physiological and/or behavioral temperature regulation. As a final example, consider the sense of taste. Under most ordinary conditions of tasting, it would appear that sapid substances are spread more or less uniformly over the tongue. Under such circumstances, there seems little need for any great capacity to localize gustatory sensations over the tongue's surface. On the other hand, there might be functional advantage—as in the thermal senses—to having a sizeable degree of spatial summation.

This chapter will deal with three types of spatial variation: spatial summation, contrast or masking, and locus of stimulation on the sensory surface. Spatial summation refers to the dependence of sensitivity and of sensory magnitude on the areal size of the stimulus. Dip one finger in water a few degrees above skin temperature and a mild sensation of warmth ensues. Dip the entire hand and a much greater sensation is felt. Contrast (masking) refers to the effect (usually of a suppressive nature) one stimulus exerts on another. The roar of a jet plane drowns out conversation, and the glare from car headlights decreases visibility on the road. With regard to locus of stimulation, sensory surfaces often are nonuniform in their sensitivity. A spot of heat on the back may be imperceptible, but to the face may be felt as fairly warm; a dim star seen from the corner of the eye disappears when you try to look straight at it.

Spatial Summation

Nearly all of the sense organs provide surfaces, of one size or another, upon which stimuli eventually impinge. The surface may be relatively small, as in the cases of the retina of the eye and the basilar membrane of the ear, or quite large, as in the cases of the skin and of the olfactory epithelium. Regardless of whether the sensory surface is large or small, sensation magnitude often depends on how much of the sensory surface is stimulated.

VISION

Summation at Threshold. Research concerning the effect of areal size on sensitivity dates back at least to the nineteenth century, when Riccò (1877) reported an inverse relation between size and intensity at the absolute visual threshold. Figure 5.1 shows a typical example of how the visual threshold varies with field size.

Over small visual angles there exists a perfect reciprocal relation between size and intensity. That is, area (A) and intensity (ϕ) can be trad-

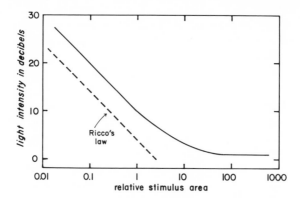

Figure 5.1. Spatial summation at the absolute threshold for vision. Over very small areas, the larger the stimulus size, the lower need be the light intensity to arouse a sensation. The dashed line has a slope of −1 in these double-logarithmic coordinates, and thereby conforms to Riccò's law of perfect areal reciprocity. Over large areas, the threshold intensity is independent of stimulus size.

ed one for the other (e.g., doubling the area and halving the intensity) to maintain a constant degree of detectability of a visual target. Within this range of area, the visual system integrates energy over the spatial dimension. The relation

$$\phi A = \text{constant} \tag{5.1}$$

is known as Riccò's law.

When area is increased beyond the region for which Riccò's law holds, the threshold intensity may continue to decline as area increases, but the rate of decline is smaller than that expressed by Eq. (5.1). Several other laws have been proposed (for example, Piper's law that $\phi A^{1/2} = $ constant) to describe partial spatial summation. When area is increased still further, it no longer has any effect on the intensity threshold, and spatial summation is said to have reached its limit.

The formal similarity between Riccò's law of spatial summation and Bloch's law of temporal summation is quite obvious. Riccò's law states that the visual system integrates luminous energy over space; Bloch's law states that the visual system integrates energy over time. Over very small areas and durations, both laws are valid: $\phi A t = $ constant. The critical areas for complete spatial summation (the analogue to critical duration for temporal summation) are about 6′ of arc in the fovea and 1° in the periphery of the dark-adapted eye (Graham, Brown, & Mote, 1939). These are the spatial limits of Riccò's law. Partial summation can extend beyond those limits. The limits on partial summation depend, however, on duration: the briefer the flash, the larger the spatial limit (Barlow, 1958).

Suprathreshold Summation. In Chapter 4 we observed how Bloch's law continues to hold when we shift from a criterion of threshold to one of suprathreshold brightness. We are now in a position to ask the question: Does Riccò's law hold for suprathreshold visual stimuli? Unfortunately, we are unable to answer that question with very much certainty. Results of several studies that have attempted to answer it fail to provide a totally satisfactory conclusion.

First, let us consider brightness matches. Willmer (1954) obtained foveal matches with very small stimuli near absolute threshold; he found complete summation (Riccò's law) for areas up to about 1', partial summation up to 3'. Ogawa, Kozaki, Takano, and Okayama (1966) also found summation at low brightnesses; subjects adjusted the luminance of test fields to match the brightness of 40' standard fields. Only partial summation ensued, but it extended out to the largest area employed—about 2°40'. Their results could be described by the equation

$$\phi \, A^\lambda = \text{constant} \qquad (5.2)$$

When $\lambda = 1$, Eq. 5.2 reduces to Riccò's law. The size of λ depended on the luminance (hence, on the brightness) of the standard field; it decreased from .6 to .35 as luminance increased from 46 to 51 dB *re* 10^{-6} cd/m². By way of contrast, Diamond (1962) found no influence of area on the brightness of small foveal stimuli. He used areas between 3.7' and 27' visual angle; even at low brightness (the 13.5' standard stimulus at 57 dB), no summation was evident. It should be pointed out, however, that the standard luminances used by Ogawa *et al.* were much lower than those used by Diamond.

Hanes (1951) worked over the range 9–144'. He had subjects produce matches for all possible pairs of stimulus sizes and at each of four luminances. At low levels of brightness (luminance of standards at 55 and 65 dB), spatial summation occurred, but only to a small degree. At 75 dB, area had no effect. But at 85 dB, there was evidence of suppression, i.e., brightness declined as area increased. Hanes's results suggest that the way brightness grows with intensity should not be the same for small and larger stimuli.

Recall that when the brightness of an extended source of light is scaled (by magnitude estimation, for example), the resulting brightness is approximately proportional to the cube root of luminance. J. C. Stevens

(1957) scaled the brightness of a small (1.5′) visual target and obtained an exponent near one-half. Mansfield (1970) has shown that the temporal locus of the Broca-Sulzer enhancement for small targets is consistent with a brightness exponent of one-half. When small stimuli are used, the stimulus duration that produces maximal brightness decreases as the square root rather than as the cube root of luminance. Such a variation in exponent would be consistent with a variation in the critical area for spatial summation; i.e., at and near threshold, the critical area for spatial summation may be greater than it is at higher levels of brightness. This variation has a direct analogy in the temporal domain where, as was outlined in Chapter 4, the critical duration for temporal summation decreases as level of brightness increases.

Although such changes in exponent and in critical area of summation are reasonably consistent with results of certain investigations (e.g., Hanes, 1951; Willmer, 1954), they are not consistent with some others (e.g., Diamond, 1962). Diamond's findings suggest no change in exponent. Similarly, Padgham and Saunders (1966) found the same exponents to

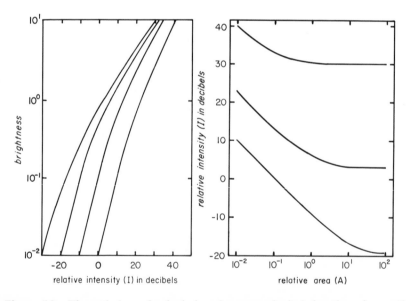

Figure 5.2. Theoretical psychophysical and sensory-physical functions for spatial summation in brightness vision. Left: psychophysical functions showing how brightness increases with intensity in decibels; each function is for a stimulus area one-tenth that of the function to its left. Right: sensory-physical functions showing how intensity and area must covary to maintain constant brightness. As level of brightness increases, the areal limit on Riccò's law decreases. [From Marks (1972).]

govern brightness functions obtained for small (1′) and extended (5°) fields.

Clearly, it is difficult to summarize what is known about suprathreshold spatial summation in brightness vision. One conclusion that appears certain is that spatial summation in foveal vision is not extensive. Even the studies that report summation find it to occur only to a small degree and, with the exception of the study of Ogawa *et al.*, only over limited ranges of area. It appears that summation is both more complete and more extensive at threshold than above it. There are two conditions that would seem to favor discovery of still greater degrees of suprathreshold summation: (a) the use of the blue light, since, in photopic vision, blue-sensitive receptors seem to display relatively extensive summation (Brindley, 1954; Wald, 1967), and/or (b) peripheral stimulation, since rods display a greater degree of summation at threshold than do cones.

Nevertheless, it remains interesting to know why summation is greater, both in magnitude and spatial extent, at threshold than at constant, suprathreshold brightness. A possibility is that the decreased summation observed above threshold results from an interaction of facilitory and inhibitory influences (Diamond, 1962; Hanes, 1951). The model of Marks (1972), outlined in Chapter 4, was extended to account for the way spatial summation changes with brightness. Extension involved hypothesizing spatial summation both of facilitory input and of inhibitory feedback. Figure 5.2 shows (a) predicted brightness functions, and (b) predicted area–luminance trading functions. The psychophysical equation underlying the curves in Figure 5.2 is

$$B(1 + B)(1 + AB) = aAL \qquad (5.3)$$

where B is brightness and L is luminance. Near threshold, when brightness is small, spatial summation is complete

$$B \approx aAL \qquad (5.4)$$

At high luminance, brightness is independent of area

$$B \approx (aL)^{1/3} \qquad (5.5)$$

Equation (5.3) does predict a decrease in critical area as brightness increases, but it does not predict a change in exponent of the psychophysical power function for brightness. If the exponent does depend on area, reformulation of the model and of Eq. (5.3) would be required.

AUDITION

The Ear as a Spatially Extended Organ. There is a fundamental sense in which the ear may be considered a spatial organ. The sensitive, receptive structure of the auditory system is the basilar membrane, and the basilar membrane is a spatially extended surface (coiled up in the cochlea). One of the most important facts about hearing is that, for much of the range of frequencies we can hear, there is a close relation between tonal frequency and the site of maximal stimulation on the basilar membrane (Békésy, 1949a & b). Stimulation by a pure tone produces a wave that travels the length of the basilar membrane. This traveling wave in turn produces a maximal displacement at some particular place along the membrane, and the site of this maximum depends on the sound frequency. Starting at low frequencies, we see (Figure 5.3) that a tone of 100 Hz stimulates maximally at the apical end of the basilar membrane, and, as frequency progressively increases, the site of maximal stimulation moves down the membrane.

There are several important consequences of this spatial representation of sound frequency: one is that a change in the width of a band of noise (i.e., its total span of frequency) or of the frequency separation between pure tones entails a change in the area of (maximal) stimulation on the basilar membrane. It therefore becomes meaningful to ask about spatial summation in audition. How does the loudness of a sound depend on the extent of the basilar membrane (maximally) stimulated?

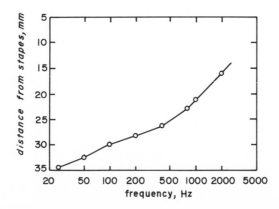

Figure 5.3. Site of maximal stimulation on the basilar membrane, as a function of the frequency of the sound stimulus. [After G. von Békésy, The vibration of the cochlear partition in anatomical preparations and in models of the inner ear. *Journal of the Acoustical Society of America,* 1949, **21,** 233–245. Courtesy of *The Journal of the Acoustical Society of America.*]

Summation at Threshold. Let us begin by examining summation at threshold; that experiment was performed by Gässler (1954). First, Gässler measured the threshold for a pure tone of 1100 Hz. Then he presented two tones, either 1095 and 1105 Hz, or 1090 and 1110 Hz, so that the average frequency remained 1100 Hz. Keeping each tone at the same intensity as the other, he found the threshold for the tone pair. The result was that the total sound pressure level remained the same; therefore, the intensity of either one of the tones was less than the threshold intensity of a single tone. Gässler added more and more tones (up to 40) and discovered that at threshold the total sound pressure of a tonal complex remained constant, but only up to a certain point. Beyond that point, the sound pressure level increased abruptly. The crucial variable is the total spread of frequencies. Up to a certain frequency interval, total sound pressure level remained constant; beyond that interval, it increased. This frequency interval is termed the *Frequenzgruppe,* or critical band. The concept of the critical band is central to much contemporary research and theory in audition.

Adding more and more frequency components to a tonal complex implies increasing the area of stimulation on the basilar membrane. Keeping a constant sound pressure level implies decreasing the sound pressure level of each component and, therefore, decreasing stimulation per unit length on the basilar membrane. It follows that at threshold the auditory system provides some form of spatial summation. There is a critical area within which sound energy is summed perfectly; thus, there is an approximation to Riccò's law for hearing.

Suprathreshold Summation. What about summation above the threshold? Once again, the important experiments have not used direct scaling procedures; rather, they have employed loudness matches. Zwicker *et al.* (1957) examined the way the loudness of a complex of pure tones depends on the frequency separation among components and the way the loudness of a band of noise depends on bandwidth. Total sound pressure level was maintained constant under both conditions. Zwicker *et al.* found that so long as the total frequency separation or bandwidth was smaller than a critical band, loudness did not change. Once the critical bandwidth was exceeded, however, loudness abruptly began to change.

The size of the critical band depends on the center frequency, but not on the sound pressure level. That is, the size of the critical band does not appear to be a level-dependent phenomenon. Nor does critical bandwidth for constant loudness seem to depend on stimulus duration. For pulses as short as 5 msec, critical bandwidth is the same as it is for longer durations (Port, 1963; Scharf, 1970). It would appear that suprathreshold

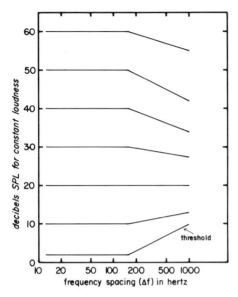

Figure 5.4. Contours of equal loudness, showing how sound pressure level and frequency spacing must covary to maintain constant loudness (center frequency, 1000 Hz). With increase in frequency spacing up to the critical bandwidth, overall SPL remains constant. When frequency spacing exceeds critical bandwidth, SPL must increase to maintain low or threshold loudness, but SPL must decrease to maintain any moderate or high loudness.

loudness obeys a rule similar to that for threshold, namely, that energy is spatially summed (within the critical band). There is, however, a significant difference. When tonal separation or bandwidth is increased beyond the critical band, sensitivity at threshold decreases, whereas suprathreshold loudness increases. This difference is made apparent in Figure 5.4, which shows constant-loudness curves as functions of frequency separation or bandwidth. The curves at and near threshold reflect Gässler's (1954) and Scharf's (1959a) findings that sound pressure level must increase when the critical band is surpassed. On the other hand, at high loudnesses the results of Zwicker *et al.* entail a decrease in sound pressure level to maintain constant loudness once the critical band is surpassed. At an intermediate sound pressure level (about 20 dB above threshold) loudness is the same for subcritical and for supercritical bands.

One implication of Figure 5.4 is that the form of the psychophysical functions for loudness depends somewhat on frequency separation or bandwidth. Narrowband noises (bandwidth less than a critical band) and pure tones should yield functions different from those produced by broadband noises (bandwidth greater than a critical band). Stevens (1972) pointed out that the loudness function for white noise departs from a power function by appearing curvilinear when plotted in log–log coordinates. Differences in the form of the loudness function reflect differences in the spacing between constant-loudness contours in Figure 5.4.

We face an interesting phenomenon: although sound energy is summed within the critical band at all levels of loudness from threshold up, spread of that energy beyond the critical band produces a decrease in sensitivity at low levels of loudness, but an increase in sensitivity at high levels. An alternative way to look at the phenomenon is the following. At low levels sensitivity is *increased* by concentrating all sound energy within a critical band rather than spreading it out more; that is, spatial summation operates within the critical band. At high levels, however, sensitivity is *reduced* by concentrating energy within a critical band; that is, some sort of mutual inhibition seems to operate within the critical band.

This point of view is suggested by the following consideration. When tones are spaced within a critical band, loudness depends on the overall sound pressure level. Neither spacing (Zwicker *et al.*, 1957) nor number of components (Scharf, 1959b) makes a difference. Now, assume the loudness of a single tone of energy E is proportional to $E^{1/3}$. The total loudness of two tones of different frequencies, but the same energy, separated by less than a critical band, equals $(2E)^{1/3}$, or $1.25E^{1/3}$. Since each of the two tones will have about the same loudness as the other, this outcome means that each tone in the pair is less loud than it would be if it were presented alone. As more and more tones are added to a complex (all falling within a critical band) the total loudness increases, but the loudness of each component tone decreases. This decrement in loudness exemplifies the phenomenon of masking, which will be considered in detail below.

It is pertinent to point out at this time the following generalization. When the frequency range of sound stimuli is limited to a critical band, it is total energy that sums. At all but low loudnesses, when sound stimuli are spaced by much more than a critical bandwidth (sufficiently to preclude mutual masking), it is loudness that sums (Fletcher, 1940; Howes, 1950; Zwicker & Feltkeller, 1955; Zwislocki, 1965).

Garner (1959), however, rejected the linear summation of loudnesses in favor of a square-root of sum of squares of loudnesses. This latter formulation is based on Garner's (1954b) lambda scale of loudness. The lambda scale, determined primarily by an equisection procedure, relates to sound energy by about a one-sixth power, rather than a one-third power. It can be shown that Garner's formulation of loudness addition follows from the nonlinear relation between loudness in lambda units and loudness in sones. Consider two pure tones, A and B, equal both in energy $(E_A = E_B = E)$ and in loudness $(L_A = L_B = L)$ and a third tone C adjusted in energy (E_C) so its loudness (L_C) equals that of A and B played together. The sensory-physical relation between E_C and E will hold regardless of which loudness scale, sone or lambda, is employed. Assume the sensory-physical relation is $E_C = 8E$. If the sone scale is employed to evaluate loudness (power-function exponent about one-third), L_A and L_B will both equal $E^{1/3}$, and L_C will equal $2E^{1/3}$. Thus, the results will be consistent with linear loudness summation. If the lambda scale is employed (power-function exponent about one-sixth), however, L_A and L_B will both equal $E^{1/6}$, and L_C will equal $\sqrt{2}\ E^{1/6}$. Thus, the results now will be consistent with root-mean-square loudness summation.

If one takes the view that energy summation in audition always constitutes spatial summation, then there exists an important difference between audition and vision. In both senses, spatial summation is most evident at threshold and near-threshold levels. But at high levels of sensation, it seems fair to say that there is little spatial summation in vision, whereas summation is still evident in hearing. Energy is summed within a critical band. The two sensory systems deal in very different ways, of course, with stimuli that affect widely separate portions of their receptor surfaces. The ear adds loudnesses but the eye does not add brightnesses.

The concept of a critical band appears time and again in relation to various auditory phenomena. The critical band is an increasing function of frequency; at low frequency, e.g., 100 Hz, the critical band is about 90-Hz wide; by 1000 Hz it has increased to a width of about 160 Hz; and by 10,000 Hz it is about 2400-Hz wide. Figure 5.5 shows how critical bandwidth depends on center frequency. Of considerable interest is the fact that the critical band seems to correspond roughly to a constant

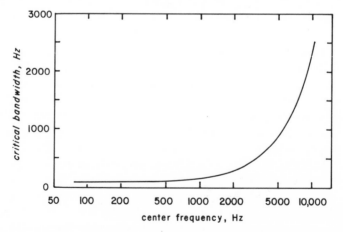

Figure 5.5. Critical bandwidth in hertz, as a function of center frequency in hertz.

length along the basilar membrane (each band about 1.3 mm), except at low frequencies (Scharf, 1970), and may even correspond to a constant number (about 1300) of primary auditory neurons (Zwislocki, 1965). Each critical band also contains approximately a constant number of units of pitch or mels (Stevens *et al.*, 1937). Critical bandwidth even manages to manifest itself in music. Tones separated by more than one critical band sound less dissonant than tones within a critical bandwidth (Plomp & Levelt, 1965; Kameoka & Kuriyagawa, 1969a & b).

TASTE

We now turn briefly to the question of spatial summation in taste. In a small experiment using sodium chloride, Bujas and Ostojčić (1941) obtained equal sensation matches at a single sensation level from one subject. The data show some evidence of summation, but less summation than that found at threshold. This result is quite similar to typical results for vision. But Smith (1971) obtained a different result. He investigated summation for sodium chloride, citric acid, quinine hydrochloride, and saccharin by means of magnitude estimation. The results demonstrated identical summation at all sensation levels, at least over the range of areas (4–126 mm²) employed. Summation was only partial; sodium chloride showed the smallest degree of summation, then citric acid, quinine, and saccharin the greatest degree. In view of the apparent conflict between results of Bujas and Ostojčić and of Smith, it is judicious at present to avoid making any strong conclusions concerning the level dependence of spatial summation in the taste sense.

SOMESTHESIS

Vibrotactile Summation. Suprathreshold spatial summation has been studied for tactile stimuli. Franzén (1969b) measured how the apparent magnitude of vibrotactile sensation grows with the amplitude of a 300-Hz vibratory stimulus applied to one, two, or three fingers. Regardless of the number of fingers stimulated, perceived vibration magnitude increased as a power function of amplitude, and the exponent in each case was a little greater than .5.

Stimulation of two fingers produced greater sensation than stimulation of one, and stimulation of three fingers greater sensation than stimulation of two. Franzén noted that the degree of summation could be described as "vectorial," i.e., sensation increased in proportion to the square root of the sum of the squares of the individual sensory magnitudes. Thus, stimulation of two fingers produced 1.4 (i.e., $\sqrt{2}$) times the magnitude produced by stimulation of one; three fingers produced 1.7($\sqrt{3}$) times the magnitude of one.

Another way of looking at Franzén's data, however, is in terms of sensory equivalence. We may ask, how much greater must be the stimulation of one finger to produce the same magnitude as stimulation of two or three fingers? The answer is that the vibration amplitude to one finger must be about 1.4 times as great as to two fingers and about 1.7 times as great as to three. These ratios correspond to differences of about 3 and 5 dB, respectively. This outcome is practically identical to that of Craig (1966), who obtained vibration matches. Craig used 1–5 vibrators (all at 60 Hz) on the fingers. In order to produce equal sensations, two vibrators required amplitudes 3 dB less than one vibrator required, three required 5 dB less than one required, and five required 7 dB less. In another experiment, he studied summation across widely disparate locations: arm, leg, and trunk. In that experiment the subjects matched the loudness of white noise to perceived vibrotactile magnitude. The finding that summation takes place as well across disparate locations as within circumscribed areas suggests that the locus of summation lies in the central nervous system, since the neural signals from disparate regions are carried over different nerve trunks. Both Franzén's and Craig's studies agree that the trading relation for vibratory sensation is

$$NP^{1/2} = \text{constant} \qquad (5.6)$$

where N is number of vibrators (area) and P is vibration amplitude. In terms of vibration intensity, assuming intensity ϕ to be proportional to the square of amplitude P, the results imply perfect spatial summation ($N\phi$ = constant). For this reason, among others, investigators such as Verrillo and Chamberlain (1972) now use units of vibration intensity instead of amplitude.

It is also possible to study vibrotactile spatial summation in the more traditional manner of spatial summation studies, namely by increasing the areal extent of a single stimulator. Summation like that expressed by Eq. (5.6) is observed at absolute threshold. Verrillo (1963, 1966a & c) found the vibratory threshold to decrease 3 dB for every doubling of stimulus area. However, certain conditions yielded no summation. The conditions were low stimulating frequency (less than about 50 Hz) and/or small stimulus size (less than about 10 mm²), or stimulation of a region of the body that contains no Pacinian corpuscles (such as the tongue). The differences suggest that the vibration sense is mediated by two different populations of receptors (Verrillo, 1966b). More recently, Verrillo and Chamberlain (1972) presented evidence to support the view that the rate of growth of sensation magnitude is related to size of threshold: the higher the threshold, the faster sensation increased with stimulus intensity. They found slightly different psychophysical power functions for 250-Hz stimulation of .28 and 2.9 cm^2 areas of the thenar eminence. The smaller area gave almost a 20% greater exponent. Smaller exponents for larger areas might mean that degree of spatial summation diminishes at higher and higher levels of vibration magnitude.

A complication to the study of areal summation in tactile perception derives from the interrelations of area, force, and pressure. Craig and Sherrick (1969) determined the influence of area on the sensory magnitude of 20-, 80-, and 250-Hz stimuli. When static pressure was held constant, they found that doubling the area was equivalent to increasing amplitude by 3 dB. Thus, the degree of summation was the same as that measured when a second stimulator is added at another location. But when force was held constant, no spatial summation resulted! Finally, when a constant area was employed, increases in force produced increases in perceived vibration magnitude. The result suggested that changes in pressure concomitant with changes in areal extent might account for the spatial summation. Somewhat similar observations were made for perception of simple pressure. In an unpublished study, Adair, J. C. Stevens, and Marks used magnitude estimation to investigate the influence of mass and area on perceived touch magnitude. Naturally, a constant weight produced a more and more intense pressure sensation as the area diminished more and more. (Imagine resting a book directly on your arm versus indirectly via an upright pencil whose point in turn is resting on your arm!) However, even constant pressure failed to produce constant sensation magnitude. It may be that the critical stimulus variable determining tactile magnitude was depth of penetration.

Thermal Summation. In marked contrast to the eye, where suprathreshold spatial summation appears to be limited in extent and difficult to measure, and the ear, where suprathreshold summation is ambiguous, the skin is a sensory organ whose surface provides clear, extensive spatial integration. The thermal senses in particular show spatial summation to a large degree. Early studies by Geblewicz (1938) and by Hardy and Oppel (1937) demonstrated that intensity and spatial extent can trade almost reciprocally and over a large range of areas at the absolute threshold for warmth. Herget *et al.* (1941), using an early version of magnitude estimation, found suprathreshold warmth to increase with increasing stimulus area. Unfortunately, their data do not permit us to evaluate quantitatively the degree of suprathreshold summation.

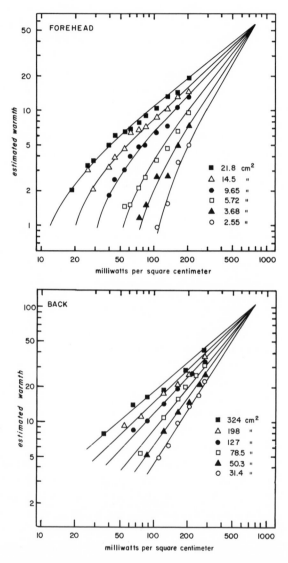

Figure 5.6. Magnitude estimates of warmth produced by thermal irradiation of the forehead (top) and of the back (bottom), as functions of stimulus irradiance in milliwatts per square centimeter. The parameter is the areal extent of stimulation. Each set of functions extrapolates to a common point at 800 mW/cm². [From J. C. Stevens and Marks (1971). Courtesy of *Perception & Psychophysics*.]

More recently, however, the problem of suprathreshold spatial summation in the warmth sense was taken up again (Marks & J. C. Stevens, 1973; J. C. Stevens & Marks, 1971). Figure 5.6 presents average magnitude estimates of warmth plotted against radiant intensity applied to the forehead and to the back.

Each set of data points corresponds to a different stimulus size. It is clear that at any given intensity level, the larger the stimulus area, the warmer the exposure felt. At low intensities the relative increase is much larger than it is at higher intensities, and, by extrapolation, the figure shows that at a very high radiant intensity (about 800 mW/cm^2) area presumably no longer would affect the magnitude of the perception. This extrapolated convergence is of special interest, since 800 mW/cm^2 approximates the threshold for pricking pain and tissue impairment under similar conditions, and other investigations (Greene & Hardy, 1958; Hardy, Wolff, & Goodell, 1940; Murgatroyd, 1964) have shown that little or no spatial summation occurs at the pain threshold. A later experiment (Marks & J. C. Stevens, 1973) compared warmth functions measured for stimulus irradiances that ranged up to 800 mW/cm^2. Results indicated convergence (disappearance of summation) at about 1000 mW/cm^2.

The warmth functions for the forehead (Figure 5.6a) curve downward at low levels of radiant intensity. This "downturn" in double logarithmic coordinates, as we have frequently seen, is typical of the behavior of psychophysical functions in the vicinity of the absolute threshold. The curves drawn in Figure 5.6 are consistent with the assumption [supported by the measurements of Hardy and Oppel (1937) and of Kenshalo et al. (1967)] that Riccò's law holds at the warmth threshold. This outcome is shown in Figure 5.7, which presents contours of equal warmth, i.e., the combinations of area and radiant intensity that produce a given constant level of warmth. At threshold, the contour is a straight line with slope equal to -1 in double-logarithmic coordinates. As warmth progresses to higher and higher levels, however, the contours tend to flatten out. This flattening reflects the fact that warmth shows decreasing spatial summation at higher and higher warmth levels.

The changing contours in Figure 5.7 look as if they are displaced from one another along the horizontal. In other words, the degree of curvature *at any irradiance* seems to be the same for different contours. That implies that the degree of spatial summation (the local slope of the trading function in log–log coordinates) depends primarily on stimulus intensity. An identical conclusion is reached with respect to summation contours derived for a shorter (.5 sec) duration (Marks & J. C. Stevens, 1973). The primary influence of duration was to increase the level of irradiance at which the psychophysical functions converge.

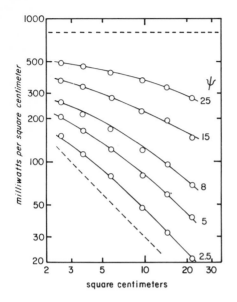

Figure 5.7. Contours of equal warmth, showing how thermal irradiance of the forehead and areal extent must covary to maintain constant warmth. The parameter is the magnitude of warmth sensation. [After J. C. Stevens and Marks (1971) . Courtesy of *Perception & Psychophysics.*]

It is almost certainly true that the thermal senses show spatial summation to a more profound degree than do any other sense modalities. We have no data at all concerning spatial summation of cold above threshold; at the absolute threshold, it appears that cold displays a somewhat smaller degree of summation than does warmth (Hardy & Oppel, 1938). (Suprathreshold summation of cold is particularly difficult to study, since there exist formidable technical difficulties to producing adequate stimulus intensities. It is much easier to produce a strong heat source than a strong heat sink.) Before concluding, however, it may prove worthwhile to speculate about possible mechanisms that might help to explain why spatial summation of warmth gradually decreases as level of intensity increases.

J. C. Stevens and Marks (1971) and Marks (1971b) sketched two possible types of underlying mechanisms. The first postulates that the degree of summation is modulated by lateral inhibition. Evidence for lateral inhibition in the warmth sense has been presented by Békésy (1962). As we shall see later in this chapter, in vision and in hearing the degree of lateral inhibiton appears to increase with increasing intensity of the inhibiting stimulus. It is, therefore, possible that the progressive diminution of summation results from the progressive augmentation of inhibition. As we saw earlier, Diamond (1962) and Hanes (1951) suggested interactions between inhibition and summation to account for spatial summation in vision. There shall be no attempt here, however, to quantify this notion. Such an attempt at quantification would require too many assumptions concerning the relation of inhibition to stimulus intensity, the spatial distribution of the effects of inhibition, and the mathematical nature of the interaction between summation and inhibition.

One aspect of the inhibitory effect, if inhibition is involved, is certain: Inhibition must take place primarily in the central, rather than in the peripheral, nervous system. That conclusion rests on the evidence that the psychophysical function for a split field (half on one side of the forehead, half on the other) is practically identical to that for a unitary field presented to one side (Marks & J. C. Stevens, 1973). Convergence of neural inputs from two sides of the midline takes place in the central nervous system. A corollary conclusion is that the variation with area of the slope of the psychophysical function for warmth also must depend on central neural factors.

An alternative to interactions between excitation and inhibition is the postulation of two independent mechanisms whose contributions add in a linear manner. According to this alternative hypothesis, one mechanism would dominate at low levels of warmth and it would display complete spatial summation; the other would come into play primarily at high levels and would display little or no summation. It turns out to be a relatively simple matter to formulate quantitatively at least one model that describes fairly accurately results for warmth. Briefly stated, the output from one mechanism (O_1) might be depicted by the hyperbolic relation

$$O_1 = \frac{A\phi}{1 + A\phi} \qquad (5.7)$$

where ϕ corresponds to irradiance (or to increase in skin temperature).

Equation (5.7) has proved useful in describing neural responses in the many sensory systems. For a discussion of its properties and possibly underlying physiological bases, see Lipetz (1971). The output from the second mechanism (O_2) would be a function of irradiance (or skin temperature) only,

$$O_2 = G(\phi) \qquad (5.8)$$

and the total output O is the sum of O_1 and O_2

$$O = \frac{A\phi}{1 + A\phi} + G(\phi) \qquad (5.9)$$

Equation (5.9) was applied to the data for the forehead (adding a small threshold constant) and for the back, under the assumption that

$$G(\phi) = \phi^2 \tag{5.10}$$

Equation (5.9) was used to provide the theoretical curves used to fit the data for the forehead when they were presented in Chapter 2 (Figure 2.4). It is especially interesting to note that a square function such as that of Eq. (5.10) is roughly consistent with data obtained from peripheral "warm" fibers serving the nasal region of the cat (Hensel & Kenshalo, 1969). Since some studies appeared to show the existence of few "warm" fibers, it had been suggested (for example, Kenshalo, 1968) that behavioral and sensory responses to warming the skin might be mediated by decreases in firing by "cold" fibers. Hensel and Kenshalo (1969) concluded their paper with the suggestion that both "warm" and "cold" fibers might be active in the range of "slightly warm" sensations.

SUMMARY

The senses of vision, hearing, taste, touch, and temperature all display spatial summation, albeit to different degrees. Summation is especially evident in somesthetic senses—vibratory sensation and warmth—where the sensory effects of stimulation can be integrated over large portions of the skin surface.

1. Warmth in particular shows spatially extensive summation, although degree of summation depends on level: at low intensities summation is nearly complete, but as intensity increases, degree of summation decreases. This level dependence reflects the variation with stimulus size in the psychophysical power function for warmth. The larger the area stimulated, the smaller the exponent. No other sensory system gives such clear evidence of level dependence in spatial summation.

2. Touch (vibratory sensation) displays complete intensity summation; as in the warmth sense the spatial integration seems to take place primarily in the central nervous system.

3. Audition also displays complete spatial summation of energy, in that there exists a spatial mapping of sound frequency onto the basilar membrane. When frequency of stimulation is limited to a critical bandwidth, it is the total sound energy that determines loudness. However, at all but low sensation levels, when frequency extends beyond critical bandwidth, even more extensive summation ensues: when tones are separated by much more than a critical band, it is their loudnesses that add.

4. Although both vision and taste display spatial summation, evidence at hand is ambiguous on the question whether summation depends on level. Taste appears to display only partial summation, i.e., concentration is more potent than area in determining taste intensity.

5. Vision displays complete summation at threshold (Riccò's law) over small visual angles (6′ in the fovea, 1° in the peripheral retina). Limits of complete suprathreshold summation are smaller. Results of different investigations range from extensive brightness summation to none. There is some evidence that the power function relating brightness to luminance has a greater exponent when visual angle is very small than when it is larger, a result that implies level-dependent variation in spatial summation.

Binocular and Binaural Summation

Closely related to the question of how sensory systems integrate spatially the effects of stimulation are the questions concerning binocular and binaural summation. These two sensory systems—visual and auditory—behave quite differently with respect to stimulation of one versus two organs. The magnitude of the difference between the functioning of the two systems can be made readily apparent by casual experimentation. Cover one ear. You can readily note a decrease in the loudness of ambient sounds. Now cover or close one eye. The brightnesses of objects seem hardly to change at all. Binaural summation of loudness is sizeable and important, both theoretically and in everyday life. Binocular summation of brightness is usually small. Although the small degree of binocular summation is of some theoretical consequence, what is important in everyday life is the *lack* of sizeable summation.

Binocular Summation

Several experiments have examined binocular and binaural summation at the absolute threshold. Most investigations of binocular summation (e.g., Matin, 1962) show a small superiority of two eyes over one; the binocular threshold usually appears to be only about 1 dB lower than the monocular (Le Grand, 1957). The binaural threshold, on the other hand, is significantly lower than the monaural, and the average difference equals about 3 dB (Chocholle, 1962; Hughes, 1938; Shaw, Newman, & Hirsh, 1947). [Caussé & Chavasse (1941, 1942) and Keys (1947) found smaller differences at very low (about 100 Hz) and very high (about 8000 Hz) stimulus frequencies.] That 3 dB difference means the threshold energy per ear in the binaural situation is one-half that in the monaural situation. But since two ears are stimulated, the *total* energy is the same in both cases. Thus, the 3 dB difference can be interpreted to reflect perfect energy summation by the two ears. Actually, this is an oversimplification. Since the monaural thresholds of any person's two ears usually differ by a few decibels, calculation of degree of binaural summation can be somewhat more complicated. In any case, summation suggests that the site in the auditory system where the threshold is determined is "central" in locus, i.e., beyond the point where neural impulses arising from the two ears converge.

J. C. Stevens (1967a) obtained magnitude estimates of the brightness of white light (3° field) presented to the left, right, or both eyes. The brightness functions covered a 60-dB range of luminances. The functions displayed a small, possibly insignificant, superiority for binocular viewing, equal approximately to 1 dB. Like the absolute threshold, suprathreshold binocular sensitivity is perhaps a little better than monocular. Assuming that brightness grows according to the cube root of luminance, the brightness aroused by stimulating two eyes was, in Stevens's experiment, about 8% greater than that aroused by stimulating one eye.

Sizeable binocular brightness interactions do, however, take place. One of the more curious binocular phenomena, known as Fechner's paradox, occurs when stimulation of the two eyes is unequal. If a weak light is presented to the second eye when the first is more strongly illuminated, the brightness can be less than it is when the second eye is not illuminated at all. Levelt (1968) has examined in detail these brightness interactions and their relation to contour and contrast. Given unequal stimulation of the two eyes (but not complete darkness to either), brightness appears to approximate an average of the brightnesses that would have resulted were each stimulated alone. The term "average" is descriptive only: Averaging may not accurately represent the nature of the underlying mechanism that is responsible (Engel, 1969).

BINAURAL SUMMATION

Measurement of suprathreshold binaural summation goes back to Docq's (1870) study in the nineteenth century. Docq varied the distances from the ears to the sound source in order to equate binaural and monaural loudnesses. The first major modern experiment was that of Fletcher and Munson (1933), who obtained monaural–binaural matches for the loudness of 125-, 1000-, and 4000-Hz tones. They found the binaural level required for equal loudness to be lower than the monaural; the decibel difference increased as the sound pressure level increased (a

100 dB SPL tone heard binaurally was as loud as a 112 dB sound heard monaurally).

Reynolds and Stevens (1960) examined binaural summation for bands of noise. Their results, shown in Figure 5.8, are based on magnitude estimations, magnitude productions, binaural–monaural matches, cross-modality matches of vibration magnitude to loudness, and binaural–monaural ratio productions (e.g., setting binaural loudness to be one-half monaural loudness). Binaural loudness grew more rapidly with sound pressure level than did monaural loudness, i.e., the binaural exponent was greater (by about 10%). Fletcher and Munson had supposed that binaural loudness is always just twice monaural. For bands of noise,

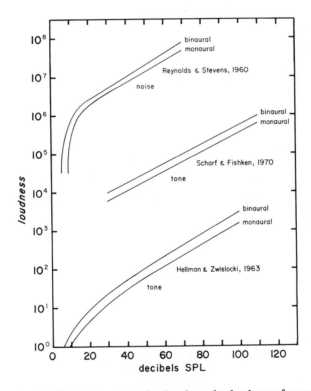

Figure 5.8. Psychophysical functions showing how the loudness of sound presented to one and to two ears increases with sound pressure level in decibels. [Results for bands of noise from Reynolds and Stevens (1960), courtesy of the authors and *The Journal of the Acoustical Society of America*. Results for 1000-Hz tones from Scharf and Fishken (1970). Copyright 1970 by the American Psychophysical Association, and reproduced by permission. Results for 1000-Hz tones from Hellman and Zwislocki (1963), courtesy of the authors and *The Journal of the Acoustical Society of America*.]

Reynolds and Stevens's results show that the 2:1 ratio occurs at only one particular sound pressure level—about 90 dB.

The loudness of pure tones presented to one or to both ears was scaled both by Hellman and Zwislocki (1963) and by Scharf and Fishken (1970). By means of magnitude estimation and magnitude production, Scharf and Fishken showed that the binaural and monaural exponents for a 1000-Hz tone are the same. That is, the degree of binaural summation of tones did not change with stimulus level (see Figure 5.8.). Their binaural and monaural loudness functions are separated by a constant 8 dB. Assuming that the loudness exponents are equal to .33 (Stevens, 1972), the results suggest that loudness of a tone presented to two ears is almost, though not quite, twice the loudness of a tone at the same sound pressure level presented to one ear. A similar conclusion was reached by Hellman and Zwislocki, who also combined results of magnitude estimation and production (a method they call "magnitude balance"). Although Hellman and Zwislocki found a slightly more rapid rate of growth for binaural, than for monaural, loudness, they concluded that binaural loudness summation is perfect: binaural loudness equals left ear plus right ear loudness. Their functions are also plotted in Figure 5.8. Instead of the constant 8-dB separation found by Scharf and Fishken, Hellman and Zwislocki found a difference that increased from about 3 dB at threshold to about 10 dB near 100 dB SPL. In this respect, their results resemble those of Reynolds and Stevens.

Scharf and Fishken also scaled monaural and binaural loudness for white noise. The psychophysical functions they obtained were curvilinear in double-logarithmic coordinates, thereby deviating from power functions. That curvature was predicted on the basis of a model for loudness summation (Zwicker & Scharf, 1970), and similar curvilinearity has appeared in other experiments (for a summary see Stevens, 1972). Nevertheless, Scharf and Fishken determined that the degree of binaural summation for white noise changes with sound pressure level. As SPL increased, the ratio of binaural to monaural loudness increased. This outcome was similar to that of Reynolds and Stevens (1960).

An alternative to linear summation of loudness between the two ears is Garner's (1959) proposal that a sound presented to two ears is 1.4 times as loud ($\sqrt{2}$ times) as the same sound presented to one ear. That hypothesis, which parallels Garner's rule for additivity of multicomponent sounds presented to a single ear, rests as we saw previously on the use of the lambda scale of loudness. The relation between loudness in lambda units and loudness in sones is approximately a square-root function.

Levelt et al. (1972) conducted an important study of binaural loudness. They examined additivity under conditions where both ears were stimulated, though not necessarily at the same sound pressure levels. Each subject's judgments produced a rank order of the loudnesses of several combinations of left ear and right ear SPLs. Application to the data of the theory of simultaneous conjoint measurement showed the rank orders to be consistent with linear additivity. That is, the effect of binural stimulation could be treated as a linear sum of effects of left ear and right ear stimulation, with no interaction. Additivity is only one of the requirements of binaural summation; the other requirement is that the additivity also hold for monaural–binaural comparisons (which they did not study).

Levelt *et al.* also attempted to decide whether it was loudness per se that displayed additivity. Treisman and Irwin (1967) had suggested that is not loudnesses, but rather what they termed "discriminal effects" of stimulation, that add. Because the data obtained by Levelt *et al.* were consistent with additivity, they applied conjoint theoretical analysis in order to derive equal-interval loudness scales. Under the assumption that the psychophysical relation between loudness and sound pressure is a power function, the average exponent of the functions turned out to equal about .5. Levelt *et al.* concluded that the exponents they obtained were close enough to exponents obtained by direct scaling experiments to deduce that it is loudnesses that add binaurally.

BINOCULAR VERSUS BINAURAL SUMMATION

It is interesting that binocular and binaural summation prove to be counterparts, in a sense, to visual and auditory spatial summation. Except perhaps for very small stimuli, suprathreshold brightness shows little or no spatial summation, and binocular brightness is only slightly greater than monocular. Auditory spatial summation, on the other hand, is extensive: energy is summed when frequencies are close together (within a critical band), and loudnesses are summed when frequencies are far apart. Binaural summation, particularly for tones, entails nearly complete loudness summation. Mention may be made that binaural summation proceeds in the same manner even when frequencies to the two ears differ (Scharf, 1969). Thus, binaural summation of loudness operates over noncorresponding regions of stimulation of the two basilar membranes.

Spatial Inhibition

Sensory inhibition has been a matter of fundamental interest and inquiry over the last number of years. The term "inhibition" often is used in several different ways. The first way inhibition is used is in the perceptual, or psychophysical, sense. For example, the presentation of an intense ring of light inhibits, i.e., decreases, the brightness of a less intense center that it surrounds. Second, there is the physiological observation that stimulation of one portion of the surface a sensory receptor can suppress the ongoing activity, spontaneous or induced, in another portion. Perhaps the best examples are Hartline's and Ratliff's measurements of the way neural firing rate of a *Limulus* ommatidium decreases with simultaneous stimulation of another ommatidium (e.g., Hartline, 1949; Ratliff & Hartline, 1959). The third way inhibition is sometimes used is as an explanatory construct. For example, earlier in this chapter it was proposed that inhibition may explain why spatial summation of brightness and of warmth diminish at high levels of stimulus intensity. Although the *construct* of inhibition is based upon psychophysical observations, the explanations themselves are not derived from observations of inhibition. There is the suggestion, but it is only a suggestion, that the same sort of process underlies inhibition as that term is used in these three ways. The present sections of the chapter will deal with the psychophysical version of inhibition, i.e., with observations that the sensory effects of some stimuli are diminished by the simultaneous presentation of other stimuli.

To be considered are examples of inhibition of sensory magnitude only, for instance, suppression of brightness and of loudness. Closely related theoretically are spatially induced changes in sensory quality. The foremost example is color contrast: A red light can induce greenness in a neighboring white, or increase the greenness of a green. It is possible, but by no means certain, that very similar mechanisms operate to produce both color and brightness contrast.

BRIGHTNESS CONTRAST

Brightness contrast is one of the most biologically important phenomena in vision. That is a strong statement, but it is supported by the role that brightness contrast plays in everyday life. Consider the following: From the smallest luminous intensity that is visible (when the retina is fully dark-adapted and is stimulated at its most sensitive site) to the intensity that produces some discomfort, there is a range of more than 100 dB, a ratio of energies more than 10 billion to 1. The visual system can respond over a greater range than even this, i.e., over a range of energies of more than a trillion to one. On the other hand, the luminances of the various objects in my living room vary, under constant illumination, only over a range of about 20 to 1. Yet that 20-to-1 ratio is sufficient to produce a wide gamut of brightnesses. Although the brightest surfaces, such as a white paper, are not extremely bright, the darkest surfaces are very dark indeed! Especially interesting are these very dark objects. My eye wanders to a shiny black rocking chair. Its surface is black, a black that is much darker than the dark gray that I see when I shut my eyes. Brightness contrast appears to be the primary mechanism which makes possible this wide range of sensory experience out of a relatively small (e.g., 20:1) range (small in terms of the capacity of the visual system) of stimulus intensity and in spite of the range-compressing effect that the visual system exerts when contrast is not present.

Look now at what happens when I increase the illumination in my room. (You cannot of course, but try it in your own room.) Some of the brightest objects do get a little brighter. Perhaps a few of the darkest ones get slightly darker. Many remain almost exactly the same. Furthermore, the relations among brightness change only slightly. The visual world remains remarkably constant despite rather large changes in overall illumination. Although the concept of *brightness constancy* has received some criticism (e.g., Freeman, 1967), mainly because perfect constancy is rarely obtained, there is nonetheless the fact that objects in complex visual scenes change less in brightness when illumination varies than isolated objects change in brightness under identical changes in illumination. No doubt, light adaptation is in some measure responsible for this relative constancy. But brightness contrast also plays a significant, probably the most significant, role.

Variables That Determine Degree of Contrast. When viewed against a dark background, the brightness of a flash of light 1 or 2 sec long grows with the cube root of luminance. That simple functional relation can be drastically altered by *simultaneous contrast*. One of the simplest configurations for the study of brightness contrast is a circular test field surrounded by an annulus. As long as the surround is less luminous than the test field, the test field's brightness is virtually unaffected. But when the surround is made more luminous than the test field, test-field brightness is markedly reduced. It is of some theoretical importance that brightness contrast is a monocular phenomenon; diminution of brightness fails to occur if the test field is presented to one eye, the surround to the other.

Figure 5.9 shows the results of magnitude estimations of the brightness of monochromatic lights of 650 nm (triangles) and 500 nm (circles) that were centered in a yellow–red surround at 80 dB *re* 10^{-6} cd/m² (Jameson & Hurvich, 1959). The surround was similar in color to the 650 nm test stimulus. Brightness of test light was judged relative to brightness of the surround, where the latter was assigned the value 100 (log brightness = 2.0). There are two major features of these data. First, it is

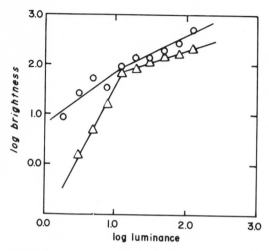

Figure 5.9. Magnitude estimates of brightness, as functions of relative luminance. The test stimuli were monochromatic lights of 650 nm (triangles) and 500 nm (circles), situated in the center of a yellow-red surround. [From D. Jameson and L. Hurvich, Perceived color and its dependence on focal, surrounding, and preceding stimulus variables. *Journal of the Optical Society of America*, 1959, **49**, 890–898. Courtesy of the authors and *The Journal of the Optical Society of America*.]

clear that when the surround is more luminous than the test field, the latter's brightness is diminished. The nature of the decrease is such that the slope of the brightness function is increased. Thus, the brightness of relatively weak stimuli is suppressed to a greater extent than is the brightness of stronger stimuli. Second, note that the degree of contrast, as exemplified by the slope of the brightness function, depends on the wavelengths of test field and surround. Contrast is greater when wavelengths are the same. Since we will not deal with spectral aspects of contrast again, it is appropriate to mention here that Jameson and Hurvich's results suggest spectral specificity of contrast. The data are consistent with the notion that there are (at least) three underlying photopic receptor systems in the eye, each with its own spectral sensitivity, and that each system has its own contrast-sensitive mechanism (Green, 1969).

One of the most important parameters in brightness contrast is the intensity of the inhibitory field. Figure 5.10 shows the results of brightness fractionations (J. C. Stevens, 1957; Stevens, 1966b). Subjects were asked to set the intensity of a 2.4° test field so that its brightness appeared to be a given fraction (one-half, one-fourth, or one-tenth) of the 5.7° surround. The dashed line shows the cube-root function presumed to operate in the dark-adapted eye when there is no contrast. Contrast produces a steepening of the brightness function; furthermore, the degree of steepening depends on the intensity of the inhibitory stimulus: the more intense the surround, the steeper the brightness function. These results are extended in Figure 5.11, which gives Stevens's (1961c) schematic representation of the relation between brightness and luminance when contrast

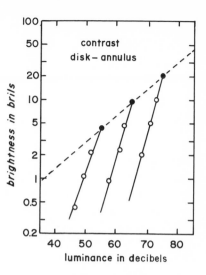

Figure 5.10. Fractional settings of brightness (open circles), as functions of luminance in decibels *re* 10^{-6} cd/m². The comparison stimuli were luminous surrounds (filled circles) that encompassed the test stimuli. [Data of J. C. Stevens (1957), presented by S. S. Stevens, Power group transformations under glare, masking, and recruitment. *Journal of the Acoustical Society of America*, 1966, **39**, 725–735. Courtesy of the author and *The Journal of the Acoustical Society of America*.]

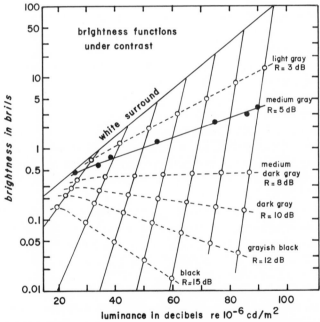

Figure 5.11. Representation of psychophysical functions (solid lines) showing how brightness increases with luminance in decibels *re* 10^{-6} cd/m² under simultaneous contrast. The line marked "white surround" gives the brightness function that obtains when no contrast is present. The dashed lines show how brightness varies when the luminances of both test and inhibiting fields are augmented in equal proportion. [From S. S. Stevens, To honor Fechner and repeal his law. *Science*, 1961, **133**, 80–86. Copyright 1961 by the American Association for the Advancement of Science. Reproduced by permission of the author and *Science*.]

173

is present. The solid line marked "white surround" gives the bril function, i.e., the cube-root function that relates the brightness of the surround to its luminance. The other solid lines represent brightness functions under contrast; their exponents are greater than one-third, and the value of the exponent increases as the luminance of the surround increases. The luminance of the surround corresponds to the point at which the contrast function intersects the bril function. At that point the luminances of surround and test field are equal; presumably, contrast no longer affects the brightness of the test field when the latter is at least as intense as the surround. (The question whether there is any small effect of less intense visual stimuli on stronger ones will be considered later in the section.) As Figures 5.9 and 5.10 show, the effects of contrast on brightness are dramatic. The psychophysical functions for brightness are much steeper (i.e., brightness grows much more rapidly with increasing luminance) when sensitivity is reduced by contrast than they are when sensitivity is reduced by light adaptation (Padgham & Saunders, 1966; Stevens & J. C. Stevens, 1960).

Each dashed line in Figure 5.11 corresponds to a constant ratio of surround luminance to test-field luminance. For example, the dashed line marked 3 dB reflects the locus of values where the test field is one-half as luminous as the surround. The implications of these dashed lines are of some consequences for our perception of the world. When overall illumination changes, absolute levels of stimulus intensity increase, but, assuming there is no change in the spectral distribution of the illumination, all of the ratios of luminances will remain the same. The dashed lines in Figure 5.11 show that the brightnesses of objects viewed under contrast change less rapidly when luminance is changed than do brightnesses viewed without contrast. At one particular contrast ratio (where the surround is 8 dB greater than test field) the brightness of an object is virtually unaffected by large changes in overall illumination. That dashed line represents brightness constancy.

One of the best everyday examples of brightness inhibition is encountered when one drives at night. The headlights from an oncoming automobile can sometimes render invisible other objects in the visual field, even objects that lie at some distance from the source of glare. Study of the reduction of brightness by glare goes back at least to the classic experiments by Schouten and Ornstein (1939), who distinguished effects of glare from effects of light adaptation. Stevens and Diamond (1965) systematically examined the effect of glare angle on the brightness function. They obtained brightness matches between 6.5′ test fields viewed with and without a small (4.5′) glare source at 113 dB *re* 10^{-6} cd/m². Most of the effect of glare on test-field brightness disappeared when separation reached 6°. In every case, the effect of glare was to steepen the brightness function, i.e., to increase the size of the brightness exponent. The smaller the glare angle, the greater the steepening. Thus, changing glare angle acts somewhat like changing intensity of an inhibiting stimulus. But there was a major difference. With glare, all of the brightness functions tended to converge at a common point, which equaled the luminance of the glare source. The results, therefore, suggest that once the luminance of the test fields equals or exceeds that of the glare source, the brightness is no longer influenced by the glare.

The magnitude of brightness contrast depends on a number of stimulus parameters. We have already examined effects of the intensity of the inhibiting field and of its separation from the test field. Other spatial parameters include configuration of test and inhibiting fields—contrast is greatest when the inhibiting field totally surrounds the test field (Horeman, 1963). Another parameter is the area of the inhibiting field. Larger inhibiting fields produce more contrast (J. C. Stevens, 1967b), although the lion's share of the inhibition is exerted by a relatively small area adjacent to the test field. Furthermore, this increase in contrast appears to be reflected by an increase in the degree to which the brightness function is steepened. It is interesting that the luminance of inhibiting field and the spatial configuration of inhibiting and test fields seem to manifest their effects primarily on the exponent of the brightness function, i.e., on the rate at which brightness of the test field increases with luminance.

Brightness in Complex Spatial Displays

Since contrast profoundly influences the way brightnesses relate to luminances in complex fields, it is not surprising that the overall brightness of a complex field does not conform to the cube-root brightness equation. Stevens and J. C. Stevens (1960) reported an experiment conducted by G. Bermant that scaled the brightness of a real-life scene—a desk covered with familiar objects—when the desk was illuminated at different levels. The result was a power function between brightness and illuminance with an exponent of .45. Given the steepening in the brightness function that contrast produces, an exponent greater than one-third is not surprising.

Bermant's experiment required subjects to gauge the overall brightness of the scene, not the brightnesses of individual elements. Thus, the sub-

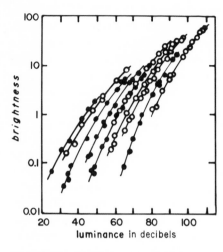

Figure 5.12. Magnitude estimates of the brightness of elements in a complex field, as functions of luminance in decibels *re* 10^{-6} cd/m². Open circles: slide projections, Filled circles: photographs. [From C. J. Bartleson and E. J. Breneman, Brightness perception in complex fields. *Journal of the Optical Society of America,* 1967, **57,** 953–957. Courtesy of the author and *The Journal of the Optical Society of America.*]

jects had somehow to integrate, or average, rather disparate brightnesses. Bartleson and Breneman (1967) and Padgham and Saunders (1966) scaled the brightness of various specified elements within complex fields. Some of the results of Bartleson and Breneman are given in Figure 5.12. Each function is for a different photograph; five were slide projections (open circles), five were prints (filled circles). The curves deviate systematically from power functions, and Bartleson and Breneman described this departure in terms of exponential decay. The equation they give may be written

$$B = kL^{\beta}\exp(-\eta L^{-\iota}) \qquad (5.11)$$

where B is the brightness of an element in the photograph, L is luminance, and η and ι depend on the luminance of the most intense element in the photograph. The meaning of the exponential term remains unclear.

A salient feature of complex visual fields is that the subjective intensities of objects fall along a dimension of whiteness–blackness, whereas isolated spots of light viewed in the dark vary along a dimension of brightness–dimness. Black can be perceived only when contrast is present. Absence of light is not the same as black. But so too is whiteness different from brightness. The qualitative difference between surface appearance (appearance as object) and aperture appearance (appearance as light) may also in large measure be due to brightness contrast.

The difference leads to the question of the perception of lightness and darkness. There appears to be some similarity here to the scales of brightness of complex fields. The lightness of gray surfaces presented in a "natural" setting grows as a positively accelerated function of intensity; the exponent of the power function reported by Stevens and Galanter (1957) equaled about 1.2. However, Warren and Poulton (1966) reported exponents as low as .5. In any case, both complex fields and grays give exponents greater than one-third, and these larger exponents probably reflect the operation of simultaneous brightness contrast. It is perhaps a perverse Pythagorean phenomenon that Munsell value (which is a scale of lightness determined by interval-scaling procedures—that is, by the setting up of equal-appearing intervals of lightness) can be described by a cube-root function relating value to reflectance (Glasser, McKenney, Reilly, & Schnelle, 1958; Saunderson & Milner, 1944; Stevens & Galanter, 1957). In fact, power functions obtained from procedures that rely on judgments of sensory intervals typically have exponents smaller than those obtained from procedures that rely on judgments of sensory ratios (e.g., Marks, 1968b). Thus, it is probably fortuitous that the same function governs both Munsell value and brightness of spots of light viewed in the dark.

Models of Brightness Contract. We come now to the question of explanation. What underlies brightness contrast? Many models of brightness contrast are based directly or indirectly on psychophysical observations of Mach bands and/or on physiological observations of lateral inhibition. Mach bands are light and dark regions that appear at points of discontinuity in spatial luminance gradients. For example, consider a spatially extended surface of low, constant luminance at one end; near the middle, luminance abruptly begins to increase, then becomes constant again, but at a higher level. At the point where luminance begins to increase, a dark band is seen; where it ceases to increase, a bright band is seen. Mach bands are prime examples of the results of lateral interactions. Békésy (1968a) noted that the bright band is more prominent than the dark band. Now, it is a reasonable assumption that the visual system imposes some sort of compressing transformation on luminance (i.e., the input–output characteristic is negatively accelerated). On the basis of the difference between bright and dark Mach bands, Békésy deduced that the site of lateral interaction must come after the site of compression.

Most notable among the physiological observations are the measurements of inhibition in the eye of the horseshoe crab, *Limulus* (Hartline, 1949; Ratliff & Hartline, 1959). The response of a given ommatidium to light is decreased by simultaneous stimulation of neighboring ommatidia. These observations would seem to make peripheral neural interactions the likely candidate for the basis of brightness contrast. However, in many respects there remains a large chasm between electrical responses in nerves of horseshoe crabs and visual perception in human beings. Quantitative prediction of brightness responses is a formidable task. A number of proposals have been made, some purely empirical, others more theoretical.

J. C. Stevens (1957), and later S. S. Stevens (1961c, 1966e), proposed that the effect of contrast (inhibition) is to impose a power transformation on the brightness function, under the restriction that the inhibiting field is more intense than the test field. Thus, the power equation for brightness

$$B = k(L - L_o)^\beta \qquad (5.12)$$

is extended by adding the qualification

$$\beta = \begin{cases} G(L_i) & \text{if } L_i > L, \\ \frac{1}{3} & \text{otherwise} \end{cases} \qquad (5.13)$$

where L_i is the luminance of the inhibiting field, B is brightness of the test field, L is its luminance, and β is the exponent. The function $G(L_i)$ increases as L_i increases; Stevens and J. C. Stevens (1960) give an expression that can be written $\beta = .33 + .04 (M - 30)$, where M is L_i in decibels. The formula applies to their conditions of measurement, namely a white inhibiting field of 5.7° that surrounds a white 2.4° test field.

Stevens and Stevens's model suggests a discontinuity in the brightness function under contrast, with a cube-root function operating when the test luminance is greater than the inhibiting luminance, and a larger exponent operating when test luminance is smaller. These discontinuities were depicted in the schematic brightness functions shown in Figure 5.11.

Horeman (1965) used the cube-root psychophysical relation between brightness and luminance of an uninhibited stimulus in order to replot brightness-matching data obtained earlier (Horeman, 1963). He concluded that the data are consistent with Stevens and Stevens's notion of a power transformation. Contrast induces a greater exponent in the power equation.

An important question concerns what happens when the luminance of a test field is the same as that of the inhibiting field. Some data, such as those of Heinemann (1955) suggest that test-field brightness is slightly reduced at that point. Stevens and J. C. Stevens (1960) and Horeman

(1965), however, noted that such an outcome would mean that the brightness of a test field will change (decrease) when its size increases, and they rejected that notion. For the most part, the evidence is in their favor. In fact, Diamond's (1962) investigation of spatial summation of brightness was conducted partly in order to answer that question. He found no evidence for a decrease in brightness when test field increased from 2.7' to 27'. However, it should be mentioned that Hanes (1951) found some evidence that intense (85 dB) lights decreased in brightness as their size increased from 9' to 36'.

A different account of the steepening of the brightness function under contrast has been given by Jameson and Hurvich (1964). Under the rubric of their opponent-process theory of visual perception, they suggested that an inhibitory field induces a "blackening" response. Mathematically, this effect appears as a subtractive term in the brightness equation. Their equation reads

$$B = k(L^\beta - qB_i) \tag{5.14}$$

The term B_i is proportional to the brightness of the inhibiting field, and the constant q depends on the spatial configuration. The Jameson-Hurvich model is continuous, in contrast to the discontinuous model proposed by Stevens and Stevens. The continuous model states that an inhibiting field affects the brightness of a test field at all test-field luminances. However, since the inhibitory effect is subtractive, and proportional to the brightness of the inhibiting field, the effect is most obvious when the test field is less intense than the inhibiting field. Thus, even though the theory assumes mutual suppressions between fields (two-way action) even when one field is much more luminous than another, the suppression is predominately one-way. The degree to which the brightness function for the test field is steepened depends, as it does in Stevens and Stevens's model, on the luminance of the inhibiting field: the greater its luminance, the greater the steepening. Figure 5.13 shows brightness functions predicted by the subtractive model of contrast proposed by Jameson and Hurvich.

This subtractive model has been extended by Marsden (1969), whose version is based upon the view that the fall-off in inhibition is a logarithmic function of the separation between stimulated points on the retina. Marsden's model specifically predicts a decrease in the brightness of a test field as its size increases.

A subtractive theory of contrast was also proposed by Treisman (1970), in a model similar to that expressed by Eq. (5.14). Since he applied the model to brightness matching data only [results obtained by

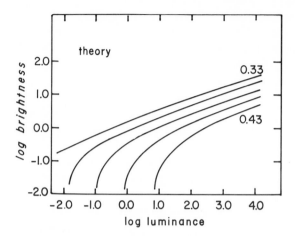

Figure 5.13. Representation of psychophysical functions showing how brightness increases with luminance under simultaneous contrast. Based on a subtractive model: the more luminous is the inhibiting stimulus, the greater is the subtraction of constant brightness from test stimuli and the larger is the function's slope in log–log coordinates. [From Jameson and Hurvich (1964). Courtesy of the authors and Microforms International Marketing Corporation.]

Heinemann (1955) and Horeman (1963)], it was not necessary for Treisman to employ a power law as the basic psychophysical relation. The mathematical nature of the transformation from brightness to luminance became a parameter of the model. Even so, Treisman found that a logarithmic transformation was not suitable, but a power transformation was. However, there are two unfortunate features of Treisman's model. First, the size of the exponent of the power relation turned out not to be constant, but to vary with the luminance of the inhibiting stimulus. Second, on the average the exponent equaled about 1.0, implying proportionality between underlying brightness and luminance. Both of these results appear to make Treisman's model limited in value.

Some other models of brightness inhibition have assumed that inhibition exerts a divisive, rather than a subtractive, effect. These other models have not, in general, attempted to account for magnitudes of perceived brightness under contrast. Instead, most divisive models have attempted to make quantitative predictions of contrast thresholds and of Mach bands. In fact, Mach (1868) proposed an equation in which the effects of excitation are divided by the effects of inhibition. A theory proposed by Diamond (1960) used divisive inhibition to account for suprathreshold contrast. Diamond's model postulated that spatial enhancement of brightness results from interactions between *on* and *off* fibers in the visual system, but that brightness depression results from inhibition of *on* fibers by *on* fibers. (*On* fibers respond with a burst of activity to stimulus onset; *off* fibers respond to stimulus offset.) The model goes on to quantify these interactions; one specific assumption was that the magnitude of neural excitation is divided by the magnitude of inhibition. Sperling's (1970) model for contrast detection also assumes that effects of excitation are divided by effects of inhibition. This type of model is of some interest because of its close relation to the model of spatial and temporal summation described in Chapter 4 and earlier in the present chapter. The latter model also assumes that inhibition (feedback) exerts a divisive effect.

Finally, we should look at the possible interpretation of brightness inhibition in terms of an adaptation-like mechanism. Whittle and Challands (1969) proposed that a gain-control (adaptational) mechanism, such as that postulated by Rushton (1965), can account for the reduction in sensitivity that is observed both when a test field is superimposed on a luminous background and when it appears in the presence of a glare source. Whittle and Challands specifically intended to cut across the usual distinction between adaptation and contrast, a distinction first clearly made by Schouten and Ornstein (1939).

The claim that adaptation can account for phenomena such as the reduction of brightness by glare has usually been based on the presumed effect of scattered light that the glare source produces. Any light source will produce entoptic scatter and therefore stimulate retinal regions in addition to the region of the source's image; under special conditions, the scattered light can even appear brighter than the image of the source (Cornsweet, 1962). The strongest arguments favoring a role for scattered light concern the capacity of scattered light to raise threshold (Holladay, 1926; Rushton & Gubisch, 1966; Stiles & Crawford, 1937). Of particular note are the observations of Holladay and of Stiles and Crawford that the increase in threshold is produced even when the glare stimulus is imaged on the blind spot of the eye. Rushton and Gubisch, however, attempted to distinguish between the capacity of a glare source to increase threshold (by scattered light) and its capacity to reduce brightness (by contrast). Schouten and Ornstein (1939) first pointed out that scattered light should add to the light of the test field and thereby increase its brightness, not decrease it. On the other hand, they also reported that a glare source did reduce the brightness of a test field even when the glare fell on the blind spot. Nevertheless, it is difficult to see how adaptational effects can account for more than a fraction of the total observed effect of brightness contrast. In particular, the perception of black, which takes place only under simultaneous contrast, is more likely to be the result of an induced effect of lateral neural inhibition than of adaptation. Certainly, blackness is phenomenologically distinct from the dimness or the depression of brightness produced by light adaptation.

Before concluding, it is worth mentioning the possibility that brightness contrast is mediated by more than one type of mechanism. Békésy (1968b) pointed out two apparently different types of lateral inhibitory effect in vision. One type, which he called "Mach-type" inhibition, acts over only small spatial extents. This inhibition produces the Mach bands that are seen near spatial discontinuities of luminance. The second type he called "Hering-type" inhibition. The latter acts over larger spatial areas and produces more gradual decrements in brightness. How different physiologically must be the processes that produce these two types of inhibition is unclear.

This concludes our discussion of brightness contrast. Contrast is surely one of the most important phenomena in vision. Yet, direct scaling procedures have been used only to a small degree in the study of contrast. Beyond the scope of this book are recent advances that have been made in the understanding of spatial inhibitory phenomena in vision. These advances have arisen to a great extent out of the use of spatially repetitive, e.g., sinusoidal, stimuli. Although most of those studies have looked at detection thresholds, the experiments of Bryngdahl (1964, 1966) and of Watanabe, Mori, Nagata, and Hiwatashi (1968) are notable exceptions.

Brightness of Spatially Sinusoidal Distributions of Luminance. Watanabe *et al.* determined curves of constant "apparent contrast." Subjects viewed a split field, each half of which contained a different spatial, sinusoidal pattern of luminance. The average luminances of the two halves were the same, but the spatial frequencies differed. The subject's task was to find two physical contrasts, one for each frequency, so that the two patterns produced the same apparent contrast. With magnitude of apparent contrast varying parametrically, the results showed that suprathreshold contrast sensitivity is a level-dependent phenomenon. The greater the apparent contrast, the less the variation in sensitivity to different frequencies. Bryngdahl (1964, 1966) attempted to determine by brightness matching how brightness varies along spatial sinusoids of different frequencies. In his experiment, by way of comparison, matches were made to the brightness maxima and minima, rather than to the overall apparent contrast. This matching paradigm is limited, however, because of the appearance of blackness, which cannot be matched by the mere absence of light. These studies, especially that of Watanabe *et al.*, are almost perfect analogues to the experiment conducted by Marks (1970) on apparent depth of temporal modulation (see Chapter 4). Use of suprathreshold, sinusoidally modulated stimuli appears likely to continue to prove valuable to the understanding of spatial interactions in brightness vision.

AUDITORY MASKING

Auditory masking proves to be in many respects a first cousin to brightness contrast, despite the fact that there are important differences between them. A major similarity surfaces at the phenomenological level. Just as an intense visual glare source or surround can depress the brightness of nearby or adjacent lights, so an intense sound suppresses the loudness of other sounds. The chirping of crickets can appear surprisingly loud against the background of a quiet summer night, but will readily dissolve into the roar of a passing jet plane. It would be easy to enumerate other daily encountered examples of auditory masking.

A terminological distinction is sometimes made between "partial masking" and "complete masking;" the latter term requires that the suppressed sound be reduced to zero loudness or to absolute threshold, whereas the former refers only to decreases in the loudness of one sound due to the presence of another (cf., Scharf, 1971). Such a terminological distinction will not be followed here. Instead, the general term, masking, will be employed to refer to *all* decreases in loudness, whether partial or complete.

Critical Bandwidth in Threshold Masking. The concept of the *critical band* takes on primary importance in auditory masking. In our examination of loudness summation (pp. 152–157), we saw how the critical band defined a range of sound frequencies over which sound energy is integrated. When energy falls totally within the critical band, loudness is independent of energy distribution over frequency. In actual fact, the concept of the critical band grew out of studies in auditory masking; it was Fletcher's (1940) assumption that when a wide band of noise masks a tone, the sound energy that is effective in producing the masking is limited to a relatively small frequency band around the masked tone—the critical band. [It was known much earlier that degree of masking of tones by tones depended on the frequency proximity of the tones: nearby tones produce more masking (Wegel & Lane, 1924).]

One way to measure the size of the critical band is to increase the bandwidth of a masking noise and measure the threshold of a test tone. As bandwidth increases, sound pressure/hertz remaining constant, threshold also increases (since there is more masking energy within the

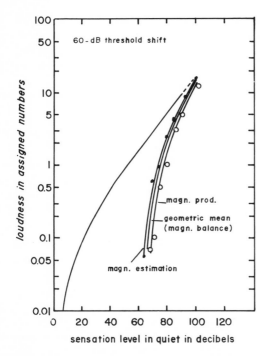

Figure 5.14. Psychophysical functions showing how the loudness of a 1000-Hz tone increases with sensation level in decibels, heard in the quiet and in the presence of masking noise. Masking is greatest at low sensation levels. [From R. P. Hellman and J. J. Zwislocki, Monaural loudness function at 1000 cps and interaural summation. *Journal of the Acoustical Society of America*, 1963, 35, 856–865. Courtesy of the authors and *The Journal of the Acoustical Society of America*.]

critical band). This increase in threshold continues up to a point, beyond which the added energy has no effect (e.g., Greenwood, 1961; Hamilton, 1957). The bandwidth or frequency range at which the threshold stops increasing is the critical band. An alternative procedure is that of Zwicker (1954), who masked a noise by two flanking tones. As the frequency separation of the tones is increased, the threshold for the noise remains constant until the critical band is reached; with further separation, the threshold then decreases.

Variables That Determine Degree of Masking. What about suprathreshold loudness? The effects of masking on loudness have been explored in a large number of studies, some of which employed loudness matching and others direct scaling procedures. Among the earliest studies was that of Steinberg and Gardner (1937), who measured how the loudness of a 1000-Hz tone decreases in the presence of white noise. They suggested that the effect of the masking noise is to subtract a constant amount of loudness from the tone. The same proposal was later made by Lochner and Burger (1961). This hypothesis resembles the formulations of Jameson and Hurvich (1964) and of Marsden (1969) for the explication of brightness contrast. Steinberg and Gardner noted that the loudness-intensity relations determined under masking are similar to psychophysical relations found, without masking, in the hearing of people who have nerve deafness (neural hearing loss), and they suggested that a similar subtractive process may also be involved in the latter case.

Hellman and Zwislocki (1964) reported a careful study of the effects of an octave (600–1200 Hz) band of noise on the loudness of a 1000-Hz tone. They compared the results of two procedures: loudness matching and "numerical loudness balance." (The latter procedure is a combination of magnitude estimation and magnitude production.) The matching and estimation data were in very close agreement. Figure 5.14 gives some data they obtained by numerical judgment. The masking noise was set to an intensity that produced an increase in threshold of about 60 dB. These data depict the basic features of loudness masking, namely that masking is greatest at low levels of the test stimulus. As the sensation level of the tone increases higher and higher above threshold, however, there is a decrease in the decibel difference between sound pressures of unmasked and masked tones that yielded equal loudness; that is, the horizontal separation between the curves decreases. Similarly, as SL increases the ratio of unmasked to masked loudness decreases; that is, the vertical separation between the curves also decreases. At high enough sensation levels, the loudness of the 1000-Hz tone is hardly changed by the introduction of the masking noise. Clearly, the loudness function as plotted in log–log coordinates is steeper under masking than it is in the quiet. Here we see a basic similarity between auditory masking and visual contrast. Both masking and contrast lead to steepening of the psychophysical functions. Hellman and Zwislocki also obtained data for a masking noise of lower intensity, and the results were similar, except that the masked loudness function was not steepened so much as was the one depicted in Figure 5.14.

The similarity between loudness under masking and brightness under contrast is striking. The significance of that similarity has been espoused with force by Stevens (1966e; Stevens & Guirao, 1967). Some of the loudness-matching data obtained by Stevens and Guirao are shown in Figure 5.15. The graph shows the sound pressure level of a tone heard in the quiet versus the sound pressure level of an equally loud tone heard in the presence of a masking noise. The dashed line at 45° shows the relation expected if the masker exerted no effect on loudness. The slope of the matching function gives the ratio between the exponent of the power function under masking and the exponent in the quiet. Slopes greater than unity (45°) mean that the masked exponent is greater. Two other aspects of the results merit note. First, these data show similar growth of

Figure 5.15. Sound pressure levels of 1000-Hz tones in the quiet required to produce the same loudnesses as 1000-Hz tones in the presence of masking noise. The parameter is the SPL of the noise. The dashed line shows the result that would be expected if the noise had no effect on the tone's loudness. [From Stevens and Guirao (1967). Courtesy of the authors and *Perception & Psychophysics*.]

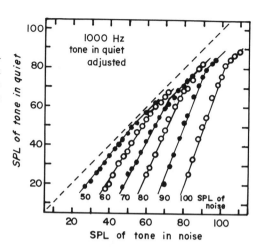

masked loudness with sound pressure to those of Hellman and Zwislocki: As the sound pressure increases, the difference between masked and unmasked loudness decreases. Second, these data too demonstrate that the steepness of the loudness function under masking depends on the intensity of the masking noise: the more intense the noise, the steeper the loudness function. That these loudness curves under masking mimic brightness curves under contrast need not be stressed any further. One interesting difference is that the masked loudness functions, especially those obtained with very intense masking noises, never quite reach the unmasked levels of loudness (dashed line), even when the test tone is 10–20 dB more intense than the noise. Stevens and Guirao suggested that this result is due to mutual masking between tone and noise.

Nevertheless, beginning at the point where noise and tone sound pressures are the same, the slopes of the functions change fairly abruptly to unity. When the tone is more intense than the noise, the exponent of the loudness function is the same as it is in the quiet. Results similar to these were obtained by Richards (1968), who used a fractionation procedure. In particular, he noted the increase in the steepness of the log loudness–log sound pressure function as the intensity of the masking noise increased. Richards's data supported Stevens's (1966e) tentative hypothesis that the exponent of the loudness function under masking increases with the .16 power of the sound pressure of the noise.

The mutual masking of tone and noise presents an important problem for analysis. Hellman (1972) measured the masking of broad and narrow bands of noise by 1000-Hz and 1400-Hz tones. Interestingly, the masking effectiveness of tone on noise is substantially less than the effectiveness of noise on tone. Equally loud tones and noises do not mask equally.

The degree of depression exerted by a masking sound depends on several parameters of the masking and masked stimuli. These include the relative intensities of masker and masked sounds and the absolute intensity of the masker, parameters that have already been considered. Studies of critical bands show that the bandwidth of a masking noise is an important variable. Although there do not seem to be any studies that systematically varied subcritical bandwidth, Hellman (1970) used loudness matching to examine the effects of variation in supercritical bandwidth. She determined how the loudness of a 1000-Hz tone changes with sound pressure in the presence of noise. The four bands varied from 925–1080 Hz (approximately one critical band) to 75–9600 Hz. When different bands were equated for overall SPL, the narrower band produced greater masking. When these different bands were adjusted in SPL to produce the same threshold masking (equal in SPL/hertz), however, they also produce identical suprathreshold masking when the loudness of the masked tone was low. When masked loudness was moderate or high, though, the narrow-band noise produced less masking than did the broader bands of noise. That is, the bandwidth had an effect on the steepness of the masked loudness function. When the bandwidth of the masking noise decreases, the steepness of the loudness function for tone increases.

Direct scaling has been used to determine the influence of a masking noise presented to one ear on the loudness of a 1000-Hz tone presented to the other (Rowley & Studebaker, 1971). Simultaneous presentation of contralateral noise yielded masking (depression of loudness), but only at low sensation levels of the tone (less than about 20 dB SL). At higher tone levels (30–80 dB), the contralateral noise produced an increase in the loudness of the tone. And at much higher levels (more than 80 dB), the tone appeared the same in loudness with and without simultaneous contralateral noise. The magnitude of these effects depended strongly on the sensation level of the masking noise; their magnitude increased as its SL increased. With the most intense noise used (100 dB SL), the loudness of weak test tones decreased as much as threefold, and the loudness of moderate tones increased by the same factor. Rowley and Studebaker suggested an interaction of summation and masking to explain the results: binaural summation of loudnesses of the noise and tone, plus masking, either by bone conduction or at some central (binaural) locus.

Loudness of Complex Sounds. One of the most important applications of our knowledge concerning auditory masking and summation is to calculations of the loudness of complex sounds. From a practical point of view, there exist many situations in which such calculations are most desirable. In industrial settings, on the streets of cities, beneath the pathways of jet planes, complex sounds and noise inundate our ears. Recent years have seen the development and refinement of a number of different schemes for calculating loudness (Stevens, 1956a, 1961a, 1972; Zwicker, 1958; Zwicker & Scharf, 1970). In its most recent version, Stevens's (1972) method makes use of: a set of equal-loudness contours (frequency-weighting functions), a cube-root power relation between loudness and sound energy, and a rule for summing loudnesses across frequency bands. This rule states that total loudness equals the loudness of the loudest component band plus a weighted sum of the loudnesses of other bands. The reduced contribution of these other bands reflects the masking action of the loudest band. Stevens gives procedures for making the computations when the sound spectrum is divided either into octave or one-third octave bands. Thus, by means of a graph showing how sound pressure level varies with frequency to maintain constant loudness, the power equation, and a summation equation plus an associated table, it is possible to calculate closely the loudness of any physical specified sound. Of special interest is Stevens's extrapolation of the equal-loudness contours to very low frequencies (1 Hz) and to high intensities (160 dB SPL). The addition of contributions of intense low frequencies aids in the precise computation of the loudness of impulsive sounds like sonic booms.

A somewhat different type of model for calculating loudness was proposed by Zwicker and Scharf (1970). They employ three intervening variables in their system: one is the "excitation pattern," which is the assumed pattern of stimulation on the basilar membrane; a second is termed "tonalness," and it is a function of sound frequency; the third is "specific loudness," which is analogous to Stevens's loudness in an individual band. Tonalness is determined by the function that relates critical bandwidth to sound frequency. Thus, Zwicker and Scharf's model specifically incorporates the concept of critical bandwidth. (Stevens's model does this in effect by dividing the sound spectrum into a large number of bands. There is a trade-off between the degree of subdivision—into octave or one-third octave bands—and the specific summation rule.) The unit of tonalness is the Bark, and the specific loudness per Bark is given as a power function of sound energy. Total loudness is the integral over tonalness of specific loudness. The Zwicker-Scharf model, like that of Stevens, seems to be a good predictor of loudness of complex sounds.

Models of Auditory Masking. Let us now turn to those models that attempt to account for auditory masking. Here we find that analogues abound to visual brightness contrast. We begin with the report that the auditory system can produce effects reminiscent of Mach bands; these auditory Mach bands appeared in measurements of masked thresholds, i.e., as enhancements in sensitivity to frequencies near the edges of a band of masking noise (Carterette, Friedman, & Lovell, 1969). Although the existence of auditory Mach bands has been questioned (Rainbolt & Small, 1972), their occurrence, if confirmed, would strongly suggest the existence of lateral inhibitory neural interactions, perhaps at a peripheral locus (e.g., on the basilar membrane). A subsequent step would be to hypothesize that auditory masking, like brightness contrast, reflects peripheral lateral interactions. It is unlikely, though, that lateral inhibition is the whole answer—at least, not inhibition solely at a peripheral level. There is a relatively small but nevertheless noticeable component to masking that takes place under binaural stimulation (tone presented to one ear, noise to the other), that is not due to bone conduction, and must therefore be central in origin (e.g., Zwislocki, Buining, & Glantz, 1968).

Just as there is reason to suspect similar underlying neural bases for masking and contrast, so similar psychophysical models have been proposed. One view is that masking involves a change in the exponent of the loudness function (Stevens, 1966e). Like his and J. C. Stevens's model for brightness contrast, Stevens's model for masking implies that loudness grows more rapidly with sound pressure level under masking than it does when no masking sound is present. This change in exponent takes place only so long as the sound pressure level of the test sound is smaller than that of the masking sound. When the test sound increases above the intensity of the masker, the masker theoretically no longer exerts an influence on loudness. This last postulate finds support in results that Stevens reviewed [data of Lochner and Burger (1961) on masking of tones and of Pollack (1949) on masking of speech]. But not all data support that postulate. Figure 5.15 shows that loudness can be suppressed even when the test tone is 10–20 dB greater than the masker. A study by Hellman (1972) also suggests the need for some extension of Stevens's model. Masking of noise by tones occurs even when the sound pressure of the noise is much greater than that of the tone. What seems to be important is not the overall sound pressure of the noise, but the sound pressure in a small band of noise frequencies around the frequency of the tone.

More important to the model, however, is that near the point of equal sound pressure levels, there is a relatively abrupt change in the exponent of the loudness function. When the test tone is more intense than the masker, the exponent of the loudness function is reduced to the same value (about one-third in terms of sound energy) as the exponent of the loudness function in the quiet.

A later elaboration of the model (Stevens & Guirao, 1967) included additional quantitative restrictions. For the case of a tone masked by white noise, it was assumed that signal-to-noise ratio at threshold is constant, i.e., that the masked threshold is proportional to the SPL of the masking noise. In particular, Stevens and Guirao showed the tone at threshold to be about 30 dB below the SPL of the noise. Since the masked loudness function shows an inflection when the SPL of the tone equals the SPL of the noise, the extended model predicts that the exponent of the masked loudness function increases linearly with the SPL (in decibels) of the masker.

Stevens extended his notion of *power transformation* to describe two other auditory phenomena. One is the relatively rapid growth of loudness with sound pressure that occurs in nerve deafness: recruitment hearing. Although weak sounds that are detectable by a normal ear may be inaudible to the recruiting ear, strong sounds may appear quite loud. The second phenomenon is the growth of loudness with sound pressure at low frequencies (less than about 400 Hz), where the log–log slope of the loudness function is greater than it is at higher frequencies. Stevens calls this phenomenon *low-frequency recruitment.*

Where masking does affect the loudness exponent, the size of the exponent depends on the sound pressure level of the masker: more intense maskers yield higher exponents. Richards (1968) provided data to support Stevens's (1966e) original suggestion that the exponent grows as the .16 power of the sound pressure of the masker.

A different interpretation of loudness under masking is that a masking sound produces a subtraction of loudness. Steinberg and Gardner (1937) were the first to suggest this possibility, and Lochner and Burger (1961) reiterated it. Subtraction of constant loudness was also postulated by Garner (1959) for use in conjunction with his lambda scale of loudness. The subtractive model is, of course, a counterpart to the model of brightness contrast put forth by Jameson and Hurvich (1964) and by Marsden (1969).

Still another explanation has been given by Zwislocki (1965). His model differs from the others in that it postulates no special process for masking. The effect that a noise exerts on the loudness of a tone, for example, is the result, according to Zwislocki, of energy summation within the critical band. The decrease in loudness comes about because of the nonlinear relation between loudness and sound energy. The model states that the total loudness (of noise and tone, in the example) within a critical band centered around the frequency of the tone is given by a power (θ) of the total energy within the band $[(E_t + E_n)^\theta]$. The loudness of the tone (L) is that total loudness minus the loudness of the noise alone (E_n^θ)

$$L = (E_t + E_n)^\theta - E_n^\theta \qquad (5.15)$$

In fact, the loudness of noise alone should not equal E_n^θ, but should be masked somewhat by the tone. Nevertheless, Eq. 5.15 was shown by Zwislocki to give a very good account of Hellman and Zwislocki's (1964) data on the masking by noise of a 1000-Hz tone (Figure 5.14).

Chocholle and Greenbaum (1966) measured the masking of pure tones by pure tones. Their results could not easily be described in terms of a power transformation, nor did they seem compatible with a model that predicts subtraction of constant loudness. However, the results would appear reasonably well fit by Zwislocki's hypothesis.

Differences between Auditory Masking and Brightness Contrast. Although the similarities between auditory masking and brightness contrast are striking, there are important differences. One is the absence of an auditory equivalent to "blackness." Absolute silence corresponds, in one sense, to the darker-than-no-light black produced by contrast. Another important difference is that loudnesses add—tone plus noise sound louder together than either alone—whereas spatially separated brightnesses usually do not. These important differences in psychophysical behaviors probably reflect fundamental differences in underlying mechanisms.

TOUCH

It is interesting, in this connection, to note that masking of tactile vibration appears to follow rules similar to those that govern masking of loudness. Sherrick (1960) found that the psychophysical function for 50-Hz vibration on the finger, in the presence of a 20–100-Hz vibratory masking stimulus presented to the palm, was steeper than the corresponding psychophysical function obtained in the "quiet." The functional similarities between audition and tactile vibration (Békésy, 1959) suggest that similar mehanisms, perhaps lateral inhibition, underile masking in both modalities.

SUMMARY

Békésy (1967) proposed that the concept of inhibition unites most or all of the senses. Detailed consideration in this section of the effects of contrast on brightness and of masking on loudness suggests remarkable correspondence in these two modalities at least. We may enumerate six principles common to the ways that inhibiting stimuli depress loudness and brightness. Some or all may also apply to other senses, such as touch (vibratory sensation).

1. Contrast and masking both relate to spatial properties of stimulation: lights are most readily dimmed by spatially contiguous inducing fields; sounds are most readily softened by contiguous frequencies (which stimulate spatially contiguous regions on the basilar membrane).

2. Depression of sensory magnitude is greatest when the intensity of the inhibited stimulus is low. In other words, the log–log slope of the inhibited loudness function or brightness function is greater than the slope of the corresponding uninhibited function.

3. Psychophysical functions measured under inhibition usually change their log–log slope only when intensity of the inhibited stimulus is less than that of the inhibiting stimulus. Higher inhibited-stimulus intensities yield "normal" rate of growth of sensation magnitude.

4. The more intense is the inhibiting stimulus, the greater is the log–log slope of the psychophysical function.

5. Contrast and masking are most extensive when inhibiting and inhibited stimuli appear qualitatively similar (are alike in wavelength or frequency composition).

6. Similar models have been proposed to describe the effects of contrast and masking on the form of the psychophysical function. One model states that both types of inhibiting stimulus induce a power transformation on the functions, i.e., alter the exponents of power functions. Another model states that both contrast and masking involve subtraction of constant sensory magnitudes from the psychophysical functions.

There also are important functional differences between contrast and masking.

1. One difference is the way inhibition depends on area of stimulation of the inhibiting stimulus. As the areal size of a visual inhibiting field (constant luminance) increases, degree of contrast also increases, albeit more and more slowly: the larger the inhibiting area, the steeper the log–log slope of the inhibited brightness function. As the (supercritical) bandwidth of a masking noise (constant SPL/Hz) increases, however, degree of masking increases only at high loudness: the wider the noise band, the less steep the log–log slope of the inhibited loudness function.

2. Loudness in complex sounds and brightness in complex fields both depend on mutual inhibitory interactions among elements. But the partially masked component loudnesses sum to form a total loudness, whereas individual component brightnesses fail to add in such a manner.

Locus of Stimulation on the Sensory Surface

For hundreds, if not thousands, of years it has been known that the best way to see a dim star at night is not to look directly at it. To mariners and astronomers this was, and is, an especially useful piece of knowledge. In general, the periphery of the dark-adapted retina is much more sensitive to light then is the retinal center or fovea. But even the peripheral retina itself is not uniform in its sensitivity. (Nor is the fovea.)

Other sensory systems also show variations in sensitivity over the sensory surface. Bitter solutions are strongest when tasted at the back of the tongue; sweet, strongest at the tip; and sour, on the sides. Sensitivity to touch, to warmth, to cold vary over the surface of the body. By way of contrast, sensitivity to thermally induced pain does not seem to vary from one body locus to another (although sensitivity to mechanical pain does): Rather, the threshold for thermal pain is reached when the skin temperature is brought to 45° C (Hardy, 1956).

VISION

Two primary questions come under consideration in this section. First, how does the sensitivity, particularly suprathreshold sensitivity, vary over the receptor surface? Second, can the sensitivity of any part of the receptor surface for a particular sensory modality be specified by a single value at all levels of sensation? For example, if a light of low luminance that is presented to one retinal locus appears n times as bright as the same luminance presented to another locus, will a high-luminance light also produce an n-to-1 ratio of brightnesses? As we shall now see, the answer is: sometimes yes, sometimes no.

Jameson (1965) and Marks (1966) examined by magnitude estimation the sensitivity of the dark-adapted eye to wide-spectrum visual targets of 1° visual angle. Their results concur in showing that a light of constant luminance appears brighter when presented to the peripheral retina than when presented to the fovea. Furthermore, the ratio of peripheral to foveal brightness depends upon level of luminance: as luminance level increases, the ratio of peripheral to foveal brightness decreases. This outcome is depicted in Figure 5.16, which was derived from the magnitude-estimation functions obtained by Marks. When luminance is near the absolute threshold of the fovea, peripheral brightness is many times foveal brightness; when luminance is high, the ratio is smaller. The brightness of an intense stimulus 20° from the fovea (where brightness is maximal) is only 25% greater than foveal brightness.

It is critical to note that these results were obtained with wide-spec-

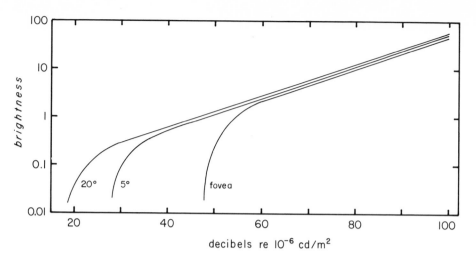

Figure 5.16. Psychophysical functions showing how the brightness of a 1° white stimulus increases with luminance in decibels *re* 10⁻⁶ cd/m² at three retinal loci: the fovea, 5°, and 20° from the fovea. Stimuli presented to the dark-adapted eye. [Derived from data in Marks (1966).]

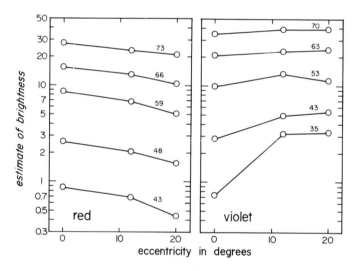

Figure 5.17. Magnitude estimates of the brightness of red light (left) and violet light (right), as functions of distance from the fovea. The parameter is the luminance of the light (1° visual angle) in decibels *re* 10⁻⁶ cd/m². Stimuli presented to the dark-adapted eye. [From Marks (1971a). Courtesy of *Perception & Psychophysics*.]

194

trum lights. It turns out that the spectral composition of the light is a significant determinant of the relation between brightness and retinal locus (Marks, 1971a). This statement is verified in Figure 5.17, which shows how brightness varies with retinal locus at a number of different luminances of violet and of red light. With violet light, as with white light, stimulation of the peripheral retina produces greater brightness than does stimulation of the fovea. Again, we see that the ratio of peripheral to foveal brightness is greatest when the luminance is low; high luminances produce a smaller peripheral–foveal brightness ratio. Use of red light, however, results in a totally different picture. Now the fovea is most sensitive, and brightness decreases continuously when the red stimulus is moved from the fovea farther and farther into the periphery. Note also that the way brightness changes with locus of stimulation does not seem to depend on luminance. Brightness changes to the same relative extent when the luminance is great as when it is small.

All of the results just described are explicable in terms of the well-known duplicity theory of vision. The eyes of most vertebrates (including man) contain two types of retinal receptor, rods and cones. In addition to being morphologically distinct, these receptors display a number of other significant anatomical and physiological differences, such as in their relative distributions over the retinal surface, their spectral sensitivities, and their neural attachments. First of all, the fovea contains only cones, the periphery, both rods and cones. The differences in local sensitivity depicted by Figures 5.16 and 5.17 seem related to the function that relates density of rods to retinal locus (Østerberg, 1935). Where the rods are most numerous (about 20° from the fovea) suprathreshold brightness is greatest. The larger relative variation in brightness at low luminances, as compared to the variation at higher luminances, may be due to the relative convergence of numbers of rods on single bipolar cells in the periphery versus one-to-one connections between cones and bipolars in the fovea.

Now for red light. There is good evidence (e.g., Wald, 1945; Walters & Wright, 1943) that cones are more sensitive on an absolute basis to long-wavelength light than are rods. If the brightness of red light is mediated primarily by cones, then the data of Figure 5.17 are readily explicable, since the density of cones is maximal in the fovea and decreases continuously with increasing distance from the fovea. Furthermore, if one-to-one connections between cones and bipolar cells predominate in the periphery as well as in the fovea, then we might not expect any dependence upon level of the ratio of peripheral to foveal brightness for red light.

Exposure to moderately intense levels of light is well known to diminish markedly the sensitivity of rods, but to leave the sensitivity of cones relatively unchanged. Then we may ask, what is the relation between brightness and retinal locus for a light-adapted eye? The answer is provided by Figure 5.18, which shows how the brightness of white light varies from one locus on the retina to another when the eye is adapted to white light (Marks, 1968a). As we might have predicted, brightness follows the pattern of density of cones: brightness is greatest in the fovea and decreases continuously from there. Again we see that the ratio of peripheral to foveal brightness does not depend upon luminance level.

[The differences between relative sensitivities of the dark-adapted and light-adapted retina imply a difference in the course of light adaptation

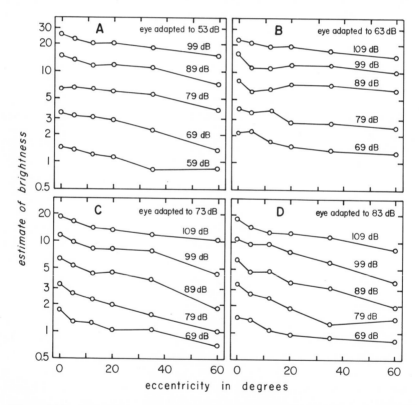

Figure 5.18. Magnitude estimates of the brightness of white light, as functions of distance from the fovea. The eye was preadapted to white light of 53, 63, 73, or 83 dB *re* 10^{-6} cd/m². The parameter within each section is the luminance of the stimulus (1° visual angle) in decibels. [From Marks (1968a). Courtesy of Microforms International Marketing Corporation.]

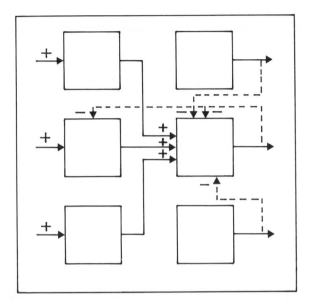

Figure 5.19. Diagram of a theoretical two-stage, feedback model of brightness. Part of the output from the second stage acts as inhibitory input to both the first and second stages. The model postulates spatial summation of both excitatory inputs and inhibitory feedback at the second stage. [From Marks (1972).]

in fovea and periphery. (For discussion of the time course of visual sensation, see Chapter 4.) A flash of constant-luminance white light will produce greater peripheal than foveal brightness in the dark-adapted eye. If the same luminance is used both to light adapt the eye and also as a subsequent test flash, it will produce greater foveal than peripheral brightness. It follows that decay in brightness (adaptation) is more extensive in the peripheral retina than in the fovea.]

In the discussion of spatial summation of brightness was sketched a model that included spatially extended inhibition as well as spatially extended summation. It turns out that the same model can account for differences in sensitivity between the peripheral and foveal retina (Marks, 1972). Figure 5.19 gives a schematic representation of one of two alternative models. The core of the model is the two-stage feedback filter outlined in Chapter 4, but extended to include convergence of excitatory (+) and feedback inhibitory (−) inputs at the second stage. Prior to such elaboration, the input–output relation of the model is

$$B(1 + B)(1 + B) = kL \qquad (5.16)$$

If the degree of inhibitory and excitatory convergence is given by ρ, then Eq. (5.16) is modified, after elaboration of the model, to

$$B(1 + B)(1 + \rho B) = k\rho L \qquad (5.17)$$

When luminance is low, and brightness (B) is small, Eq. (5.17) yields

$$B \approx k\rho L \qquad (5.18)$$

When luminance is high, and brightness relatively large

$$B \approx (kL)^{1/3} \qquad (5.19)$$

Equation (5.16) describes the condition of foveal stimulation, since convergence is minimal for the cone system. Under stimulation of retinal rods (peripheral stimulation) convergence is great, and Eq. 5.17 describes the input–output relation, assuming a suitably large value for ρ. Low levels of luminance produce much greater brightness in the periphery than in the fovea, but as luminance increases, the ratio of peripheral to foveal brightness decreases. There is no difficulty accounting for the variation in brightness across the retina produced by red light. Since retinal cones, primarily, are stimulated, there is no need to invoke convergence. Differences in brightness probably result from the variation in density of retinal cones and, it appears, this variation manifests itself as a change in sensitivity that is independent of level of brightness.

AUDITION

Because of the close relationship between sound frequency and site of maximal stimulation on the basilar membrane, the problems associated with varying locus of stimulation translate into problems associated with varying sound frequency. Chapter 3 dealt in greater detail with the spectral sensitivity of the ear. Equally intense sounds at different frequencies do not appear equally loud. Over the range 2000–5000 Hz, sounds are louder than they are at lower and at higher frequencies. The contours of threshold sensitivity (Figure 3.4) and of equal loudness (Figure 3.5) define the sensitivities of the auditory system to different sound frequencies, and, therefore, to some extent they may also define the relative sensitivities of different parts of the basilar membrane. This last relation must be qualified because of the fact that a good deal of the nonuniform sensitivity of the auditory system derives from variation from one frequency to another in the transmission of sound through the middle ear. When middle ear nonuniformities are eliminated, the sensitivity of the high-frequency portion of the basilar membrane appears greater than the sensitivity of the low-frequency portion (Zwislocki, 1965).

A second major feature of the ear's local sensitivity concerns the variation in perceived qualities. When locus of maximal stimulation is varied by changing sound frequency, the most obvious phenomenal variation is one of pitch. The close dependence of pitch on frequency has been known since the time of Galileo. Scales of the pitch magnitude of pure tones (Stevens *et al.* 1937; Stevens & Volkmann, 1940) show one specially interesting feature. Pitch is related to frequency in about the same manner that locus of stimulation on the basilar membrane is related to frequency. In other words, there is a very intimate connection between pitch and the locus of maximal stimulation on the basilar membrane. That relation is considered again in Chapter 6.

SOMESTHESIS

This section concludes with some consideration of locus of stimulation of the skin. The section concerned with spatial summation noted that the psychophysical function for warmth depends on the size of the stimulus: warmth grows faster with irradiance when area is small than when it is large, and this is true for stimulation both of the forehead and the back. Of special interest here is the fact that the rate of growth of warmth was the same when the area of the back stimulated was more than 20 times that of the forehead (J. C. Stevens & Marks, 1971). For example, stimulation of 7 cm^2 of the forehead and of 160 cm^2 of the back both yielded psychophysical functions with exponents equal to 1.0. Alternatively stated, for the same stimulus size warmth grows more rapidly on the back than on the forehead. The magnitudes of warmth for these two loci are not directly comparable, since the forehead and back were examined in separate experiments. However, it would not seem unreasonable to make the assumption that at the threshold of thermal pain the sensations produced on the forehead and back are about the same. It follows from that assumption, and from the data of Figure 5.6, that for a given area and irradiance, warmth on the forehead exceeds that on the back. The ratio of forehead-to-back warmth diminishes, however, as intensity level increases toward the threshold of thermal pain.

The more rapid growth of warmth with intensity on the back than forehead may be correlated with the back's higher warmth threshold (again, given the same areal extent of stimulation). That correlation seems in turn to relate to differences in density of neural innervation—the forehead probably has many more warmth receptors per unit surface area. The notion that the rate of growth of sensory magnitude is related to neural innervation is not new. Békésy (1955, 1958) propounded the idea with respect to sensations aroused by vibratory stimuli. More recently, Verrillo and Chamberlain (1972) confirmed Békésy's observations. They obtained magnitude balances (averages of production and estimation) for vibratory stimuli presented to the finger, thenar eminence, and forearm. The exponents of the power functions that they obtained were directly related to measured absolute thresholds. Both exponent and threshold were in turn related to neural innervation. When the vibratory stimuli were presented together with a static surround (which prevented spread of vibration), thresholds, and therefore exponents, were inverse functions of the *density of innervation.* When no surround was present, thresholds and exponents were inverse functions of the *number of stimulated neural units.*

In the warmth sense as well as in the vibration sense, the critical variable determining sensitivity—threshold and suprathreshold—seems to be the relative number of stimulated receptors. This generalization is strengthened by the knowledge of the nature of spatial summation in both senses (see pp. 157–164). When the number of stimulated receptors is small (either because stimulus size is small or because a poorly innervated region of the body is stimulated), the sensitivity to weak stimuli is poor (the absolute threshold is high). But the growth of sensation with increasing stimulus intensity is relatively rapid. Thus, differences in sensitivity are much smaller at high levels of sensory magnitude than at low. At high sensation levels, differences in sensitivity are relatively small, both over different body regions and for different stimulus sizes presented to the same region. A similar and possibly related phenomenon appears to be recruitment in auditory nerve deafness. Threshold is increased abnormally, but loudness increases more rapidly than normal, so intense sounds may appear about as loud to a recruiting as to a normal ear. Furthermore, Steinberg and Gardner (1937) and Stevens (1966e) pointed out functional similarities between recruitment and auditory masking. And Stevens added "low-frequency recruitment" to the group. Whether the relatively large exponents measured under all of these conditions depend on similar underlying mechanisms remains to be determined.

SUMMARY

Sensory systems, e.g., visual, auditory, somesthetic, often display nonuniform sensitivity to stimulation of different portions of the sensory surface. In most of these systems, nonuniformity is level dependent; that is, the variations in sensitivity are not identical when measured at all levels of sensation. Typically, the magnitude of the variation diminishes as sensation level increases. Examples are:

1. Relative sensitivity of the dark-adapted peripheral retina is greater than that of the fovea (so long as the stimulating light is not red). Variations in sensitivity are on the order of 20 dB at threshold, but decrease to 2–3 dB at high brightness. Relative sensitivity of the light-adapted retina (or of the dark-adapted retina stimulated with red light), however, is independent of brightness level. It appears that relative sensitivity depends on level when both rods and cones mediate brightness, but does not depend on level when only cones mediate brightness.

2. Relative sensitivity of the basilar membrane is greater to high than to low frequencies, especially at threshold and at low loudness. Low-frequency recruitment (relatively rapid rate of growth in the loudness of low-frequency tones), however, leads to reduction in the differences between sensitivity to high and to low frequencies when loudness is great.

3. The skin's surface varies from place to place in its sensitivity to thermal and to vibratory stimulation. But absolute sensitivity and rate of growth are inversely related, so the local variations in sensitivity are greatest at low, smaller at high sensation levels.

This concludes the chapter on the dependence of sensory magnitudes on spatial parameters of stimulation. In comparison to the rich spatial complexity of our perceptions, it is clear that we have dealt with what, relatively speaking, are effects of extremely simple spatial variations. Yet, the ramifications of even these admittedly simple types of variation often lead to complex questions and problems. It is difficult to emphasize enough the functional importance of sensory responses to spatial patterns of stimulation. Spatial variations are important not only with regard to the perceptual information that they provide for the organism, but also to sensory scientists who endeavor to understand better the nature of the mechanisms underlying the perceptions. One important feature of some of the models that attempt to account for the sensory responses, and in particular of the model of visual brightness, is their reliance on data obtained by direct evaluations of sensory magnitudes. Many models of sensory processes attempt to account only for invariances, i.e., for constant-sensation functions; they predict only how two or more stimulus parameters must vary in order to maintain invariant some sensory attribute or attributes. But these functions say nothing about how sensory magnitudes might vary with stimulus parameters. The model of visual brightness assumes that brightness (not very near threshold) grows as the cube root of luminance. An underlying assumption, therefore, is that when a cube-root function between brightness and luminance is obtained—by the methods of magnitude estimation and magnitude production, for example —that the function is relatively uninfluenced by response biases, but rather exemplifies a "pure" psychophysical relation. In fact, it takes more than one experiment, and more than one procedure to ascertain what is the prototypical psychophysical function for any modality. Even then, there is no guarantee that an average or prototype is unbiased. There are few sensory modalities for which the experimentation has been extensive enough to permit more than a rough guess at pure psychophysical relations.

Chapter 6

QUALITATIVE AND HEDONIC
ATTRIBUTES OF SENSATION

All of the chapters thus far have considered only the intensive aspect of sensations and the ways in which sensory magnitude depends on physical attributes of the stimulus: intensity, frequency or wavelength, duration, spatial extent, and so on. But sensations have other dimensions and attributes besides intensity. Not only do sensations vary from dim to bright, from soft to loud, from warm to hot. They also have distinctive qualities: lights have hues that vary from blue to yellow and from green to red; sounds have pitches that vary from low to high; tastes may be sweet, salty, sour, bitter, or combinations of these. And all of these qualities vary not only with the composition of the stimulus (wavelength of light, frequency of sound, chemical composition), but also with stimulus intensity, duration, and spatial extent and configuration.

Attributes of Sensation

The present chapter deals with some of these attributes other than sensory intensity, and in particular it deals with what we may generally call *sensory quality*. No attempt will be made, however, to relate the qualitative dimensions of sensory experience to all of the relevant stimulus parameters. Instead, the chapter will provide an overview of the measurement of qualitative variables by means of direct scaling procedures.

SENSORY QUALITY

It should be stated at the outset that the term "quality" is being used quite broadly; a few of the sensory dimensions that will be considered in

203

this chapter are not usually termed qualitative. In vision, for example, we shall consider both hue and saturation. Although hue may properly be termed qualitative, saturation is, phenomenologically speaking, clearly more quantitative in character. Saturation is more of a "how much" than a "what kind." The same sort of distinction may be made among auditory attributes. The pitch of a sound is usually referred to as qualitative, its loudness as quantitative. But loudness is not the only auditory dimension that is clearly quantitative in nature; so also are the auditory attributes of volume (the apparent size of a sound) and density (the apparent compactness of a sound). Phenomenologically considered, both volume and density are dimensions that vary in how much, rather than in what kind.

The distinction between qualitative and quantitative dimensions does not necessarily lead to differences in quantifiability. Qualitative variables such as pitch in theory may be quantified (scaled) just as readily as loudness, volume, and density may be. But even though qualitative dimensions may be quantified, it is not clear that the quantification is of the same sort as it is for quantitative dimensions. Instead of reflecting how much, scales for qualitative attributes may instead reflect how much different. Such a dichotomy between scales for qualitative and for quantitative dimensions might underlie the suggestion made by Stevens and Galanter (1957) that scales for pitch may provide at best only interval scales, but not ratio scales.

SENSORY AFFECT

Beyond the purely qualitative, there is another aspect of much of our sensory experience that is of prime importance. This aspect is the affective, or hedonic, value of our sensations. At one end of the scale of affective responses are esthetic responses—responses to faces and figures, to sculpture and painting, to literature and music. We know relatively little, and understand even less, about esthetic responses to these complex stimuli. At a much simpler level, however, there are responses, responses to much less complex stimuli, that are clearly hedonic (although we may wish to refrain from calling them esthetic). Sensory hedonics goes beyond the obvious point that stimuli from nearly all modalities become unpleasant when their intensities are great enough. To be sure, intense light, sound, heat, and cold can be unpleasant, distressing, and even biologically threatening. But in addition, even moderate levels of stimulation may be pleasing, as a sweet taste often is, or unpleasant, as a bitter taste can be.

Pfaffmann (1960) wrote of the reinforcing value of certain stimuli. Pleasant or unpleasant, reinforcing or punishing, are characteristics particularly peculiar to many stimuli from the chemical senses. Consider the extent to which the industries of food and drink are oriented toward hedonic responses. Nor are the hedonics of olfaction ignored by those involved in the manufacture and sale of perfumes and deodorants. Affect is prominent also in the realm of thermal responses. Distinct pleasantness or unpleasantness may accompany heating or cooling the skin, especially when large regions of the skin surface are involved. Only for vision, hearing, and to some extent, touch, are hedonic responses the exception rather than the rule. Vision and hearing especially are more informational, less hedonic. They lead us to deduce and induce; the chemical and thermal senses, and often touch, tend to seduce.

MULTIDIMENSIONAL SCALING

In addition to the study of various individual sensory dimensions, many investigators have concerned themselves with how our perceptions integrate variations over more than one sensory dimension. This is the realm of multidimensional scaling. Rarely is it possible to stimulate any sensory modality so as to produce sensations that vary along only one dimension. Typically, any change in stimulation brings about variation along several subjective dimensions. When we set about to study a single sensory function, such as brightness *re* energy and wavelength of light, we ask the subject to ignore variations in hue and saturation, to abstract the single dimension of brightness, and respond only to variation in it. Determination of the relations between brightness and energy entails two sets of operations: first, the measurement of brightness, i.e., the determination of scale values for each stimulus; and second, the determination of the psychophysical function, i.e., the relation between scale values of brightness and scale values of energy.

A somewhat different approach is required in multidimensional scaling. First of all, the major concern of multidimensional scaling is with all (or at least many) of the sensory dimensions of the modality. This means that the subject must respond to more than one sensory dimension; in some cases he is told what the dimensions are, but, more often, no dimensions are specified. Examples of the latter approach are experiments in which subjects are asked to judge stimuli on the basis of their similarity. Fundamentally, then, multidimensional scaling is concerned with *psychological* relations, the relations of various sensory (psychological) attributes or dimensions to each other.

The primary goal of multidimensional scaling is to construct or abstract a finite number of sensory dimensions and to measure simultaneously the magnitudes of the variations from one stimulus to another on all of these dimensions. A major task of multidimensional scaling, is, therefore, to find a single psychological unit that can serve to measure all of the dimensions. In the example of vision, one aim is to determine a unit that can be used to describe variations in brightness, hue, and saturation. Obviously this cannot be a unit merely of any one of them alone, i.e., not just a unit of brightness, of hue, or of saturation. It must, instead, be some superordinate unit of sensory difference. In that sense and for that reason, multidimensional scaling often is applied to subjects' judgments of difference or of similarity among stimuli.

Multidimensional scaling (of similarity, for instance) involves two processes. One consists of determining unit differences along each of the sensory dimensions—e.g., how much change in hue is equal to how much change in saturation. The second consists of determining how differences across dimensions combine—e.g., if differences add linearly. Tversky and Krantz (1970) have given mathematical formulation to the principles that underlie two such processes.

Several procedures for multidimensional scaling have been devised. Some of these procedures are based primarily on the frequencies that stimuli are confused, e.g., scaling based on just noticeable differences (JNDs). A common assumption is that equal degrees of confusability imply equal sensory differences. Some other procedures employ ratings or categorical judgments, but process the results in accordance with models that, like the JND methods, center themselves around confusability. With the advent of direct ratio-scaling procedures such as magnitude estimation, there has also been application of these procedures to multidimensional situations. One of the most common of these procedures is the numerical estimation of similarity or difference. This procedure actually has a relatively long history. Newhall (1939), for example, described methods for the ratio estimation of differences in color. Presented two pairs of stimuli, a subject is asked to estimate the ratio of the similarity of the members that comprise one pair to the similarity of those that comprise the other. The judgment may be a verbal estimate of the ratio, but more often consists of setting physical distances to the appropriate ratio. Sometimes it is convenient to permit the subject to place the stimuli (if they are samples of colored paper, for instance) so that their physical separation corresponds to their psychological difference. Methods of ratio estimation were employed to determine the psychological spacing of color samples from the Munsell system (Newhall, Nickerson, & Judd, 1943).

One possible difficulty with these procedures stems from the fact that they require judgments of difference. In spite of the fact that the judgments themselves—numerical or distance estimates—can have ratio properties, the properties are those of sensory or perceptual intervals. And there is evidence that ratio judgments of intervals can often deviate widely from intervals based on ratio judgments of magnitudes, even when only one sensory dimension varies (Beck & Shaw, 1967; Dawson, 1971; Marks & Cain, 1972). This subject is discussed in more detail in Chapter 7.

Ekman (1963) in particular was critical of the use of estimates of difference. He devised an alternative method of similarity rating that is of great potential value. In Ekman's similarity-estimation procedure, a subject judges how much of stimulus A appears to be (psychologically) in stimulus B and how much of B appears to be in A. Thus, for each pair of stimuli the subject makes two judgments of sensory ratios. Ekman's use of multidimensional similarity, therefore, does not rely on the estimation of difference or distance among sensations, but rather reflects ratio relations among them (scalar products).

This procedure of similarity estimation Ekman (1970) characterized as a *content method*, in contrast to the estimation of sensory difference, which is a *distance method*. He applied both methods to four domains, including odor mixtures and colors. In general, results obtained using Ekman's content procedure provided clearer structure to the underlying dimensions. A major conclusion was that distance methods may yield simplest descriptions of perceptual responses, but content methods may reveal more of the fundamental psychological components. Ekman's procedure could prove to be of special value for modalities that contain a large number of sensory dimensions, but where different subsets of the dimenstions are elicited by different stimuli. Olfaction may be such a modality.

Often the dimensions revealed by multidimensional scaling can be represented in some multidimensional space. Determining that space entails finding out how the several dimensions combine, e.g., how changes in hue, brightness, and saturation add up. A common assumption is that the space is Euclidean. In Euclidean space, total distance is given as the square root of the sum of the squared distances on component dimensions; total distance between two stimuli embedded in a two-dimensional space is measured, therefore, as the crow flies. Another type of space—one that is no more complex than Euclidean space—would give the resultant distance as the simple sum of the individual, component distances. Such a *city-block* metric may be encountered when the salience of the individual dimensions is very great (Shepard, 1964).

Both Euclidean and city-block spaces are examples of Minkowski-n space; this general metric is defined by the equation

$$D = (\Sigma d_i{}^n)^{1/n} \tag{6.1}$$

where D is total distance, d_i is distance on component dimension i, and $n \geqq$ 1. When $n = 1$, the metric is city-block; when $n = 2$, the metric is

Euclidean. Sometimes, results of multidimensional-scaling experiments suggest values of n equal neither to 1 nor 2. The Euclidean and city-block metrics are much easier to conceptualize than are the other Minkowski-n metrics; frequently, it is assumed that psychological space is either Euclidean or city-block. In any case, note that Eq. (6.1), when applied to psychological (sensory) space, is an example of a psychosensory relation. It defines a structure among psychological dimensions.

Dimensions of Visual Experience

Hue

In previous chapters, our concern with vision was limited exclusively to brightness. Variations in brightness among objects in the visual field (which can be related to corresponding luminances and to spatial and temporal variables) provide much information about ourselves and the world around us that is useful to us. Even if brightness information were the sole type of visual information available (as it is for the small number of people who are totally color blind), we could still do quite well in getting about in the world. (Many totally color-blind people cannot; the reason, however, is that they also suffer some defect in visual acuity or in brightness vision.) After all, we have no trouble following events in a monochrome motion picture. But in fact most of us are fortunate also to enjoy color vision. And color is a salient feature of visual perception, even if it is in a sense disposable. Although color vision is hardly necessary, differences in color are readily noticed and remembered. We often tend to sort out and to categorize objects on the basis of color; it is both interesting and important that the categories used in color classification can themselves vary widely from culture to culture (Conklin, 1955).

Two psychological dimensions comprise the general category of color: hue and saturation. The primary stimulus variable that determines hue is wavelength. A simple means to measure the hue of a monochromatic stimulus is the following. First, equate lights of different wavelengths so that they are equal in brightness. Then, estimate the amount of each primary color, e.g., blue, green, yellow, and red—that comprises the hue at each wavelength. In order to simplify such an experiment, the identities and number of primary colors would probably be preselected, and they might consist of the four hues just mentioned. It turns out that there are some limitations on possible responses; these limitations are intrinsic to the nature of our color vision. It is not possible to have a light that appears yellow-blue or green-red. The two members of each pair occur in opposition. This opposition is the basis of one type of theory of color vision (see Hurvich & Jameson, 1957), a type that originated with Hering (1878) in the nineteenth century.

The procedure just described was employed by Jameson and Hurvich (1959) to scale the hue of stimuli viewed in colored surrounds. The spectral composition of the surround affects the perceived hue of the test field in accordance with predictions of their opponent-process theory: for example, changing the surround from blue-green to yellow-red increases the blueness of stimuli of short wavelengths.

A similar procedure was employed by Yager and Taylor (1970) to assess how hue varies with the luminance of a visual stimulus. With a small (17′) stimulus at 550 nm, estimates of yellow and of green increased as power functions of luminance; the exponents were almost identical (.27 for green, .29 for yellow). At low luminances, the functions departed from power-law behavior as greenness and yellowness approached their respective thresholds; the threshold for yellow was quite a bit higher than that for green. This difference between thresholds, and the small difference between exponents of the two psychophysical functions, helped to account for another result, namely that the reported percentage of yellow to yellow plus green increased as luminance increased.

Two variants of this general hue-scaling procedure have been used in experiments conducted by Boynton and his colleagues. Both are related to the method of constant sum (Metfessel, 1947; Comrey, 1950). In one version subjects are asked to give only color names (e.g., blue, green, yellow, red) as responses, with point scores assigned by the experimenter (2 points to the first hue named and 1 to the second, or 3 points if only one hue is named) (Boynton, Schafer, & Neun, 1964; Boynton & Gordon, 1965). In the second version the subject is permitted to divide up a constant total (100 points) however he desires between color names (Sternheim & Boynton, 1966). Both methods appear to give reliable and consistent results.

An example of color-naming functions determined by the first method (Boynton & Gordon, 1965) is shown in Figure 6.1. The test field subtended a visual angle of 3°; stimulus exposures lasted .3 sec. Blueness was maximal at about 460–470 nm, greenness at 510 nm, yellowness at 580–590 nm, and redness increased steadily at very long wavelengths. In addition, Figure 6.1 shows how perceived hue changes with brightness, i.e., as retinal illuminance increases from 100 to 1000 trolands. (One troland is defined as the retinal illuminance produced by viewing a luminance of 1 candela per square meter through a pupil 1 square millimeter in area.) As brightness increases, short wavelengths appear bluer, long

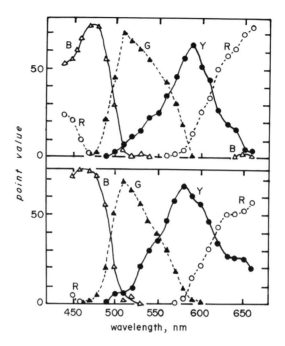

Figure 6.1. Color names as functions of stimulus wavelength. Total point scores for blue (B), green (G), yellow (Y), and red (R). Top: retinal illuminance of 100 trolands. Bottom: retinal illuminance of 1000 trolands. [Data of subject TY. From R. M. Boynton and J. Gordon, Bezold-Brücke hue shift measured by color-naming technique. *Journal of the Optical Society of America,* 1965, 55, 78–86. Courtesy of the authors and *The Journal of the Optical Society of America.*]

wavelengths, yellower. These changes are consistent with other measurements of the Bezold-Brücke shift, i.e., the change in wavelength required when intensity changes in order to maintain hue constant; the concomitant change in wavelength serves to offset the change in hue that would otherwise take place. Measurements of the Bezold-Brücke shift by direct hue matching at different luminances (Boynton & Gordon, 1965; Purdy, 1931) show that as luminance varies, contours of constant perceived hues shift toward three loci, located at about 470, 510, and 570 nm. These are loci of invariant hues, ones that do not change with luminance.

Jacobs and Gaylord (1967) used the color-naming procedure of Boynton, Schafer, and Neun, and of Boynton and Gordon to show how chromatic adaptation (adaptation to spectral lights) changes the perceived hue of subsequent spectral stimuli. Adaptation to green light decreases greenness of subsequent stimuli, adaptation to red decreases redness, and so on.

The same procedure was used by Kaiser (1968), who modified it in two ways. First, subjects were permitted also to give achromatic (white) responses; second, they could specify whether the achromatic response equaled about one-fourth, one-half, or all of the total color of the stimulus. Using a small (12') test field and stimuli that varied in wavelength, 565–590 nm, Kaiser showed that Bloch's law holds for color naming. That is, the color-naming response remained identical as long as the total energy (the product of intensity and duration) stayed the same. (See Chapter 4 for a discussion of Bloch's law of temporal summation in brightness vision.)

Ishak, Bouma, and van Bussel (1970) allowed their subjects to use four color names (blue, green, yellow, red) and employed a method of constant sum, identical to that of Sternheim and Boynton (1966). They applied the procedure to scale the hue of 60 samples of Munsell colors (visual angle of 10 × 4.5°) viewed against black, gray, and light backgrounds. The proportion of the constant sum assigned to each color name was not quite linearly related to scale values of Munsell hue, but the degree of nonlinearity was not large. Ishak *et al.* also noted, in line with results of Jameson and Hurvich (1959), that colored surrounds induce complementary color changes in the perceived hue of test stimuli.

The experiments described in this section utilized a classificatory scheme comprising four primary hues. Such a scheme seems quite adequate for systematic investigations of how wavelength composition, luminance, contrast, adaptation, etc. determine perceived hue. Yet the question arises whether the four so-called primaries—red, green, yellow, blue— are truly fundamental to color vision, or alternatively, are categories based to a large degree on learning, that is, derived from a particular cultural tradition. In the Hanunóo culture of Polynesia, for example, colors are classified quite differently (Conklin, 1955). The primary level of distinction is a four-category scheme: darkness (including black, dark colors, violet, and blue), lightness (e.g., white), redness (including orange and yellow), and greenness (including light brown).

That the four primary hues—and, by implication, color-naming functions like those of Figure 6.1—are in fact related to fundamental properties of the visual system is suggested by data on wavelength discrimination; changes in wavelength are most discriminable around 490 and 580 nm; changes are least discriminable at very long (>650 nm) and very short (<450 nm) wavelengths, with a local minimum in discriminability around 550 nm (Judd, 1932). Wright (1947) pointed out that wavelength discrimination is greatest where hue changes rapidly (wavelengths where color-naming functions cross), least where hue changes slowly (wavelengths near the center of a single color category). These relations imply that there are indeed fundamental, primary hues intrinsic to the operation of the visual system.

SATURATION

A second visual dimension, saturation, refers to the amount of color perceived in a visual stimulus, that is, the proportion of color to the total sensation. Quantitative judgments of saturation go back to studies in the 1920s conducted by Richardson (1929) and by Maxwell (1929), who obtained estimates of the redness of stimuli of different chromatic composition. Their subjects viewed a line whose ends signified "white" and "red" and were asked to mark the point on the line that corresponded to saturation of test stimuli. Richardson found saturation to increase as a positively accelerated function of the chromatic proportion of the stimulus.

More recent studies by Galifret (1959), Onley, Klingberg, Dainoff, and Rollman (1963), Panek and Stevens (1966), Indow and Stevens (1966), and Indow (1967) examined how saturation depends on the colorimetric purity of the stimulus, i.e., on the percentage of pure spectral component in a mixture of spectral light with white. The general finding has been that saturation increases as a positively accelerated function of purity. An exception is the results of Galifret (1959), who scaled saturation of red by the methods of magnitude estimation and ratio production; he found saturation to relate almost linearly to colorimetric purity. It is difficult to know why Galifret obtained a linear psychophysical relation, since nonlinearity, and in particular a positively accelerated function, appears to be the rule for saturation.

Onley *et al.* (1963) compared saturation functions, obtained by the method of magnitude estimation, for red and green lights that varied in luminance and stimulus size, as well as in purity. All of the conditions produced power functions with an exponent range 1.8–2.3. The size of the exponent (average about 2.0) did not seem to depend in any systematic way upon luminance or size. The exponents were systematically smaller (average about 1.6), however, when subjects were presented heterochromatic comparisons (e.g., red standard stimulus for judgments of greenness). Heterochromatic comparison is quite difficult, and difficult tasks often produce low estimates of exponents (Stevens & Greenbaum, 1966; see Chapter 1, pp. 26–28), so the result of Onley *et al.* is not surprising.

Panek and Stevens (1966) and Indow (1967) also studied saturation of red, the former by the method of magnitude estimation and the latter by magnitude estimation and by ratio production. Both of these studies reported power functions with exponents in the range 1.5–2.0; these are close to the values reported by Onley *et al.* A much lower exponent (e.g., .5), however, may be found when subjects' responses are limited to an interval, rather than a ratio, scale (Warren, 1967).

The most extensive study of the relation between saturation and colorimetric purity is that of Indow and Stevens (1966). By means of magnitude estimation and heterochromatic matching, they determined how the psychophysical function for saturation depends on wavelength and on luminance (they studied four primary hues—blue, green, yellow, and red —plus two intermediate hues—orange and yellow-green). The spectral components of their stimuli were small bands of light (not monochromatic). A summary of the results is depicted by Figure 6.2, which shows log–log plots of saturation versus purity for each of the four primary hues at each of four levels of luminance.

First, note that luminance has little effect on the saturation of red (dominant wavelength 632 nm). The exponent of the power function for saturation of red increases only slightly as luminance increases. The saturation of a low-purity (e.g., 10%) stimulus increases somewhat with luminance; the saturation of a high-purity (100%) stimulus first increases, then decreases. But luminance has an enormous effect on saturation of yellow (dominant wavelength 579 nm). The exponent increases from 1.6 at a luminance of 66 dB (re 10^{-6} cd/m²) to 5.0 at 86 dB! Luminance also has a sizeable, though not nearly so large, effect on saturation of green (537 nm), and a still smaller effect on blue (478 nm).

Why should luminance modify rate of growth of saturation in so varied a way? One possibility is that the different ways that saturation exponents increase with luminance relate to different ways that chromatic components increase with luminance. More likely, or in addition, the different relations between exponent and luminance may reflect some sort of interactions among chromatic mechanisms. There is probably a close connection between (*a*) the relatively small increases in exponent with luminance found for red and blue and (*b*) the desaturation of these colors at high luminance.

The functions depicted in Figure 6.2 also show how saturation varies from one hue to another when stimuli are at the same level of luminance and have the same degree of purity. At the top of each function appears the value of saturation of 100% purity. For example, at a luminance of

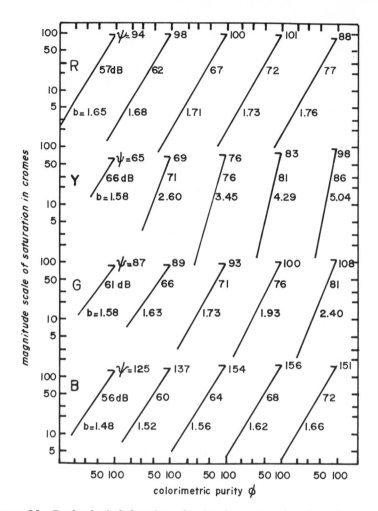

Figure 6.2. Psychophysical functions showing how saturation depends on colorimetric purity of red (R), yellow (Y), green (G), and blue (B) light. The values of ψ at the top of each function indicate the maximal saturations (at 100% purity); the values of b indicate the exponents of the power functions. The luminance of the stimulating light is given in decibels re 10^{-6} cd/m². [From Indow and Stevens (1966). Courtesy of the authors and *Perception & Psychophysics.*]

about 71–72 dB and 100% purity, blue appears the most saturated of the colors, next red, then green, and finally, yellow. Similar order of saturation holds for almost all combinations of luminance and purity.

That various wavelengths differ in the degrees of maximal saturation has been known for a long time. Psychophysical relations between saturation and wavelength for monochromatic light were determined by Jameson and Hurvich (1959) and by Jacobs (1967). Jameson and Hurvich showed that the spectral composition of a surround modulates the degree of saturation of a chromatic test field; thus a yellow-red surround decreases the saturation of a long-wavelength stimulus. Comparable results were obtained when subjects estimated numerically the saturation of Munsell samples (Ishak *et al.*, 1970). Jacobs also showed how saturation depends on the nature of prior adaptation: After "neutral" (incandescent light) adaptation, saturation was greatest with stimuli at very long wavelengths, i.e., near 650 nm (red), and least near 570 nm (yellow). Prior spectral adaptation has the effect of reducing the saturation of the corresponding portion of the spectrum. Thus, "red" adaptation reduced the saturation of long-wavelength stimuli, and "blue" adaptation reduced the saturation of short-wavelength stimuli.

MULTIDIMENSIONAL SCALING AND VISUAL COLOR SPACE

We turn now to the use of direct scaling procedures in attempts to measure simultaneously several visual dimensions, all with a common psychological unit. Attempts to integrate visual dimensions via multidimensional scaling usually have employed Munsell color samples as stimuli. One reason is that psychological scales for Munsell colors have been well worked out for each of the three individual dimensions. For example, a subcommittee of the Optical Society of America concerned with the spacing of Munsell colors based its 1943 report on a total of 3 million ratio judgments of color difference (Newhall *et al.*, 1943). The Munsell system specifies units on each of three dimensions: lightness (Munsell *value*), hue (Munsell *hue*), and saturation (Munsell *chroma*). The units for these three demensions are, however, independent of one another. Multidimensional scaling may provide the link to enable all colors to be put into a visual space that contains a single common unit.

The method of matching physical distance to psychological or perceived distance was employed by Indow and Uchizono (1960) to scale differences in hue and chroma, and by Indow and Kanazawa (1960) to scale differences in hue, value, and chroma. Munsell samples that varied in hue and chroma could be described by a two-dimensional metric, and samples that varied in hue, value, and chroma by a three-dimensional metric. The metrics were Euclidean, that is, n equaled 2 in Eq. (6.1). Colors that form complementary pairs all fell nearly opposite one another, and lines for different chroma all converged at a single (achromatic) point. However, there were some disagreements between the spacing obtained along the hue dimension and spacing defined by Munsell hue.

In both experiments it sometimes happened that distances in the derived multidimensional space were not quite linearly related to the original estimates of distance. That is, the mapping of results onto a Euclidean space required a distortion (nonlinear transformation) of the estimates. Alternatively speaking, if we assume that the derived, Euclidean space is a valid representation of color space, we may conclude that the original distances given by some of the subjects were distorted (nonlinear) functions of distances in the underlying space. Such nonlinear relations between multidimensional distance and the original estimates of distance are not always found, however. Ramsay (1968) obtained ratio estimates of color differences, from which he derived a color space. He found no significant nonlinearity between original ratio estimates and ratios calculated from the multidimensional space. Ramsay's two-dimensional Euclidean space was a close replica of Munsell spacing of his stimuli. The fact that noticeable nonlineariites can occur between estimates and scale values may reflect some response bias intrinsic to judgments of sensory intervals or differences.

The method of matching physical distance to similarity was employed by Helm (1964), who used as stimuli Munsell samples that varied in hue and chroma. The results of the multidimensional scaling gave a two-dimensional Euclidean space that agreed closely with the spacing among samples given by the Munsell notation. Color-blind subjects gave distorted versions of the space, and their results implied additional psychological dimensions. Of some interest is Helm's finding that the interstimulus distances he obtained for the same stimuli by the method of successive intervals (in which interval sizes depend on variability in the use of response categories) were nonlinearly (logarithmically) related to distances obtained by the method of direct (distance) scaling. The interstimulus distances derived from successive intervals would produce a simple two-dimensional space only after they had been transformed according to an exponential function. Because of this outcome, Helm concluded that the distance-matching procedure is a more satisfactory means for multidimensional scaling than is successive intervals.

For an example of direct similarity scaling of stimuli, rather than scaling of distances, we turn to Ekman's (1963) examination of similarity of spectral lights. He employed monochromatic stimuli that varied from 593 to 674 nm and that were adjusted in radiance to give the same brightness. For every pair of lights (A and B), each subject estimated the proportion of A that appeared in B and the proportion of B that appeared in A. From the matrix of responses Ekman derived two factors: one of them increased with increasing wavelength, and he called it a "red" factor; the other reached a maximum at about 590–600 nm, and he called it a "yellow" factor. A similar result obtained for an experiment (Ekman, 1970) on spectral lights between 522 and 580 nm. Here the two factors obtained were "green" and "yellow".

Multidimensional color space, by its nature, contains a metric that is different from the metric underlying most of the other direct scales that have been considered. A multidimensional space gives *distances* between stimuli, distances that are inversely related to similarity. But distance is not necessarily proportional to *difference in sensory magnitude.* Lights of 30 and 40 units in brightness may not be as similar as lights of 20 and 30 units. Eisler and Ekman (1959), for instance, proposed that similarity is proportional to the difference between sensory magnitudes divided by their average. Whether Eisler and Ekman's equation is or is not of general applicability, it is nevertheless important to bear in mind that similarity or distance is not the same thing as sensory magnitude.

The dichotomy between magnitude and distance becomes especially important in certain experimental tasks. In the method of category rating, in the method of bisection, and in other methods whereby subjects are asked (or assumed) to equate or indicate the sizes of sensory intervals between stimuli, subjects may tend to judge distance or similarity, rather than difference in magnitude. Given a sensory dimension on which distance is not proportional to difference between magnitudes, then, the results of interval- and ratio-scaling experiments will disagree. One interpretation of the nonlinear relation that interval scales often bear to the corresponding ratio scales rests, therefore, on the assumption that the two types of procedure tap into different aspects of the sensory experience. Under this interpretation, neither type of scale, interval or ratio scale, is necessarily more correct than the other. One, however, may be more appropriate to a particular situation. The answer one gets frequently depends on the question one asks.

Dimensions of Auditory Experience

Whereas simple visual stimuli are known to vary along three psychological dimensions—brightness, hue, and saturation—simple auditory stimuli are known to vary along at least four. In addition to loudness and pitch, we can order sounds along dimensions of tonal volume and density. Pitch—the high–low dimension of sounds—is usually taken to be the qualitative dimension of hearing; there is a sense, therefore, in which pitch may be considered an analogue of hue. Both volume and density, on the other hand, are phenomenally more quantitative than pitch, i.e., more like loudness. Volume refers to the apparent size of sounds and density to their apparent compactness. Earlier in the twentieth century, it was common to find reference to another auditory dimension, namely brightness (e.g., Abraham, 1920; Rich, 1919). Some, like Hartshorne (1934) and Rich (1919), noted a close similarity between brightness and pitch, and one suggestion was that brightness and pitch represented the same auditory attribute. After all, high-pitch sounds are bright and low-pitched sounds are dull. On the other hand, there is some evidence that brightness may be the same as density (Boring & Stevens, 1936).

Complex auditory stimuli, of course, can vary along still other dimensions. There are tonal consonance (see Chapter 3, pp. 93–94), vocality (the attribute that distinguishes vowels), and musical timbre. No attempt will be made here to describe evaluations of these attributes that pertain primarily to more complex sounds.

PITCH

The pitch of a pure tone is first and foremost a function of sound frequency. Pitch also seems to change slightly when sound pressure changes (Snow, 1936; Stevens, 1935), but that effect is relatively small. It is frequency that primarily determines pitch.

It is instructive to examine briefly the nature of pitch by comparing it to hue. Both are qualitative dimensions of their respective sensory modalities. To be sure, they behave quite differently in mixtures: colors blend to form new color qualities, whereas pitches often remain distinguishable (except when masked). Of course, when tones very close in frequency produce the waxing and waning in loudness that is known as beats (or intense tones produce combination tones) can we say that a new quality emerges. But more to the point, visual sensations can be hueless (gray, black, or white). What about pitch? Is tone quality (pitch) something added to the other qualities, or does pitch necessarily inhere in auditory experience? Can there be pitchless sound?

On the negative side is the statement by Stevens et al. (1937) that some subjects who attempted to fractionate pitch often found it very difficult to know what "zero" pitch was. Stevens and Galanter (1957) suggested that there may be no true zero point on a pitch scale, and if so, then the mel scale of pitch, to be discussed shortly, would be only an interval scale, not a ratio scale. This point of view is supported by Beck and Shaw's (1962) finding that a change in the numerical value (modulus) assigned to the pitch of a standard stimulus had the effect of producing a constant addition to magnitude estimates of the pitch of other stimuli.

On the positive side is the evidence that subjects generally concur with one another and are reasonably consistent from one experiment to the next in ratio judgments of pitch. It is not unlikely, then, that subjects do have some zero pitch in mind when they judge pitch numerically. Some noises, and pure tones of very low frequencies, can often appear to be almost pitchless. To decide whether some sounds are completely pitchless, however, may be a difficult or impossible question to answer.

Formulation of a psychophysical function relating pitch to frequency was one of the earliest successes of psychophysical scaling. Stevens *et al.* (1937) used the method of fractionation to scale the pitch of pure tones. Subjects adjusted the frequency of a variable tone so that its pitch appeared half as great as that of a standard. All tones appeared at 60 dB loudness level. The pitch function that Stevens *et al.* derived is shown in Figure 6.3. The unit of pitch was called the *mel;* the pitch of a 1000-Hz tone was given the value of 1000 mels. Siegel (1965) obtained very simi-

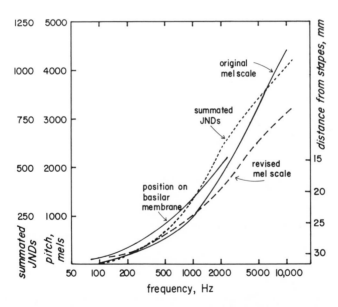

Figure 6.3. The original mel scale of pitch, the revised mel scale, summated JNDs for discrimination between sound frequencies, and site of maximal stimulation on the basilar membrane, all as functions of sound frequency. [The original mel scale is from Stevens, Volkmann, and Newman (1937), courtesy of the authors and *The Journal of the Acoustical Society of America.* The revised mel scale is from Stevens and Volkmann (1940), courtesy of the authors and the University of Illinois Press. Summated JNDs are from Shower and Biddulph (1931). Site of maximal stimulation on the basilar membrane is from Békésy (1949b).]

lar results using the identical procedure. Only for frequencies greater than 4000 Hz was there any sizeable difference between his data and those of Stevens *et al.*; Siegel's data showed relatively greater values of pitch at those high frequencies.

In a second experiment, Stevens and Volkmann (1940) derived a somewhat different psychophysical function for pitch. This latter function was based on results obtained by an equisection procedure, in which the subjects adjusted the frequencies of several pure tones (all at 55 dB loudness level) in order to mark off equal sense-distances. The "revised" pitch function is shown as the dashed line in Figure 6.3. Although the two functions are similar, there are some systematic differences: the revised mel function falls above the original function at low frequencies (20–1000 Hz), but below it at high frequencies (>1000 Hz). Actually, the synchrony of the two functions at 1000 Hz is a matter of definition; the important point is that the revised mel function increases less rapidly with frequency than does the original function. It may well be that this difference is another example of the less rapid rate of growth that seems to be found, in many sensory domains, when interval-scaling, rather than ratio-scaling, procedures are employed (see Chapter 7).

Stevens and Volkmann also obtained fractionations of pitch, and these results were in agreement with the revised mel scale. The difference between these later fractionations and the earlier ones obtained by Stevens, Volkmann, and Newman stemmed from the fact that in the later experiments, subjects could also listen to a low-frequency (40-Hz) tone at any time. This tone, presumed to lie near zero pitch, was made available because, as was said, some of the subjects in the earlier experiment reported difficulty in imagining what zero pitch was like. However, it is important to note that availability of the 40-Hz stimulus may have led subjects to bisect, rather than to fractionate.

Like all other sensory dimensions, pitch depends upon a multiplicity of stimulus variables. Of these variables frequency may be the most important, but it is not the only variable. Mention has already been made that pitch also depends on sound pressure. It may be presumed that the mel function would be slightly different if measured at a very low or very high loudness level. When loudness level increases, the pitch of tones high in frequency increases, whereas that of tones low in frequency decreases (Snow, 1936; Stevens, 1935). Another important variable is stimulus duration: as duration increases, pitch at high frequencies increases, and pitch at low frequencies decreases (Pollack, 1968). Thus, increases in both duration and intensity act in a similar manner.

There is no doubt that variations in procedure can affect the results of scaling experiments, and the scaling of pitch is no exception. Beck and Shaw (1961) obtained magnitude estimates of pitch. They found that when the standard tone was near the middle of the frequency range, the results resembled the original mel scale, but when the standard was near the bottom of the range, the results resembled the revised scale. Changes in the number assigned to the standard tone can also affect the form of the mel scale (Beck & Shaw, 1962). In spite of these variations, judgments of pitch seem somewhat less sensitive to procedural variables than do judgments of many other sensory dimensions, such as loudness. Also plotted in Figure 6.3 against sound frequency are summated JNDs for frequency (from Shower & Biddulph, 1931), and position of maximal stimulation on the basilar membrane (Békésy, 1949b). The similarity among all of these functions is remarkable. To a rough first approximation, the pitch of a pure tone seems to be nearly linearly related to position of maximal stimulation on the basilar membrane, and this is true whether pitch is determined by a ratio-scaling procedure, such as fractionation, an interval-scaling procedure, such as equisection, or a discriminability-scaling procedure, such as summation of JNDs.

Because very similar pitch functions were obtained from quite different scaling procedures, Stevens (1957) classified pitch as an example of a *metathetic* continuum, and thereby distinguished it from examples of prothetic continua like loudness and brightness (see Chapter 1, pp. 19–20). It is important to keep in mind that the relative invariance of the relations between pitch and frequency appeared for the most part from experiments employing pure tones. Although there appears to be no systematic difference between results obtained for pure tones and for somewhat more complex stimuli, e.g., piano notes (Beck & Shaw, 1962), nevertheless, pitch may relate differently to the frequency of other complex sounds. (Even in the case of piano notes, much of the sound energy is contained in the fundamental frequency component.)

It is well known that the pitch of certain complex tones (e.g., a complex consisting of frequencies 200, 300, 400, 500, and 600 Hz) is not so simply related to the frequencies of the stimulus. The dominant pitch in that example would correspond to that of a pure tone at 100 Hz. Experiments such as those of Schouten (1938) and Plomp (1967) support the view that pitch of complex sounds depends on the overall periodicity of the stimulus waveform. Houtsma and Goldstein (1972) present evidence that periodicity pitch can appear with dichotic stimulation (frequencies different to the two ears); thus, periodicity pitch seems determined somewhere in the central, rather than in the peripheral, auditory nervous system.

The simple relation between pitch of pure tones and locus of maximal stimulation on the basilar membrane fails at low frequencies. Zwislocki (1965) suggested that this failure relates to variation along the basilar membrane in the density of neural innervation. He concluded that pitch actually corresponds most closely not with distance along the membrane but with number of primary auditory neurons: 1 mel equals approximately 12 neurons. That there may be so close a relation between pitch and cochlear anatomy is interesting and remarkable.

But pitch cannot always be related to properties of the cochlea: as was just mentioned, periodicity pitch seems to arise further "upstream." Furthermore, in the perception of melody we use the dimension of pitch in a manner quite unrelated to the mel scale (either original or revised). Both musical and nonmusical individuals respond to simple patterns of tones in a manner quite different from that predicted by the mel scale (Attneave & Olson, 1971). The way subjects treat melody—in terms of transposition—implies an underlying scale for pitch that is closely related to the logarithm of sound frequency (like the common musical scale).

One possibility is that the melodic aspect of pitch bears the same sort of relationship to the mel scale that multidimensional similarity of color dimensions bear to scaled magnitudes of hue, saturation, and brightness. This possibility—that melodic pitch reflects similarity—is supported by the high similarity between pure tones separated by octaves (frequencies in the relation 2:1). Octave similarity may be considered a type of logarithmic psychophysical function. Alternatively (or perhaps concomitantly), musical pitch can be treated not as a linear dimension (like the mel scales), but rather as a two-dimensional spiral in which tones one octave apart reside directly above and below each other. Important questions concern the relations between the two types of pitch: mel and musical. Is one derived from the other? Or are they unrelated, merely reflections of independent modes of processing auditory information?

TONAL VOLUME AND DENSITY

Even simple auditory stimuli such as pure tones produce sensations that vary along several psychological dimensions. In addition to loudness and pitch, these dimensions include tonal volume and density. Tonal volume is an attribute of sounds that corresponds to their apparent size; some sounds are big, others small. Sound frequency is an important, though not the sole, determinant of volume: high frequencies yield small sounds, low frequencies, large ones (Rich, 1919; Thomas, 1949, 1952). Lest it be assumed that volume depends solely on learned associations (e.g., small objects produce high-frequency sounds), it is worthwhile to point out that a congenitally blind subject had no special difficulty judging tonal volume; the judgments were not essentially different from those of sighted subjects (Stevens, 1934).

How volume depends on wide ranges of both frequency and sound pressure was determined by Terrace and Stevens (1962). Magnitude estimates of volume increase with increasing sound pressure, but decrease with increasing sound frequency. At any one frequency, volume grows as a power function of sound intensity, but with an exponent (rate of growth) that systematically increases with frequency, from a low of about .05 (in terms of sound energy) at 100 Hz, to a high of .4 at 5000 Hz. This means that the change in volume with frequency is greatest at low sound pressure levels; at high levels, where volume is already great, it changes less with frequency. Given constant sound pressure levels, how-

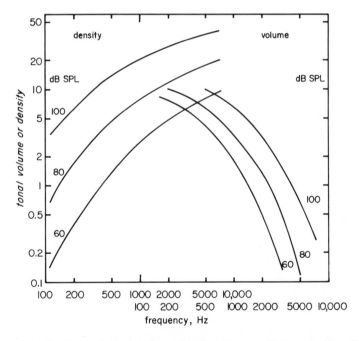

Figure 6.4. Psychophysical functions showing how auditory density of bands of noise (left) and auditory volume of pure tones (right) depend on sound frequency. The curves were derived at three constant sound pressure levels (60, 80, and 100 dB). [Contours for volume from Terrace and Stevens (1962), courtesy of the authors and the University of Illinois Press. Contours for density from Guirao and Stevens (1964), courtesy of the authors and *The Journal of Acoustical Society of America*.]

ever, volume is not a power function of frequency; in double-logarithmic coordinates the functions are concave downward. This is shown in Figure 6.4.

Similar measurements were made for auditory density (Guirao & Stevens, 1964). Density is the attribute of sounds that describes their apparent compactness. As frequency of sound increases, density increases. However, density is not just the reverse of volume, for both density and volume increase with sound pressure level. Magnitude estimates of the density of pure tones and of bands of noise, obtained by Guirao and Stevens, show that density, like volume, grows as a power function of sound intensity. As in the case of volume, the exponent systematically depends on frequency. In the case of density, however, the size of the exponent decreases as sound frequency increases. As Figure 6.4 shows, density also is curvilinearly related to sound frequency in logarithmic coordinates.

Note that when volume and density are related to sound frequency at constant sound pressure levels, loudness is not constant. Loudness varies systematically with sound frequency (see Chapter 3). For example, along the density contours shown in Figure 6.4, the loudness will be much less at low frequencies than elsewhere, but along the volume contours, loudness will be least at high frequencies. The shapes of the contours suggest that if density and volume were determined at constant loudness, their re-

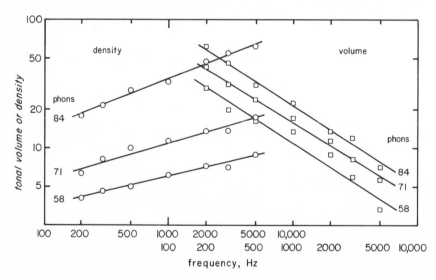

Figure 6.5. Magnitude estimates of the density (left) and the volume (right) of pure tones, as functions of sound frequency. The determinations were made at three constant levels of loudness (58, 71, and 84 dB loudness level).

lations to sound frequency might be linear in log–log coordinates. Figure 6.5 shows previously unpublished data obtained by the author. The functions were determined by a series of experiments that involved magnitude estimation of loudness, volume, and density. When volume and density are measured at constant loudness, they are power functions of sound frequency.

We now have looked at four attributes of sound: loudness, pitch, volume, and density, all of which relate in different ways to sound intensity and frequency. What are the interrelations among these attributes? Stevens, Guirao, and Slawson (1965) found that estimated loudness (L) of a band of noise is proportional to, and, given appropriate choice of units, equal to the product of its volume (V) and its density (D)

$$L = V \cdot D \qquad (6.2)$$

It follows from Eq. (6.2) that volume and density are the inverse of one another, under the restriction that loudness is constant. If two sounds are equated for loudness, but one is twice as dense as the other, then it must also be half as voluminous. Constant loudness implies a constant product of volume and density. Alternatively, given constant volume (or density), loudness and density (or volume) vary proportionately.

Equation (6.2) is an example of a psychosensory law. It describes relations among psychological dimensions only, and it contains no reference to physical variables. Unpublished results of the author imply that Eq. (6.2) holds also for pairs of tones sounded simultaneously. When two tones are well separated in frequency, so that their loudnesses sum approximately linearly, then their volumes also add linearly. Densities, however, add reciprocally, i.e., total density equals the reciprocal of the sum of the reciprocals of the individual densities. These psychosensory laws give structure to sensory dimensions. If other such interrelations can be discovered, one outcome might be to help answer the question of whether sensory measurement is like fundamental physical measurement. Luce (1972) has pointed out that units of physical measurement (length, mass, time, etc.) interrelate when they enter into simple physical formulas. Now, there is clearly an analogy between Eq. (6.2) and the physical formula that defines density as the ratio of mass to volume. If the formulation of simple laws is a criterion for deciding whether some other sort of measurement is like physical measurement, then elucidation of psychosensory laws producing a system of interrelating sensory dimensions, would suggest that sensory and physical measurement are alike.

One meaning of Eq. (6.2) is that loudness, volume, and density are not all mutually independent variables. Given knowledge of any two of them, it is possible to compute the value of the third. In fact, this mutual dependence can be extended to include pitch. Because of the ways that pitch, loudness, volume, and density relate to sound intensity and frequency (in particular, because most of the functions are single-valued, i.e., monotonic), it is also the case that, given a knowledge of the values on any two psychological dimensions, it is possible to calculate the values on both the other two. For example, specification of loudness and pitch of a sound suffice in theory to determine also what must be the volume and density. This reduction of the system to two degrees of freedom does not necessarily diminish the psychological reality of any two of the four auditory attributes. It does suggest, however, that there need not be a search for four different (independent) neural codes that underlie them.

An implication of the nonindependence of the four auditory attributes, and particularly of the possible reduction of them to two, is that a similarity space for simple sounds such as pure tones may be two-dimensional, as sketched in Figure 6.6. Either of two pairs—pitch and loudness, or volume and density—could serve as orthogonal dimensions. To be determined is whether such a space could be Euclidean [$n = 2$ in Eq. (6.1)]. Probably it could not. As a matter of fact, one interesting possibility emerges if the dimensions shown in Figure 6.6 correspond to the logarithms of the variables, loudness, density, and volume. Logarithmic transformation of Eq. (6.2) gives log loudness as the linear sum of log density and log volume. A city-block metric [$n = 1$ in Eq. (6.1)] of logarithmically transformed variables would, therefore, be consistent with Eq. (6.2).

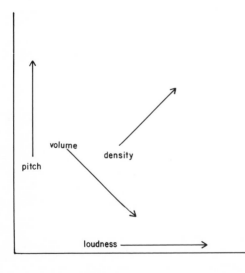

Figure 6.6. A possible two-dimensional representation for the auditory similarity of pure tones.

After this last section was written, there appeared a study by Carvellas and Schneider (1972), who obtained numerical estimates of the dissimilarities among pure tones that varied both in frequency and loudness level. The estimates were consistent with a two-dimensional space; one dimension corresponded to loudness, the other to pitch. Furthermore, the metric of the space was city-block. It would be interesting to know whether the results might also be consistent with some other pair of dimensions, such as volume and density.

Olfactory Quality

The question of what comprises odor quality is especially perplexing. Whereas there is reasonably good agreement concerning psychologically primary hues in vision and primary qualities in taste, there is no universal agreement concerning primary odor qualities. Several schemes have been proposed, but each has met with only limited acceptance. The fundamental dimensions of odor quality, if indeed fundamental dimensions exist, remain to be discovered, or, if they have been discovered, to be verified.

Prior to the twentieth century, several attempts were made to classify odors according to their qualities. The number of primary odors that was postulated varied widely, from von Haller's three through Linnaeus's seven to Zwaardemaker's nine. One of the best known schemes of the twentieth century is Henning's (1916a) six independent qualities: ethereal, fragrant, putrid, spicy, resinous, and burnt. Henning arranged the six primaries at the points of a prism, and he believed that all odors lay either on the prism's edge or surface. Another contemporary scheme is the four-component system of Crocker and Henderson (1927), consisting of the qualities fragrant, acid, burnt, and caprylic. Every odor, according to this classification, can be given a score that may range 0–8 on each of the four dimensions. A more recent classification is Amoore's (1965) seven primaries, five of which he claimed relate to molecular shape. The notion that molecular shape is a critical stimulus variable for the chemical senses dates back to Democritus, 23 centuries ago, who theorized such a relation for taste and, by implication, also for olfaction. In contrast to these schemes, all of which contain relatively small numbers of primary odor qualities, Harper, Bate-Smith, Land, and Griffiths (1968) proposed 44 primaries, and Amoore (1969) more recently estimated that there may be as many as 20–30; the latter estimate was based on the possibility that primaries might be determined from specific anosmias (odor blindnesses), as Guillot (1948) previously suggested.

Theoretically, if we knew what were the primary dimensions of odor quality, it would be possible to use the set of qualities (especially if the number were relatively small) in a procedure to obtain an "odor profile," much like the taste profile described in Chapter 4, p. 135, and also in the next section. That is, it might be possible to ask subjects to estimate the magnitude of each primary subjective component of various test stimuli. Thus far, this has not been done. And perhaps that is to the best, given the present state of knowledge about odor quality.

A few attempts have been made to apply direct scaling procedures to the problem of odor quality. Engen (1962, 1964) applied two procedures, ratio estimation and Ekman's (1963) similarity estimation, to determine the nature of the perception of undiluted aliphatic alcohols that contained from three (propanol) to eight (octanol) carbon atoms in their chains. Odorants that were close to one another in chain length appeared most similar. [It is interesting that patterns of excitation and inhibition measured electrophysiologically in the olfactory bulb of the frog also show greatest similarity for neighboring chain lengths (Døving, 1966).] Factor analysis of the resulting matrices of estimates showed three prominent dimensions. One of these dimensions increased with increasing molecular chain length, whereas another dimension decreased with chain length. The third was not related to chain length. Subjects reported that the stimuli became more "musty," less "fresh" smelling, and less pleasant with longer and longer carbon chains. Ratings given by another group of subjects confirmed the reports that the longer chains appeared less pleasant. It appears that pleasantness was a significant determinant of similarity. Further consideration of the hedonic aspects of odors appears in the next section.

Another attribute of the odors of aliphatic alcohols that seems to correlate with chain length is "oiliness." Henion (1970b) found that magnitude estimates of oiliness increased as chain length increased from 2 (ethanol) to 8 carbon atoms (octanol), although oiliness becomes smaller again with 9 (nonanol) and 10 atoms (decanol). Perhaps a valuable way to get at an answer to the question of odor quality is to start with a set of homologous odorants such as those used by Engen and Henion, and to try to tease apart the components.

Gustatory Quality

Taste Quality and the Taste Profile

The number of qualities of sapid substances is usually put at four: sweet, sour, salty, and bitter. These four "classical" qualities were for-

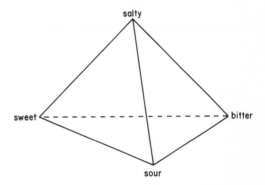

Figure 6.7. A three-dimensional representation of taste quality: Henning's taste tetrahedron. The four vertices of the tetrahedron represent pure sweet, salty, bitter, and sour tastes. [After Henning (1916b).]

malized somewhat by Henning (1916b), who set up a taste tetrahedron. A diagram of Henning's tetrahedron is reproduced in Figure 6.7. It is presumed that the four, pure, simple taste qualities could serve as constituents of any more complex taste. A related proposal is that, given four substances which produce the four simple tastes, any other taste can be compounded from suitable mixture of the four. Some attempts were described by von Skramlik (1926) to duplicate complex tastes by mixtures of a few primaries. Sucrose and sodium chloride are typical examples of pure sweet and salty tastes. Acids, such as tartaric, and quinine salts, such as quinine sulfate, provide fairly pure sour and bitter tastes. The hypothesis of quadrisapidity (mixtures of four primaries necessary and sufficient to match any taste) provides a parallel to the trichromacy of vision, i.e., to the visual system's ability to produce any hue from appropriate combination of three suitably selected primaries.

The two proposals—one that there are four psychologically primary tastes, the other that any taste can be constructed from four physical primaries—are not, however, the same, but instead reflect two different usages of the term "primary." The one is a statement of *psychological primaries,* the second a statement of *mixture primaries.* Vision, for example, is usually considered to have four psychological primary hues (blue, yellow, green, and red, organized into opponent pairs), yet normal color vision is trichromatic (only three mixture primaries are needed to match any color).

There is a major difference between taste and vision that makes the analogy concerning mixtures somewhat tenuous. This difference is that visual quality (hue) is, phenomenally, superposed on visual intensity (brightness). It is perfectly possible to have a "neutral" (white) light, but there probably is no "neutral" (white) taste, except perhaps at low concentrations between thresholds for detection and recognition. Taste quality is inextricably tied up with taste intensity: the taste qualities themselves have intensity, and overall taste intensity seems to be a sum of the intensities of the constituent qualities.

A direct procedure for evaluating the four taste qualities is the *taste profile*. Determination of the taste profile for a substance entails the simultaneous evaluation, by magnitude estimation, for example, of the intensities of all four constituent taste qualities. McBurney and Shick (1971) used this procedure to profile the taste of 26 compounds, and also to profile the taste of distilled water presented just after each compound (water taste). The various salts tested (e.g., chlorides and sulfates of sodium, potassium, and ammonium) tended to have salty and bitter tastes, and to yield water tastes that were sweet and sour. Acids tasted sour and yielded water tastes that were primarily sweet. Bitter substances, such as quinine sulfate and caffeine, yielded water tastes that also were primarily sweet. McBurney and Shick looked for evidence of "opponent processes" operating to code taste quality, but found little evidence to support that concept. A more extensive study was made by Bartoshuk (1968) of the water tastes produced by many substances. Adaptation to a particular concentration of a compound yielded water tastes for all lower concentrations of the same compound; the water taste was greatest in magnitude when the subsequent stimulus was distilled water.

The tastes of a dozen salts, sampled after adaptation to distilled water and after adaptation to sodium chloride, were determined according to a taste profile (Smith & McBurney, 1969). The saltiness of all of the salts diminished after adaptation to sodium chloride, although the sweet, sour, and bitter qualities present in some salts occasionally increased. Adaptation to sucrose and to saccharin decreased the sweetness of normally sweet compounds (compounds that taste sweet when presented after distilled water) (McBurney, 1972), and adaptation to citric acid decreased thes sourness of normally sour compounds (McBurney et al., 1972). These results suggested that each of saltiness, sweetness, and sourness is coded by a single mechanism. However, adaptation to quinine hydrochloride decreased the bitterness of most substances, but not all (McBurney et al., 1972). The bitterness of urea and citric acid were unaffected. Furthermore, adaptation to urea and citric acid affected bitterness of quinine little. It may be that the coding of the bitter taste is more complex than that of the three other taste primaries.

The possibility that the four taste qualities might show some fundamental interactions (in addition to mutual suppression) is an old view. Pikler (1922) and Hartshorne (1934) propounded, primarily on phenomenological grounds, opponent-process theories of taste. Hartshorne's theory was an avowedly affective one. That is, the essential differences among qualities were theorized to emanate from differences in affect. This is perhaps most clearly seen in the opposition of sweet and bitter, qualities whose affective aspects are pleasantness and unpleasantness. Although evidence for an opponent-process mechanism in the mediation of taste quality is scant, nevertheless the importance of affective dimensions of taste is major. Hedonic aspects of taste quality will be considered later in this chapter.

Experiments that employ procedures of taste profiling provide data that enable us to place the tastes of various compounds onto points, either on the surface of, or within, Henning's prism. On the basis of data from McBurney and Shick (1971), for example, we would place .1 M sodium chloride near the *salty* apex, .18 M sucrose near the *sweet* apex, .05 M ammonium bromide on the *salty-bitter* edge, .4 M sodium sulfate in the middle of the *salty-sour-bitter* face, and .026 M calcium chloride in the center of the prism. It is important to point out that no single quality profile (or point on the surface of the prism) can be used to describe the quality of a substance at all of its concentrations. For example, weak levels of sodium chloride can appear predominately sweet, and strong concentrations of saccharin can appear bitter. However, the taste profile gives an adequate description of the quality of any particular concentration of a substance, and a series of profiles can describe how quality changes with concentration.

MULTIDIMENSIONAL SCALING OF TASTE QUALITY

There remains an arbitrariness to the structure of Henning's tetrahedron. In addition to its assumption that four qualities suffice to describe all tastes, the tetrahedron specifies the psychological distances that separate pure tastes. In particular, it assumes that each primary quality is equally distant from all three others. More adequate representation of the space for taste quality probably cannot come about from quality profiles alone. Actually, there exists an indefinitely large number of quality spaces. Any one space provides a representation of the distances among substances that are equally strong subjectively, and for every level of overall subjective intensity there exists another space. The same compound can change spatial position dramatically if its taste quality depends strongly on concentration (as, for example, when sodium chloride tastes sweet when weak, salty when strong).

Schiffman and Erickson (1971) obtained similarity ratings of pairs of taste substances taken from a set of 19, all of which were equated for total taste intensity. Although the procedure produced metric data (the

subjects marked distances on a line to indicate perceived difference between members of a pair), only the nonmetric properties (rank orders)
were employed in determination of the multidimensional space. The results indicated a nearly Euclidean, three-dimensional space. In contrast
to Henning's equidistant model, Schiffman and Erickson's data suggest
that pure sour substances are located somewhat nearer to salty and bitter
ones than sweet, salty, and bitter ones are to each other. In its basic
structure, however, the space resembles that postulated by Henning. But
Schiffman and Erickson avoided interpreting their results in terms of the
usual four primary tastes. (If they had, it would probably have been necessary to add a fifth primary: alkaline, as exemplified by sodium hydrox-
ide.) Instead, they related the space to three presumed underlying dimensions, namely hedonic quality of the taste, molecular weight of the substance, and pH.

An interesting approach to multidimensional scaling of taste quality is
the cross-modality matching procedure of Gregson (1965, 1966). Each
subject judged complex taste stimuli by rating his degree of certainty that
each of a set of pictures (which code, by color, combinations of four primary tastes) described the stimuli. The nonmetric data were analyzed
into multidimensional spaces, either of three dimensions (salty, sweet,
and sour) when sucrose, sodium chloride, and citric acid were employed,
or of four (bitter also) when quinine was added. Thus, Gregson found a
separate dimension corresponding to each taste quality. Of some interest
was the fact that the best mathematical solutions were consistently non-
Euclidean. Schiffman and Erickson also found some deviation from Euclidean space. In all of these studies, the best spatial representations involved values of $n > 2$ in Eq. (6.1). Thus, the best representations were
much more like Euclidean than like city-block space.

A similarity space that closely approximated Henning's tetrahedron
was obtained by Yoshida (1963) from numerical judgments of similarity
among 11 substances. However, two of the substances he tested fell outside the prism: monosodium glutamate and alum potassium.

There is an interesting and curious aspect to Henning's tetrahedron as
it is considered with regard to multidimensional space for gustatory sensations. Although Henning's tetrahedron specifies four primary tastes, it
can be embedded in a three-dimensional space. Any taste substance can,
therefore, be specified by a set of three, rather than four, numbers. In
fact, Henning's classification can itself be modified in a similar fashion. If
values of the four primaries are normalized so that (a) every taste substance receives a numerical score for each quality, and (b) the sum of
the scores is always the same (e.g., 100%), then scores on only three di-

mensions (qualities) are needed to provide complete specification of the taste of each substance. The logic underlying this system is identical to that used to specify the three chromaticity coordinates of any spectrum of light with only two numbers.

Hedonic Responses to Sensory Stimuli

Often neglected, or at least underrepresented, in the study of the senses are the hedonic or affective responses to sensory stimuli. [An excellent survey of early work on hedonics is Beebe-Center's (1932) *The Psychology of Pleasantness and Unpleasantness*.] Not only do our sensations vary in strength and in quality, but also in the nature and degree of affect, whether positive or negative. In a sense, the frequent neglect of affective responses is understandable. Our knowledge of how the senses operate is most complete in the modalities of vision and hearing, and it seems fair to say that those two senses are intrinsically the least affective. This is not to say that affect is totally absent from those modalities. Given even relatively simple stimuli, experimental subjects (and ordinary people) may report that they prefer some stimuli to others. Thus, sounds that are low in frequency and weak in intensity seem to be preferred to sounds high in frequency and strong in intensity (Singer & Young, 1941). There may be a maximum of preference or pleasantness in the region 500–700 Hz and 50 dB SPL (Vitz, 1972). Guilford (1954) applied a Thurstonian model (by which scaled differences depend on variability in responses) to the rating data obtained by Singer and Young, and he was able to generate isohedonic (constant-preference) contours for auditory pure tones. In a similar vein, Helson and Lansford (1970) examined extensively how pleasantness of color samples varies with the colors of the samples, the colors of their backgrounds, and the spectral distribution of the light that illuminates them.

In spite of the affective differences found among simple visual and auditory stimuli, it is, nevertheless, true that the nature of affect in vision and in hearing is heavily dependent on contextual variables and on past experience. Music provides a good example. A host of factors are involved in determining whether, at some particular moment, I prefer Bach, Beethoven, or the Beatles; and experiential factors determine whether at any time I will like any of them at all. To be sure, there is some extent to which certain frequency intervals of tonal combinations are more or less pleasurable. Plomp and Levelt (1965) showed there to be a close relation between perception of consonance and dissonance and the critical bandwidth of the auditory system. But even consonance and dissonance, regardless of the degree to which they may be intrinsically related to the operation of the auditory system, belong to a plane of affect that lies somewhat remote from the positive and negative affect that is associated with the tastes, for example, of sucrose and of quinine. For taste, for olfaction, and for the somesthetic senses (especially pain and temperature), the hedonic dimension of sensation reigns high, even if not quite supreme.

The hedonic aspects of sensation at times are recognized, and occasionally attempts have been made to relate hedonics to schemes for classification of the senses. Sherrington (1906) pointed out that the dichotomy between "distance" receptors (vision, hearing) on the one hand and "contact" receptors (touch, taste) on the other also entails an affective distinction. Affect is often a strong component of the sensations aroused by stimulation of contact receptors. Another scheme is that of Troland (1928), who divided the sensory qualities into the categories of "beneceptive," "nociceptive," and "neutroceptive." Under the heading beneceptive, came the sweet and weak salty tastes, flowery and fruity odors, and mild warmth; under nociceptive, came pain, putrid odors, intense sour and salty tastes, cold, and intense warmth; under neutroceptive, came all of sight, sound, and touch. Troland's system was based on the notion of biological utility, not directly on affect—yet the relation between affective perceptions on the one hand and utility and danger on the other is often very close. A more recent proponent of this sort of relation is Cabanac (1971). He argues that whether a stimulus is perceived as pleasant or unpleasant often depends on internal factors, factors that also determine biological utility of the stimulation. For example, an external thermal stimulus that lowers previously normal skin temperature can be pleasant when internal body temperature is higher than normal, but unpleasant when internal temperature is lower. External stimuli that tend to maintain internal homeostasis are usually perceived as pleasant, those that disrupt homeostasis are usually perceived as unpleasant. In any case, there is an important connection between affect and motivation (Young, 1959).

The affect or hedonic tone of a sensation depends on all of the same stimulus variables that are involved in determining sensory intensity and quality. Foremost among these stimulus variables are its intensity and qualitative structure. It is obvious that very strong stimulation in every modality can be unpleasant: intense lights and sounds, extremes of temperature, perhaps even very high concentrations of otherwise pleasant substances, such as sucrose. One of the first attempts at quantification of the rules relating hedonic tone to intensity was given by Wundt (1874, 1907), who claimed that stimuli just above threshold are pleasant, that they increase in pleasantness up to a maximum as intensity increases, and that they then decrease through indifference (affective neutrality) to unpleasantness at very high intensities. A contemporary proponent of this sort of view is Berlyne (1971), who incorporated the Wundtian scheme into a more general theory of affect and esthetics. Others, such as Stout (1915), point out that some stimuli are naturally unpleasant, that they increase monotonically in unpleasantness as their intensity increases. In fact, there is probably no single rule that holds without exception. Relations of hedonic value and magnitude to stimulus intensity are complex and varied; in some cases, affect may be more simply and directly related to other sensory properties (magnitude and quality) than to stimulus properties.

OLFACTORY HEDONICS

The hedonic aspect of olfactory sensations is often profound. Henning (1916a) claimed that odors that lay along the burnt–putrid edge of his odor prism are unpleasant, and those on the ethereal–spicy–fragrant–resinous side are pleasant. A plane through the prism divides it into pleasant and unpleasant portions, with indifferent odors along the plane.

In at least one experiment concerned with olfactory similarity (Woskow, 1968), the primary factor obtained seemed to be pleasantness. The hedonic value of odor sensation may be modified by other variables such as internal state; ingestive factors, for instance, influence pleasantness and unpleasantness (Duclaux & Cabanac, 1970). After ingestion of glucose, a previously pleasant odor (orange syrup) can become unpleasant.

Engen (1962, 1964) found that of the aliphatic alcohols, those with long chains of carbon atoms are the least pleasant. His finding was confirmed by Henion (1971d), who scaled pleasantness of undiluted alcohols. Except for a few reversals, pleasantness declined as chain length increased. The variation in pleasantness was not great, however: decanol (C_{10}) was judged about half as pleasant as ethanol (C_2). Cain (1966) also found a general decline in pleasantness as chain length increased. However, he also found evidence of local disturbance of that otherwise monotonic relation. Pentanol (C_5) was judged especially unpleasant.

Only a small range of pleasantness resulted when the pleasantness of amyl acetate was scaled at several concentrations (Henion, 1971a). An increase of 64:1 in concentration led to less than a 2:1 decrease in pleasantness. Henion pointed out that the range of pleasantness he obtained was smaller than the range of odor intensity obtained by Engen and Lindström (1963) for the identical range of concentrations. It is difficult, however, to accept without question the validity of such a difference in range when it is based upon results obtained in different laboratories with different groups of subjects.

Henion's observation must be considered in light of the study conducted by Engen and McBurney (1964) on pleasantness of 18 different odorants. They found that the range of magnitude estimates of pleasantness can be very large—as great as 200:1 from one odor to another. The simuli examined were diverse, ranging from caprylic acid to jasmin to camphor. It was Engen and McBurney's conclusion that the range of hedonic response is probably larger than the range of odor intensity. Henion (1971b) compared directly ratio judgments (magnitude and ratio estimates) of pleasantness and of intensity of a series of concentrations of amyl acetate. Results showed a slightly greater range when odor intensity was judged than when pleasantness was judged. Because pleasantness and intensity were almost reciprocally related, Henion concluded that both were really only aspects of a single dimension. That subjective intensity and pleasantness will not always be so simply related, however, appears certain. It is possible that, for some subset of odorous stimuli, pleasantness and intensity comprise a single dimension. In order to demonstrate convincingly that such a rule is valid, it would be necessary to show in addition that two or more stimuli (perhaps of different concentrations and exposure durations) adjusted to appear equally strong also appear equally pleasant or unpleasant.

TASTE HEDONICS

To repeat, there is often a close relation between biological utility and hedonic tone. The frequently encountered relation between the sweet taste and nutrition and between the bitter taste and poison is clear enough. Yet the mild salty taste also, according to Troland (1928), can be classified as beneceptive, since sodium chloride is vital to well-being. It turns out that all four of the primary taste qualities fall into Troland's categories of beneceptive and nociceptive. There are few, if any, biologically neutral tastes.

Ekman and Åkesson (1965) compared magnitude estimates of sensory intensity (saltiness and sweetness) to estimates of pleasantness (or, as they termed it, preference) of solutions of sodium chloride and of sucrose. For both substances, sensory intensity increased monotonically with increases in concentration (power functions with exponents 1.5–1.6), whereas pleasantness decreased with increases in concentration of sodium chloride and varied nonmonotonically with concentration of sucrose. Figure 6.8 plots the psychosensory relations: it shows how pleasantness (preference) related, on the average, to saltiness and to sweetness. Even though the average relation between pleasantness and saltiness is monotonic, there were large variations between subjects: three of the eight subjects gave nonmonotonic relations between preference and saltiness, and one gave a monotonically increasing one. For sucrose, the results were more uniform from subject to subject, all showing the same sort of Wundtian increase and decrease. It is interesting that, as in olfaction, pleasantness in taste can be modified by internal factors. The same concentration of sucrose may be pleasant when the solution is not ingested, but unpleasant when it is (Cabanac, Minaire, & Adair, 1968). Taste pleasantness depends on more than just gustatory stimulation.

Results similar to those of Ekman and Åkesson were obtained by Kocher and Fisher (1969), who scaled, by magnitude estimation, pleasantness of (preference for) sodium chloride, sucrose, citric acid, and quinine hydrochloride. With the first three substances, pleasantness hardly varied as concentration increased up to about .03 M; thereafter, pleasantness increased with concentration for sucrose, but decreased slowly for sodium chloride and more rapidly for citric acid. Quinine hydrochloride showed a steady decrease in pleasantness as concentration increased. The failure of Kocher and Fisher to find a nonmonotonic function for sucrose, such as that found by Ekman and Åkesson, is interesting. Ekman and Åkesson found the peak in pleasantness to occur at a sucrose concentration of .1 M. Moskowitz (1971c) found the peak to occur at a much

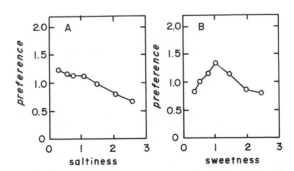

Figure 6.8. Magnitude estimates of pleasantness (preference), as functions of the corresponding magnitude estimates of saltiness and of sweetness. The molar concentrations range: .00086–.011 M for sodium chloride and .015–.22 M for sucrose. [From Ekman and Åkesson (1965). Courtesy of the author and *Scandanavian Journal of Psychology*.]

higher concentration (2.0 M). This concentration fell beyond the range used by Kocher and Fisher.

Moskowitz (1971c) determined the pleasantness and sweetness of 32 sugars by magnitude estimation. In all cases, pleasantness increased less rapidly with concentration than did sweetness, and for eight of the sugars, pleasantness was not monotonically related to concentration. These latter substances included sucrose, which reached its maximum in pleasantness at 2.0 M. At such a concentration, a sucrose solution is extremely viscous. The several sugars varied greatly from one to another in their pleasantness: the least pleasant of those examined was mannose, whose quality has a large bitter component as well as a sweet one.

The number of questions about sensory hedonics that direct scaling can help to answer is great. How does hedonic tone vary with concentration for other substances? For bitter substances, for example, is unpleasantness proportional to taste intensity? What are isohedonic points for different substances (e.g., when are quinines and acids all equally unpleasant)? Will multidimensional spaces for taste, determined from judgments of similarity, automatically contain a hedonic dimension? Or will judgments of hedonic similarity yield quite different relations among sapid substances from relations yielded by judgments of general taste similarity?

HEDONICS OF THE SKIN SENSES

Affect in the skin senses is by no means limited to the displeasure produced by strong mechanical, chemical, or thermal stimulation. Affect may be positive as well, to wit:

> The sensation of Warmth, on emerging from cold, is one of the greatest of physical enjoyments. It may be acute, as in drinking warm liquid, or massive, as in the bath, or other warm surrounding. Of passive physical pleasure, it is perhaps the typical form; other modes may be, and constantly are, illustrated by comparison with it; as are also the genial passive emotions—love, beauty & c [Bain, 1868, p. 34].

It is interesting that Bain chose as his example of positive affect warming after cooling. Phenomenally considered, that order of events may produce greater affect than cooling after warming. Recall that Troland (1928) classified mild warmth (but not mild cold) as beneceptive. There is even some cross-cultural evidence (Jeddi, 1970) that warmth is considered to be intrinsically positive in affect (pleasant), whereas cool is considered negative (unpleasant). And newborn monkeys prefer a metallic, but warmed, artificial mother to a cloth or furry, but cooled, one (Harlow, Harlow, & Suomi, 1971; Jeddi, 1970).

In the realm of the skin senses, only the surface of the hedonic and affective iceberg has been touched by direct scaling. Ekman and his colleagues looked at the unpleasantness of electrical stimulation. Unpleasantness grows quite rapidly with increasing electrical current to the fingers; the psychophysical relation is approximately a square function (Ekman et al., 1964). It might be argued that this highly accelerated psychophysical function serves the useful biological role of warning us of impending danger. Unpleasantness of electric shock also increases with increasing stimulus duration. The increase in unpleasantness with duration is not so rapid, however, as it is with increasing current (Ekman et al., 1966). The former relation is more nearly logarithmic, i.e., negatively accelerated. Ekman et al. (1968) found the logarithmic relation to hold up to the longest duration they examined, namely 3 sec. It would not be surprising if unpleasantness continues to grow when stimulus duration extends even longer.

In the thermal realm, it seems that temporal aspects of pleasure and displeasure may be somewhat complex. On the positive side, it is probably clear from experience (at least from frequent winter experiences in much of the world) that heating one's over-cooled body rapidly produces a feeling of pleasure. On the other hand, the original discomfort that was produced by cooling the skin may have taken a minute or two to develop. When I walk outside on a cold day without a coat, the *sensation* of cold comes on quickly, but *discomfort* appears only after some time elapses. Discomfort also typically requires stimulation of large areas of skin. Stimulation of small areas if intense enough may produce pain, but still not the generalized discomfort produced by stimulation of large areas.

The usual stimulus for temperature sensation, and for affect as well, is a change in temperature of the skin from the neutral, or physiological zero. (It is also possible to change the state of thermal comfort by manipulation of internal body temperature.) Small changes (about 1° C) in skin temperature from its neutral level (about 33° C) are decidedly more pleasant than is no change at all, but are less pleasant than very large changes (changes more than 3° C) (Haber, 1958). Both increases and decreases in skin temperature are perceived as pleasant. However, if the baseline skin temperature is raised from 33 to 36–40° C, no upward change in temperature is perceived as pleasant.

Marked heating and cooling of the body produce discomforts, but the two types of discomfort (warm and cold) do not bear symmetrical relations to the magnitude of stimulation. J. C. Stevens, Marks, and Gagge

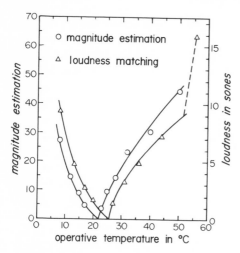

Figure 6.9. Magnitude estimates of (circles) and loudness matches to (triangles) the discomfort produced by stimulation of the front surface of the body, as functions of operative temperature (which denotes the combined effects of air temperature and thermal irradiation). Discomfort increases when operative temperature changes either down or up from the neutral range, 22–24°C. [From J. C. Stevens, Marks, and Gagge (1969). Courtesy of Academic Press.]

(1969) measured how discomfort depends on "operative temperature" when a large portion (most of the front surface) of the body is exposed to the stimulus. In that experiment, air temperature was maintained at 4° C and the level of infrared radiation was adjusted so that the effective temperature of the environment could vary from 4 to 55° C. Figure 6.9 shows the results obtained in two experiments, one by magnitude estimation, the other by cross-modality matching, i.e., when numbers and when loudnesses were matched to the discomfort produced by 2-min stimulations.

In the warm, discomfort increased as a negatively accelerated function of operative temperature (power function with exponent of .7), but in the cold as a positively accelerated function (exponent of 1.7). That difference in exponent—a more rapidly growing response to cooling—may reflect the body's greater capacity to make adequate physiological responses to heating than to cooling. Because physiologically regulatory responses to cooling are meager, when the skin is cooled, behavioral responses are needed, and discomfort presumably provides the motivation for behavioral temperature-regulation.

In the experiment just described, the subjects were in a normal thermal state at the start of each session (none had fever). Internal body temperature is another important factor in determining the state of thermal comfort; had internal body temperature deviated sizeably from 37° C, the results might well have been different from those shown in Figure 6.9. Cabanac (1969) demonstrated that hyperthermia and hypothermia (deviant internal temperatures) interact with superficial (skin) temperature to determine thermal pleasure and displeasure. The same thermal stimulus can be either pleasurable or unpleasant, depending on internal body temperature. When internal temperature is raised above normal, a lowered skin temperature is perceived as pleasant, but a raised one is perceived as unpleasant. Again, the interaction would tend to lead to homeostasis (Cabanac, 1971). The hedonic iceberg is certainly not shallow.

Summary

Qualitative attributes of sensation have generally received much less study by direct scaling procedures than have quantitative attributes. It is also true that qualitative dimensions appear much less analogous from one sense to another. Certainly the dimensions of brightness, loudness, odor intensity, taste intensity, etc., bear much more similarity to each other than do hue, pitch, odor quality, and taste quality. But, of course, this is precisely what difference among sense modalities is all about. These qualitative variations from sense to sense are in large measure what make vision to be vision and not audition, audition to be audition and not olfaction, olfaction to be olfaction and not taste.

But it is not only in the phenomenological domain that the several senses appear similar with respect to sensory intensity, dissimilar with respect to sensory quality. In their psychophysical relations, too, there is a difference. Sensory intensity depends primarily on stimulus intensity, and, furthermore, there appears to be a general psychophysical power law applicable to the growth of sensory magnitude on most if not all continua. Quite the opposite seems true for sensory quality, which shows little generality in its relation to its primary stimulus variable, namely composition.

The four primary hues are blue, green, yellow, and red. The magnitude or proportion of each hue relates nonmonotonically to the wavelength of monochromatic light. Blueness reaches its maximum near 470 nm, greenness its maximum near 510 nm, yellowness its maximum near 580 nm, and redness increases continuously at long wavelengths. The relative proportions of colors aroused at any wavelength vary also with luminance (Bezold-Brücke effect).

Taste, too, shows four psychological primaries: sweet, bitter, salty, and sour. Fairly pure versions of these tastes can be elicited by single compounds, e.g., sucrose, quinine, sodium chloride, tartaric acid. Also like vision, change in stimulus intensity (here, concentration) often modifies perceived quality. Unlike vision, however, in taste there is no known metric for stimulus composition. Thus, it is not known if there exists some continuous dimension on the physical side that is analogous to wavelength of light.

Audition, like vision, does have a continuous metric for stimulus composition, namely sound frequency. But unlike vision, the variation in sound quality (pitch) is basically unidimensional: pitch increases from low to high as frequency increases. Equal units of pitch in mels correspond roughly to equal distances along the basilar membrane. Pitch varies also with sound intensity, but only to a small degree.

The psychophysics of olfactory quality is understood only to a small extent. Not known clearly is the nature of psychologically primary odors, whether primary odors exist, or how odor quality depends on stimulus composition.

There exist some other perceptual dimensions that may be referred to loosely as qualitative. Saturation in vision depends foremost on spectral purity: the psychophysical relation is positively accelerated, a power function with exponents greater than unity. The precise values of the exponents depend on parameters such as wavelength composition and luminance. Auditory volume (apparent size of sounds) and density (apparent compactness) depend strongly on both sound frequency and intensity. The four attributes of simple sounds—pitch, loudness, volume, and density—are not independent of one another. The last three interrelate by the simple rule: loudness equals the product of volume and density.

Often it is possible to scale simultaneously variations on several dimensions of some modality and thereby to determine a multidimensional sensory space. Distances between stimuli in the space reflect psychological difference or dissimilarity. Given a constant visual intensity (i.e., lightness), color space is two-dimensional and Euclidean. Pure colors fall close to a circle, with complementary colors (e.g., red and green, blue and yellow) opposite each other. Lines that connect constant hues meet at an achromatic point near the center; distances outward from the achromatic point reflect saturations. The dimension of sensory intensity (lightness) can be added to make the space three-dimensional.

A three-dimensional, approximately Euclidean, space is also applicable to taste. Gustatory space contains four nodal points, one for each of the primary taste qualities.

The metric for auditory space appears to be city-block rather than Euclidean. Dissimilarities among pure tones can be described in terms of a linear sum of differences in two dimensions: loudness and pitch.

There is another dimension besides sensory intensity common to all modalities, namely affect. Unpleasantness is a quality that often accompanies high-intensity stimulation. But even at low stimulus intensities, variations in affect are noticeable, especially in the chemical and somesthetic senses. Wundt proposed a general psychophysical function for affect: as stimulus intensity increases, hedonic intensity (positive in value) first increases, reaches a maximum of pleasantness, then decreases through indifference (neutral value), and finally increases in unpleasantness (negative value). But not all stimuli obey that rule; under normal conditions, sucrose, for instance, may never become unpleasant, even at high concentrations, and quinine is unpleasant at very low concentrations. Pleasantness does vary nonmonotonically with sucrose concentration (first increasing, then decreasing), but unpleasantness increases monotonically with increasing concentration of quinine, sodium chloride, and citric acid. Pleasantness decreases monotonically with increasing odorant concentration of amyl acetate. Unpleasantness increases with increase in number of carbon atoms of aliphatic alcohols. Not surprisingly, unpleasantness grows continuously when electrical stimulation increases in current and when environmental temperature increases its deviation from neutral. Wundtian nonmonotonicity does take place in the thermal realm, however, when only a small area of skin is stimulated and initial skin temperature is normal.

The affective dimensions of sensation often correlate closely with homeostatic mechanisms and biological utility. There appears to be a close connection between sensory affect and the motivational aspect of stimulation.

Chapter 7

INTERPRETATIONS OF SENSORY SCALES

In Chapters 3–6 we examined the application of direct scaling to the study of sensory processes. The present chapter deals with some persistent issues and unresolved questions in the realm of sensory scaling. Let us first briefly consider the concept *sensory magnitude*, as it has been used above, in terms of the language of intervening variables and hypothetical constructs. There is a fundamental sense in which sensory magnitude has been employed as a type of intervening variable. We saw, for example, how constants of the psychophysical power law change systematically with parametric variations in stimulation. In a basic sense, the notion of sensory magnitude served to relate to one another the several stimulus parameters which affect sensation. The psychophysical equations appeared as intermediaries to sensory-physical equations—equations that relate to one another physical variables for constant sensation levels. Bloch's law of temporal summation is an example.

A significant question is whether the concept of sensory magnitude can contain any additional denotative meaning. If sensory magnitude is to serve only as an intervening variable, then any monotone transformation of direct ratio scales of sensory magnitude would serve just as well as the original scales. For example, logarithmic transformation of direct scales would not change the interrelations among stimulus parameters that define any particular constant level of response. In other words, we can ask whether the magnitude scales of sensation that we have examined turn out to serve merely as heuristic devices, useful primarily for organizational value, but are devoid of any particular meaning in their numerial assignments, i.e., in the relations of these numbers to one another. This is a fundamental issue that will arise throughout Chapter 7.

The Nature of Scale Types

Previously, reference was made to different types of scale, in particular to ratio scales and interval scales. Typically, type of scale was described in terms of the scaling procedure used. Thus, methods like magnitude estimation and fractionation were said to yield ratio scales—scales that contain meaningful zero points; methods like category rating and bisection were said to yield interval scales—scales that contain arbitrary zero points. However, other types of scale also exist. Furthermore, sensory measurements that yield one type of scale do not always relate linearly to measurements that yield another type of scale.

Although it is appropriate at this point to examine the relations between scales of sensory magnitude obtained by what we termed "ratio scaling" procedures and scales obtained by other procedures, it is first necessary to describe in greater detail some of the major differences among types of scales. Perhaps the most common division of scale types is that of Stevens (1946, 1958). He described four major types of scale: ratio, interval, ordinal, and nominal. To this list may be added five others: hyperordinal, log interval, power, difference, and absolute scales. Since we will not be concerned with nominal, ordinal, or hyperordinal scales, we shall omit further consideration of them. For a more detailed discussion of scale types, see Suppes and Zinnes (1963).

CHARACTERISTICS OF SCALES

Before proceeding to consider the different types of scale themselves, it is necessary to look more carefully at the defining characteristics of a scale. Determination of scale values for some attributes or objects entails two major factors: first, a set of *empirical operations* (experimental procedures); second, a means for determining all of the sets of possible *numerical representations* for the attributes or objects. This second factor concerns what is termed the *uniqueness* properties of the scale. In developing a system that leads to a ratio scale (for example, a ratio scale of length), we would find that any of an infinite number of numerical representations are possible, but that the types of representation are highly restricted. In a particular instance, three line segments might have possible scale values of 1, 2, and 5, or 2, 4, and 10, or 1000, 2000, and 5000, among others. The uniqueness of the scale—in this case a ratio scale—

TABLE 7.1

RELATIONS AMONG SCALE VALUES THAT REMAIN INVARIANT UNDER ADMISSIBLE
TRANSFORMATION FOR VARIOUS TYPES OF SCALE

Scale type	Invariance	Transformation
Interval	$x_1 - x_2 = a(x_3 - x_4)$	$x' = bx + c$
Logarithmic interval	$\dfrac{x_1}{x_2} = \left(\dfrac{x_3}{x_4}\right)^a$	$x' = bx^c$
Difference	$x_1 = x_2 + a$	$x' = x + b$
Power	$x_1 = x_2{}^a, x > 1$	$x' = x^b$
Ratio	$x_1 = ax_2$	$x' = bx$
Absolute	All relations invariant	$x' = x$

lies in the fact that each of the possible scale assignments is related to every other possible assignment in a particular, simple manner, i.e., by multiplication by a positive constant. Multiplication is the means for going from inches to feet or centimeters to meters. The term *ratio scale* implies that what is left invariant by any change from one scale assignment to another is the ratios of values between any two objects on the scale.

Mathematical transformations that leave invariant the appropriate relations are often termed *admissible transformations*. In fact, perhaps the easiest way to distinguish among scale types is in terms of invariances or admissible transformations. Table 7.1 and Figure 7.1, which are extensions of the scheme proposed by Stevens (1946), give examples of scale types in terms of their invariances (the relations among scale values that are unchanged) and admissible transformations (the types of mathematical changes that maintain the invariances).

The second column in Table 7.1 shows the relations among scale values that remain unchanged (invariant) under admissible transformation. Let us take the relation $x_1 - x_2 = a(x_3 - x_4)$, which is invariant under interval scales, and apply it to temperature measured in degrees Celsius and degrees Fahrenheit. The difference between 100° C and 130° C is three times the difference between 10° C and 20° C. If we change to the Fahrenheit system, the corresponding temperatures are 212° F and 266° F in the first example, 50° F and 68° F in the second. The difference between the first two temperatures is still three times the difference between the second two.

The third column of Table 7.1 shows the mathematical types of admissible transformation that characterize each type of scale. These are also diagrammed in Figure 7.1, which shows the interrelations among restrictions on admissible transformations. The invariances follow from the mathematical nature of the admissible transformations.

Basic Scale Types

The most powerful (most restrictive) type of scale is an *absolute scale*. On this type of scale, each number corresponds in an absolute manner to one object or attribute measured (Suppes & Zinnes, 1963; von Neumann

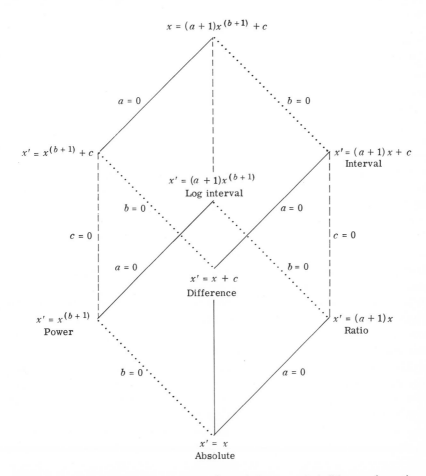

Figure 7.1. Scale types given in terms of restrictions on admissible transformations. The most general transformation permits addition of a constant, multiplication by a constant, and exponentiation to a constant. If addition is not permitted $(c = 0)$, it becomes a logarithmic interval scale; if exponentiation is not permitted $(b = 0)$, it becomes an interval scale; if addition and exponentiation are not permitted $(c = 0, b = 0)$, it becomes a ratio scale; if addition and multiplication are not permitted $(c = 0, a = 0)$, it becomes a power scale; if multiplication and exponentiation are not permitted $(a = 0, b = 0)$, it becomes a difference scale; if addition, exponentiation, and multiplication are not permitted $(c = 0, b = 0, a = 0)$, it becomes an absolute scale.

& Morgenstern, 1947). An example is a counting scale, a scale that refers to particular numbers of attributes, objects, or events. Clearly, if any transformation were made, information would be lost. Thus, an absolute scale permits no transformation other than identity.

The absolute scale may be thought of as a restricted case of a ratio scale. Examples of ratio scales abound: mass in kilograms, length in meters, temperature in degrees Kelvin, etc. Ratio scales contain meaningful zero points. They are totally restricted except in terms of their unit of measurement, in that, as was mentioned above, multiplication by a positive constant constitutes an admissible transformation (e.g., change of unit from inches to feet).

Ratio scales thus have one degree of freedom that is, one unspecified parameter (the unit). If the number of unspecified parameters is taken as a measure of the power of a scale, then the ratio scale is weaker than the absolute scale, since the latter has no unspecified parameters and no admissible transformations (other than identity). Equal to the ratio scale in power are two other scales: the *difference scale* and the *power scale*. The difference scale has a specified unit, but no true zero point. Thus, addition of a constant constitutes the only admissible transformation. An example of a class of difference scales is the common decibel scale of sound intensity or pressure. The unit of the decibel scale (the decibel itself) is fixed. The zero point is arbitrary, however, so we may change, for instance, from a scale of sound pressure level to a scale of sensation level.

For the power scale, admissible transformation consists of raising the scale values to some (constant) power. Because of the equivalence among the three types of scale (ratio, difference, and power) in terms of the number of unspecified parameters, it may be possible to convert a scale of one type into another with no concomitant loss of information. For example, let us assume that by means of a procedure such as magnitude estimation we have obtained a scale for some attribute, and we believe we have obtained a ratio scale for this attribute. It is perfectly possible to transform this scale: by taking logarithms into a difference scale; by exponentiating into a power scale. Psychophysical relations between scale values and stimulus intensity would, therefore, change from a power function to a logarithmic function in the first example, or into a complex exponential function in the second. All of the transformed scales might then be considered scales of sensory magnitude, and any one of them could play the same mathematical role as any other (e.g., in predicting the nature and exponent for cross-modality matching experiments between modalities or sensory-physical laws within a single modality).

Two types of scale remain to be considered: *interval* and *logarithmic interval scales*. Again, these two types of scale are similar in that each has the same number of unspecified parameters, in the present case two. In the case of the logarithmic interval scale, admissible transformation consists of both multiplication by a constant and of exponentiation to a constant. Although few attempts have been made to devise procedures with the specific purpose of generating such scales, they are of some theoretical importance. One interesting use is in the interpretation of direct sensory scales. The so-called ratio procedures, like magnitude estimation, yield, under given conditions, psychophysical power functions whose exponents often vary from one experiment to another. To the extent that the exponents of the functions do vary (but *not* because of differences in stimulus configuration, duration, etc.), the sensory scales may be considered to be invariant up to exponentiation as well as multiplication: that is, they may be considered more properly logarithmic interval, rather than ratio scales.

A generally more important scale is the interval scale; familiar examples are the commonly used temperature scales, Celsius and Fahrenheit. Interval scales have specified neither the unit nor the origin (zero point). Some of the earliest attempts to measure sensory magnitudes involved procedures aimed at producing interval scales. Results of some of these attempts were discussed in Chapter 1. Psychophysical procedures that lead to interval scales may utilize either direct or indirect approaches. Direct approaches include equisection and bisection, method of category rating, and method of equal-appearing intervals. The indirect approach is exemplified by discriminability scales, integrated JND scales, and scales produced by the method of successive intervals. It is desirable at this point to turn to the results of experiments that have used various scaling procedures and to examine the relations among scales derived by the different procedures.

Scales of Sensation

Direct Interval Scales

Bisection Scales. First for consideration will be scales generated from direct interval procedures. One of the earliest examples is that of Plateau (1872), who asked eight artists to paint a gray that appeared to be halfway between a white and a black. The method of *bisection* is seemingly quite simple, although some important variables, such as the order in which stimuli are presented, can significantly affect the results.

In a modification of the method of bisection, the subject is given two end points and asked to set values of two or more additional stimuli, so as to mark off three or more equal-appearing intervals; the modified procedure is called *equisection*. The bisection procedure was especially popular in the nineteenth century. Titchener (1905) reviewed many of these early studies. A number of studies using bisection had as their aim an investigation of the validity of Fechner's logarithmic law, which predicts that the point of bisection should lie at a stimulus value equal to the geometric mean of the values of the end stimuli. This prediction sometimes has, (e.g., Angell, 1891; Newman, Volkmann, & Stevens, 1937) but more often has not, been confirmed (e.g., Cross, 1965; Merkel, 1888, 1889, 1894; Stevens, 1955, 1961c). Typically, the bisection points are larger than the geometric mean.

It is also possible to inquire whether the results of bisection experiments are consistent with a power relation between stimulus and sensation. Thus, if the interval between the stimuli ϕ_a and ϕ_b is to be bisected, then the bisection point ϕ_c is predicted

$$\phi_c = \left(\frac{\phi_a^{\alpha} - \phi_b^{\alpha}}{2} \right)^{1/\alpha} \tag{7.1}$$

where α is the exponent of the (presumed) underlying psychophysical power law. In order for the results of bisection to be consistent with a power relation, it is necessary that bisections between different pairs of stimuli all predict the same exponent α. This outcome has occurred on some occasions [for example, a set of bisections of brightness intervals reported by Stevens (1961c)]. On the other hand, bisection results often do not consistently predict a single power relation (see Stevens, 1955; Stewart, Fagot, & Eskildsen, 1967). A possible reason might be that bisection points reflect something different from sensory magnitude. One possibility is that subjects adjust stimuli in bisection in order to produce sensations that appear equally similar to the sensations produced by the two end stimuli.

When results of bisection experiments are interpreted in terms of power functions, the exponents of the functions [values of α in Eq. (7.1)] can be compared to the exponents obtained by ratio-scaling procedures such as magnitude estimation. Most often, the bisection exponents are smaller than the corresponding ratio exponents [e.g., results for loudness by Stevens (1955)]. However, recall from Chapter 6 that although pitch is not related to sound frequency by a power function, the mel scale derived by equisection is approximately linearly related to the mel scale derived by ratio procedures.

Gage (1934a & b) attempted to examine the consistency of the bisection procedure for both loudness and brightness. The subject's task was first to bisect the loudness (or brightness) interval AB, yielding a bisection point C. Then AC was bisected to yield point D and CB was bisected to yield point E. Finally, the interval DE was bisected and the resulting point F compared to the original bisection point C. The series of operations is depicted in Figure 7.2. Gage reported significant deviations of the final bisection point F from the original point C both when loudness and when brightness intervals were examined. Presumably, the initial and final bisection points should agree if the bisection procedure is internally consistent. Gage concluded that the bisection procedure cannot produce a scale of sensory magnitude, and, furthermore, he questioned whether any scale of sensory magnitude could be devised.

Although Gage's results indicate some sort of systematic violation of a

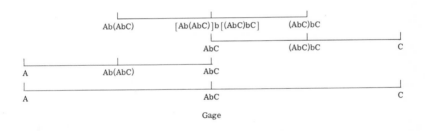

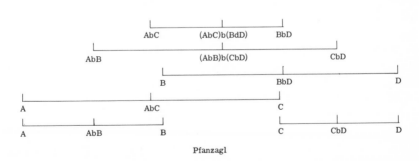

Figure 7.2. Tests of consistency for the bisection (b) of sensory intervals. Top: Gage's (1934a & b) experiments. The bisecton of A and C gives AbC. The bisection of A and AbC gives Ab (AbC); the bisection of C and AbC gives (AbC)bC. The bisection of Ab (AbC) and (AbC)bC gives [Ab (AbC)]b[(AbC)bC], which should equal AbC. Bottom: Test of Pfanzagl's (1959) axiom of bisymmetry. The bisection of A and B gives AbB; the bisection of C and D gives CbD. The bisection of A and C gives AbC; the bisection of B and D gives BbD. The bisection of AbB and CbD should equal the bisection of AbC and BbD.

rule in bisection, there are several types of rule that may have been violated. An interesting axiomatization applicable to bisection has been presented by Pfanzagl (1959). The crux of Pfanzagl's system is the notion of bisymmetry. An interval scale can be produced from the results of bisection procedures if the axiom of bisymmetry is not empirically violated. The empirical test of bisymmetry is related to Gage's procedure, but is somewhat different. If we use the symbol b for the bisection operation, then Gage's experiments tested the equivalence between AbC (the initial bisection) and $[Ab(AbC)]b[AbC(bC)]$ (The bisection of the "25%" and "75%" points). The bisymmetry axiom can be represented as the equivalence between $(AbB)b(CbD)$ and $(AbC)b(BbD)$. (See Figure 7.2.) What is particularly interesting is that, as Pfanzagl showed, the bisymmetry axiom can be validated (or at least not violated), yet Gage's results still obtain. This outcome can occur if the bisection method is either not reflexive (identical, e.g., $A = A$ for all A) or not commutative (independent of order, e.g., $AbB = BbA$). Failure of reflexivity is unlikely, since that would entail lack of identity between some point and itself. But failure of commutativity appears to be quite likely (Fagot & Stewart, 1970).

Failure of commutativity means that the order that the stimuli are presented makes a difference for results of bisection. That fact has been known for a long time; Stevens (1957) termed "hysteresis" the result that bisection points fall higher when stimuli are presented in an ascending order of intensity than they fall when stimuli are presented in a descending order. Newman et al. (1937) repeated Gage's experiment on loudness, but used both ascending and descending orders of stimulus presentation. Averaging the results they found a much smaller deviation from consistency than Gage. Wolff (1935) conducted an experiment of even greater complexity, but the core of the experiment contained the same test as that of Gage and of Newman et al. Wolff found reasonable agreement between initial and final bisection points.

Fagot and Stewart (1970) concluded from the results of bisections of brightness intervals that lack of consistency in Gage-type experiments results from failure of commutativity, not failure of bisymmetry. Their conclusion might have been stronger had they controlled experimentally for simultaneous brightness contrast, which may have influenced their results. Cross (1965) investigated the bisymmetry axiom directly in obtaining bisections of loudness intervals. The results for one subject appeared to confirm bisymmetry; those for the second subject were inconclusive. Bisymmetry without commutativity does not preclude the possibility of constructing an interval scale from bisection data, according to Pfanzagl.

Under such a restriction, however, it becomes necessary to generalize the interpretation of the bisection point, which may then correspond to something other than a half-way point between the ends bisected. The response bias that causes the bisection point to deviate from a mid-point is presumably the basis for the hysteresis effect described above. To repeat, though, the existence of such response bias does not preclude construction of an interval scale.

The question of possible agreement between magnitude and bisection scales of sensation awaits a more complete examination of bisymmetry and of the nature of the underlying bisection scales. A similar examination is due for the equisection procedure, a generalized form of bisection whereby more than two equal-appearing intervals are marked off at the same time. We shall deal with some scales based on equisection data for loudness below, when we examine the lambda scale proposed by Garner (1954b).

Category Rating Scales. One of the most popular types of scaling procedure makes use of ratings or categories. In *category scaling,* contiguous categories (usually numbers) are employed to mark off steps along whatever continuum is being scaled. Often, the subject is told that each step should correspond to a constant sensory interval. Rating scales certainly have seen widespread use: some examples, selected to show the variation in application, are scales of heaviness (Cowdrick, 1917), thermal discomfort (Winslow, Herrington, & Gagge, 1937), and esthetic preference (Saffir, 1937). When subjects set about to partition a prothetic sensory continuum (such as brightness or loudness), the result is usually not linearly related to direct ratio estimates of sensory magnitudes (Stevens & Galanter, 1957). Although occasionally a linear relation obtains (e.g., Warren & Poulton, 1962), the relation between category scales and magnitude scales is more typically described by a concave downward curve (see Figure 7.3). The nature of the nonlinear relationship between the two types of scale (or, equivalently, between category scales and stimulus intensity) has intrigued a number of investigators.

It has often been stated (e.g., Baird, 1970; Galanter, 1962; Helm, Messick, & Tucker, 1961; Junge, 1967; Michels & Helson, 1949; Thurlow & Melamed, 1967; Torgerson, 1961) that category scales are logarithmically related to the corresponding magnitude scales, or, equivalently, logarithmically related to stimulus intensity. Quasi-logarithmic relations were also proposed by Schneider and Lane (1963) and by Eisler (1962a & b, 1963b). Eisler based his equation on the notion that category scales mimic scales of integrated just noticeable differences (JNDs). Stevens and Galanter (1957) demonstrated clearly, however, that logarithmic functions are the exception rather than the rule. The curvature of the functions depicted in Figure 7.3, for example, is not extreme enough to be described by a logarithmic relation. A few other investigators suggested a power function to describe the relation between interval scales of sensation (such as scales derived from methods of equal-appearing intervals and bisection) and stimulus intensity (e.g., Glasser *et al.*, 1958; McGill, 1960; Stevens, 1961c). In a related vein, Ekman and Künnapas (1960)

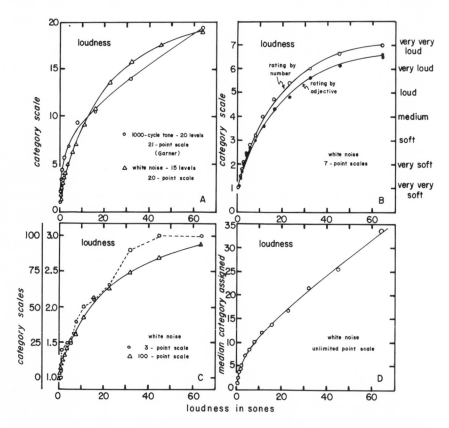

Figure 7.3. Category scales for loudness, as functions of the corresponding loudnesses in sones (i.e., loudnesses derived from ratio-scaling procedures such as magnitude estimation). All of the function are curvilinear, although the use of an unlimited number of categories (lower right) produced a function that is almost linear. [From S. S. Stevens and E. Galanter, Ratio scales and category scales for a dozen perceptual continua. *Journal of Experimental Psychology*, 1957, **54**, 377–411. Copyright 1957 by the American Psychological Association, and reproduced by permission.]

report category scales for brightness to be linearly related to the corresponding ratio scales when the latter were raised to powers less than 1.

It has been claimed that there is a general power relation between category scales and stimulus intensity on prothetic continua

$$C = a\phi^{\alpha} + c' \qquad (7.2)$$

where C is category rating; c' is an estimated parameter, necessary because a category scale is at best an interval scale (Marks, 1968b). The exponent α is usually smaller than the exponent of the corresponding magnitude function β. As the value of α approaches zero, the category scale becomes more nearly a logarithmic function of stimulus intensity (Fagot, 1966; Marks, 1968b).

Stevens (1971a) has used the term "virtual" to describe power-function exponents obtained by interval-scaling procedures. This term would apply to exponents obtained from bisection and equisection experiments, among others, as well as those obtained from category ratings. The claim that a power function describes the relation between category ratings and stimulus intensity is by itself of some importance. In addition, however, it appears that the size of the exponent α can serve as an index of the extent to which various experimental parameters systematically affect the form of the category scale (Marks, 1968b). As the stimulus range or

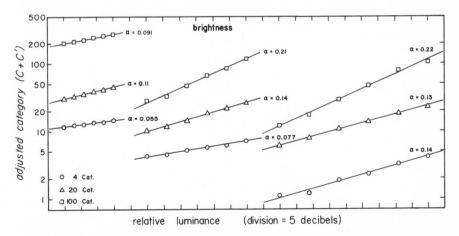

Figure 7.4. Category scales for the brightness of white light, as functions of luminance. Both coordinates are logarithmic. The exponent α of the categorical power function increases both with increase in stimulus range and with increase in number of available categories. [From Marks (1968b). Courtesy of University of Illinois Press.]

number of available response categories increased, for example, the size of α increased (Figure 7.4). In other words, increasing stimulus and response ranges appear to make category scales more nearly linearly related to corresponding magnitude scales.

There is an entire group of continua, termed metathetic by Stevens (1957), for which the relation between category and magnitude scales is quite different. More often, category and magnitude scales are linearly related on metathetic continua (Stevens & Galanter, 1957). The example of pitch has already been given: fractionation (Stevens *et al.*, 1937) and equisection (Stevens & Volkmann, 1940) yielded very similar scales of pitch (the mel scales), and category ratings of pitch were found usually to be linear functions of pitch in mels (Stevens & Galanter, 1957). The latter relation is not always linear, however, because category ratings are so very susceptible to the influence of variables such as stimulus range and stimulus distribution. J. C. Stevens (1958) found category scales to be much more susceptible to contextual influence than were magnitude estimation scales of loudness.

Estimation of Sensory Intervals. A variant of interval-scaling procedures calls upon subjects to make ratio-type judgments of intervals. This procedure has been employed extensively in the determination of intervals or differences among the visual appearances of colored surfaces (Newhall, 1939; Newhall *et al.*, 1943). In one version of this method, a given pair of stimuli is set in front of the subject as a standard; the subject is told that the physical distance between the two members of the pair represents the sensory interval or difference. The subject's task is to adjust the physical distance between members of each subsequent pair to reflect proportionally its apparent difference. In another version of the task, the subject gives numerical estimates of the interval or difference between members of each stimulus pair. Regardless of which task is used, the essence of the procedure is to evaluate directly the size of sensory intervals.

Beck and Shaw (1967) and Dawson (1971) obtained numerical estimates of intervals of loudness, and they found that estimates of intervals are not proportional to the intervals calculated from the sone function for loudness. The sone function gives loudness to be proportional to about the .3 power of sound energy. Stevens (1971a) noted that results of Beck and Shaw and of Dawson could be accounted for fairly well if a lower exponent is assumed—one equal to about .15.

Similar lack of proportionality is found between estimates of intervals and intervals calculated between estimates of magnitudes on other sensory and perceptual continua, such as heaviness of weights (Curtis, Attneave, & Harrington, 1968; Marks & Cain, 1972), brightness of lights (Curtis, 1970), and apparent area of circles (Dawson, 1971; Marks & Cain, 1972; Rule, Curtis, & Markley, 1970). The results of all of these experiments agreed in their finding of nonlinear relations between estimates of intervals and the intervals predicted from estimates of magnitudes. Beck and Shaw (1967) found estimates of loudness intervals to approximate the sizes of intervals calculated from Garner's (1954b) lambda scale. That result is not surprising, since the lambda scale was constructed in large part on the basis of equisection, itself an interval-scaling procedure. Loudness in lambda units relates to sound energy by about a .13 power. Curtis *et al.* (1968) and Marks and Cain (1972) also noted, as did Stevens, that direct judgments of intervals appear to be governed by exponents lower than those that govern estimates of sensory magnitudes. Estimates of intervals behave, therefore, in much the same manner as do bisections and category ratings. In fact, Marks and Cain found that scale values from estimates of intervals were linearly related to category ratings on two of the three continua that they examined (apparent area and apparent roughness, but not apparent heaviness).

What are interval scales of sensation all about? One view is that, on prothetic continua, they are biased versions of "true" scales; presumably, the latter obtain only by ratio-scaling procedures. Stevens (1971a) refers to the "virtual" exponents, which govern the growth of sensation on interval scales, in contrast to the "actual" exponents, which govern growth on magnitude (ratio) scales. An alternative view is that there exist prototypical interval scales that describe relations among sensations different from relations described by magnitude scales. This alternative is implicit in several models of categorical judgment: Michels and Helson (1949) described category judgments in terms of adaptation-level theory; Helm, Messick, and Tucker (1961) suggested that category scales are equiprecision scales; Eisler (1962a & b, 1963b) argued that they parallel scales of integrated JNDs; Baird (1970) claimed category (and magnitude) scales result from the ways subjects code sensory information, and they relate directly to sensory and cognitive channel capacities.

Another possibility is that interval scales are scales of *dissimilarity* rather than scales of *magnitude*. Eisler and Ekman (1959) and Ekman, Goude, and Waern (1961) showed for unidimensional stimuli, and Ekman, Engen, Künnapas, and Lindman (1964) showed for multidimensional stimuli, that similarity is closely associated with sensory and perceptual *ratios*. The general similarity equation they proposed is

$$\psi_s = \frac{2\psi_a}{\psi_a + \psi_b} \tag{7.3}$$

where ψ_s is similarity and ψ_a and ψ_b are sensory magnitudes (ψ_b greater than ψ_a). If a scale were produced in which all stimuli were spaced in steps of equal similarity (as defined by Eq. [7.3]), then that scale would consist of a logarithmic spacing among the sensory magnitudes of the stimuli, and, therefore, it would also bear a logarithmic relation to stimulus intensity (given that the sensory magnitudes follow a psychophysical power law). The same conclusion results if the similarity Eq. (7.3) is modified along the line suggested by Künnapas and Künnapas (1971). They proposed similarity to be proportional to the ratio of the sensory magnitudes raised to a power. [It may be pointed out that Junge (1967) claimed that all psychophysical scales—interval and magnitude— are based on similarity judgments.]

The relation between similarity and stimulus intensity might not be precisely logarithmic. Nonetheless, if interval scales are scales of similarity, then they probably should bear some nonlinear relation to scales of sensory magnitude. That expectation also presupposes that prototypical ("pure") interval scales can be obtained. In order to obtain pure interval scales, iterative experimental procedures are often employed so as to remove biasing effects of variables such as stimulus spacing (Pollack, 1965a & b; Stevens & Galanter, 1957). One deduction that may be made from these considerations is that bias in interval-scaling experiments can act in several directions. If we assume that interval scales reflect similarity rather than magnitude, then a given procedure and choice of stimuli might bias results of interval scaling toward, as well as away from, agreement with results of magnitude scaling.

Perhaps the strongest support for the view that direct interval scales are scales of dissimilarity comes from multidimensional scaling. Several studies have obtained ratio judgments of perceived intervals among colors, where the intervals or differences were taken to be measures of dissimilarity. In spite of the fact that it might be thought much more difficult to judge multidimensional, than unidimensional, stimuli, nevertheless data of Helm (1964), Indow and Kanazawa (1960), Indow and Uchizono (1960), and Ramsay (1968) also show the judgments of multidimensional visual stimuli to be consistent with relatively simple underlying Euclidean spaces. These similarity spaces contain one dimension (roughly circular) for hue, a second one (radial) for saturation, and a third for lightness. Because these results show a basic agreement with the spacing of stimuli given by the Munsell system (a set of three unidimensional interval scales), they also tend to support the view that similarity or dissimilarity underlies the direct judgment of sensory intervals.

INDIRECT INTERVAL SCALES

JND Scales. We turn now to a different class of interval scales. The interval scales we have considered thus far are examples of *direct scales,* direct because there is an isomorphic or one-to-one correspondence between responses made by subjects and scale values assigned by the experimenter. The new class of scale we shall consider is that of *indirect* or *processed scales.* In the derivation of processed scales, there is no one-to-one correspondence between particular responses and scale values. Rather, what is necessary is mathematical processing of data in order to obtain scale values. Processing typically centers about the variability in the data. The most famous example of a processed scale, described in some detail in Chapter 1, is the summated JND scale. Fechner (1860), of course, was the first to propose summating just noticeable differences in order to derive a scale of sensation magnitude. We saw in Chapter 1 that Fechner suggested two methods by which JND scales may be erected. The first method entails merely counting off just perceptible steps—JNDs—starting at the absolute threshold. The second method involves mathematical derivation, beginning with Weber's law, and ending with the solution of a differential equation that yields Fechner's famous logarithmic law.

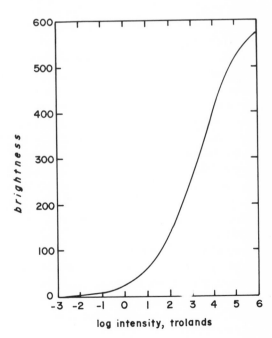

Figure 7.5. Scale of brightness derived by the summation of JNDs for luminance discrimination, as a function of retinal illuminance in trolands. [From *The Principles of Psychophysiology* by Leonard T. Troland © 1930 by Litton Educational Publishing, Inc. Reprinted by permission of Van Nostrand Reinhold Company.]

Fechner's first method, the graphical summation of JNDs, historically has been much more frequently invoked in the measurement of sensation. An early JND scale for brightness (or "brilliance," as it was then called) was published by Troland in 1930. Troland carefully distinguished between a "Fechner function," that is, an integrated JND scale (of which Figure 7.5 is an example) and Fechner's logarithmic law. It is quite clear that the JND scale for brightness that Troland calculated deviates both at low and high intensities from a logarithmic relation.

A question that has persistently agitated investigators in psychophysics is whether just noticeable differences can provide a unit of sensory magnitude. Specifically, the question has been whether all JNDs, i.e., JNDs measured at all sensation levels, are subjectively equal. One solution has been to make the subjective equality of JNDs a postulate, rather than an hypothesis (Cobb, 1932; Luce & Galanter, 1963). However, there are means by which one can attempt to test the internal consistency of the notion the JNDs can provide units of measures of sensory magnitude. We shall consider the general question of consistency in greater detail below. For the present, let us merely mention that JND scales have been shown to conflict in some instances with data obtained by the prototypical procedure of sensory-physical measurement, namely direct intrasensory matching.

This is not to say that scales of integrated JNDs are totally without use. For although it appears that JND scales cannot always provide consistent measurements of sensory magnitude, they are indeed scales of discrimination, of discriminability. They tell us something about the separations or differences among stimuli in terms of the relative degree of discriminability between them. Stimuli close together on a JND scale are poorly discriminable; stimuli farther apart are more discriminable. It is to be hoped, of course, that equal distances along a JND scale would provide a measure of equal discriminability. If they do, then JND scales help to furnish us with important information concerning the relation between sensory magnitudes and discrimination.

Consider as an example two tones at frequencies A and B, adjusted in intensity to match at one level of loudness (A_1 and B_1), then adjusted again in intensity to match at a different loudness (A_2 and B_2). The same number of JNDs may not separate A_1 and A_2 as separate B_1 and B_2. Equal changes in sensation do not always entail equal changes in discriminability. It is likely, as Luce and Edwards (1958) pointed out, that different measures of sensation, e.g., magnitude scales and JND scales, tap different sensory processes and mechanisms (or tap the same mechanisms differently).

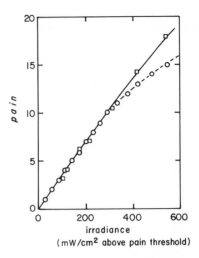

Figure 7.6. The dol scale of pain, derived by the summation of JNDs, and magnitude estimates of pain, both as functions of stimulus irradiance (milliwatts per square centimeter) above the pain threshold. [The dol scale is from Hardy, Wolff, and Goodell (1947). The magnitude estimates are from Adair, J. C. Stevens, and Marks (1968).]

Again, it is necessary to point out that these generalizations seem to hold true primarily for prothetic continua. Only for prothetic continua is it generally true that scales of integrated JNDs are approximately logarithmically related to stimulus intensity, and, therefore, approximately logarithmically related to scales of sensory magnitude generated by direct ratio-scaling procedures such as magnitude estimation. (Although Troland's Fechner function for brightness deviates from pure logarithmicity at the extremes, it is nearly logarithmic over much of the middle range.) A notable exception, however, is the scale of integrated JNDs for pain produced by radiant heat—the *dol scale* derived by Hardy *et al.* (1947). Over much of its range, the dol scale is a linear function of stimulus intensity (see Figure 7.6), and, in fact, is also linear with the magnitude-estimation scale for pain (Adair *et al.*, 1968).

Other exceptions arise within the class of metathetic continua. Metathetic continua are, in fact, characterized by a relative uniformity of the subjective size of the JND over the stimulus range. At least, the JND does not seem to increase systematically in the manner that it does for prothetic continua. Summated JNDs for frequency discrimination, for example, provide close linear agreement with magnitude estimates and category ratings of pitch (Stevens & Volkmann, 1940; Zwislocki, 1965).

Other Indirect Scales. Closely related to the JND scale is the class of discriminability scales whose development originated with Thurstone (1927a & b). Central to Thurstone's approach was the view (like Fechner's) that discriminability can provide a measure of difference in sensory magnitude. To be more specific, Thurstone propounded a *law of comparative judgment* to relate or, more properly, to define scale differences in terms of the probability that one stimulus is judged greater than another (e.g., in a paired-comparison experiment). By making various quantitative assumptions concerning the nature of variability in judgment and of the interrelations among variables, this law (or equation) could be simplified.

The similarity between Thurstonian and Fechnerian scales was pointed out by Stevens (1960). Thurstonian scales lie at a greater level of sophistication than do traditional Fechnerian scales. In constructing JND scales, the unit of the scale, the JND, is selected at a particular, and thereby automatically quite arbitrary, cut-off level of discrimination. [It is possible to increase the sophistication of the Fechnerian approach by expanding the index of discrimination from an arbitrary one—the JND —to a more general one (Falmagne, 1971).] That is, the criterion and procedure for determining the JND are basically left to the discretion of the experimenter. Furthermore, since only the preselected cut-off is used, the method is wasteful of information obtained in the discrimination experiment. Thurstonian scaling makes use of all or nearly all of the information available from the comparative judgments of greater and less. In his famous Case V, Thurstone (1927a) assumed that "equally often noticed differences are equal," an assumption directly equivalent to the Fechnerian assumption of equal subjective magnitudes of JNDs. Again, as in the Fechnerian case, the hypothesis of equality may be replaced by its postulation. Thurstone's system and his law of comparative judgment will not be discussed here. Although the Thurstonian methods have been used extensively, as in the measurement of attitudes (Frederiksen & Gulliksen, 1964), their use in studies of sensory processes has been quite limited.

We turn now to a different type of processed scale. The Fechnerian and Thurstonian methods we have discussed thus far involve processing of ordinal data (judgments of greater or less), or, more precisely, the probabilities of these judgments. The second class of scale involves processing not of simple, ordinal (discrimination) judgments, but rather rating or identification judgments. These scales have been described by several investigators (Adams & Messick, 1958; Attneave, 1949; Garner & Hake, 1951; Saffir, 1937) and have been assigned several names. The terminology we shall adopt to describe this scaling procedure is the *method of successive intervals*. In essence, the method consists of application to categorical judgments of scaling procedures analogous to the procedure of comparative judgment that Thurstone applied to ordinal data.

The most general application has been called the *law of categorical judgment* (Torgerson, 1954). This name immediately conjures an analogy to Thurstone's law of comparative judgment, to which it is a first cousin. Whereas comparative judgment refers to probabilities of judgments greater or less for paired comparisons of stimuli, categorical judgment refers to probabilities of categorical assignments (ratings) for single stimuli. Basic to the law of categorical judgment is the view that the derived scale must take into account the degree of variability involved in categorical assignments; this variability may reflect both variation in sensory effects (perceptual variability) and variation in location of category boundaries (judgmental variability). As in the case of comparative judgment, various simplifying assumptions can be made. For example, it is common to assume that category boundaries remain constant from moment to moment. It is such a simplification that leads to the well-known method of successive intervals (see Adams & Messick, 1958; Saffir, 1937; Torgerson, 1954). Essentially the same model was proposed by several other researchers (Attneave, 1949; Garner & Hake, 1951).

It is useful to compare results obtained by the method of successive intervals to those obtained by taking averages of category ratings, since the same data can be employed to produce the two scales. In the case of average ratings, it is assumed that each successive category reflects a constant unit change along a perceptual dimension. Not so in the case of successive intervals. Instead, that method assumes that the category widths may be unequal; their sizes must be estimated from the variability in the judgments, i.e., from the distributions of probabilities of categorical assignments. In the method of successive intervals, the category boundaries are adjusted in order to provide consistency with some assumption about the variability (e.g., that the distribution of categories assigned to each stimulus is normal) and to reflect constant discriminability from category to category. There is no particular reason, therefore, to expect that scales derived by successive intervals will be linearly related to the average category ratings.

Little surprise is it either, in the light of the previous discussion, that scales derived by the method of successive intervals appear nonlinearly related to magnitude scales. Galanter and Messick (1961) found that a successive-intervals scale for the loudness of a 1000-Hz tone was a logarithmic function of the magnitude scale obtained from the same subjects; the mean-category scale was, as is typical, also nonlinearly related to the magnitude scale. The degree of nonlinearity was less, however, for the category scale than for the successive-intervals scale. Garner (1952) derived an equal-discriminability scale along similar lines for a 1000-Hz tone. Garner's scale was close to a logarithmic function of sound pressure. Garner did not obtain a magnitude scale with which to compare the discriminability scale, but, presumably, if he had the two would have been logarithmically related. His equal-discriminability scale derived from categorical judgments was also found to relate linearly with a scale of summated JNDs for intensity discrimination obtained by Riesz (1928). Even for numbers themselves, an equal-discriminability scale is approximately a logarithmic function of actual numbers (Rule, 1969). It is not an unlikely generalization that processed scales, such as those produced by Garner's procedure or by the method of successive intervals, reflect quite well relative stimulus discriminability and will usually be linear functions of corresponding summated JND scales.

Validity of Sensory Scales

The question of validity (discussed briefly in Chapter 1) has probably been the most generally persistent and perplexing problem to envelop psychophysical scaling. From the time that Fechner proposed utilizing discriminability as a measure of sensory magnitude, critics have attacked, often with great justification, attempts to define scales of sensation and to relate sensory magnitudes to stimulus intensities. Often these critics did not agree with one another, heated debates developed, expelling radiation to a greater degree in the infrared region than in the visible. There were and are, of course, critics who deny the possibility of ever measuring anything as subjective as sensory magnitude. Some, like James, denied that sensations have magnitudes: ergo, nothing to measure. "Every sensation presents itself as an indivisible unit; and it is quite impossible to read any clear meaning into the notion that they are masses of units combined [James, 1892, p. 23]." James, like some others, was unable to see how sensations could be added together or divided up: therefore, they cannot be quantities. But it is now abundantly clear that empirical operations of additivity as such are not necessary in order to produce numerical scales (Luce & Tukey, 1964; Shepard, 1966; Stevens, 1951; Zinnes, 1969).

Once we reject the notion that sensations are not quantitative, it becomes necessary to face the evidence just described, namely that different scaling procedures often yield dissimilar results, and ask the questions: What procedure or procedures produce valid scales of sensory magnitude? Do any? What can possibly be meant by valid? There do not appear to be similar questions concerning validity with respect to measurement in the physical sciences. Why not? One reason is that scales for many physical attributes bear clear, invariant relationships to one another in physical laws; another is that different procedures for deriving scales yield similar, or identical, results. Let us look again at this second matter —the relations between scales derived by different procedures—to see whether these interrelations can help decide about validity.

It is generally the case, as we have now frequently seen, that different scaling procedures produce scales for prothetic sensory continua that are nonlinearly related to one another. But we have found at least one exception, namely the dol scale for pain, produced by summating JNDs for thermal pain. Over at least part of its range, the dol scale is proportional to a scale derived from magnitude estimates. Perhaps for that reason we might wish to consider the continuum of thermal pain to be not prothetic, but, rather, metathetic, since one of the defining characteristics of metathetic continua is that discriminability is relatively uniform, i.e., the Fechnerian hypothesis of constancy of the subjective size of the JND is approximately verified.

One of the most extensively examined metathetic continua is pitch. Let us reiterate the general result that the original mel scale (Stevens *et al.*, 1937), constructed on the basis of fractionations, is very similar to the revised mel scale (Stevens & Volkmann, 1940), constructed primarily on the basis of equisection data. Magnitude estimates of pitch may approximate either scale (Beck & Shaw, 1961; Siegel, 1965; Stevens & Galanter, 1957). All of these scales tend to relate in a fairly linear manner to pitch determined by summating JNDs (Stevens & Volkmann, 1940). Figure 6.3 (p. 219) shows the relation of all of these psychophysical measures of pitch, and of locus of maximum stimulation of the basilar membrane, to sound frequency. The pitch of pure tones, whether measured by direct ratio-, direct interval-, or discriminability-scaling procedures, appears to be a straightforward function of the site of maximal stimulation of the basilar membrane. For pitch, therefore, as for thermal pain, there does not seem to be a problem of validity. If any one of the scaling procedures produces a valid scale for pitch, then they all do. And given the simple relation of pitch to an underlying physiological, or at least anatomical, structure, there seems no reason to deny validity of any or all procedures. Perhaps the most important consideration is that the convergence of different scaling procedures in their results suggests strongly that each of the methods *can* produce some sort of meaningful scale, a valid measure of something. The question becomes, for any particular continuum, whether these various somethings are the same (or linear functions of one another) when different procedures are employed. The convergence of results has particular importance for the direct procedures, e.g., magnitude estimation and category rating, since the results suggest that subjects are able to make valid numerical judgments, to use the number continuum as responses in an apparently linear fashion, at least under some circumstances.

INTERNAL CONSISTENCY OF SENSORY SCALES

Of course, there does exist a minimal criterion for evaluating the validity of sensory scales. This criterion, which we shall term *internal consistency,* requires that the assignments of scale values agree with, or at least not violate, sensory equalities determined by direct matching. One of the classical approaches to the study of sensory processes is sensory-physical, via direct matching. That is, one determines the parametric variations among stimulus attributes that leave unchanged some aspect of sensation. Because matching is so fundamental an operation in sensory measurement, we consider matching to be fully a part of the system of scaling. Ergo, we use the term, internal consistency. To give some examples, the frequency and intensity of a tone can be jointly varied in order to maintain constant loudness or pitch; the areal extent and intensity of radiation can be varied to maintain constant warmth; the wavelength, duration, and intensity of light can be varied to maintain constant brightness, hue, or saturation. The criterion of internal consistency requires that all of the stimulus combinations that produce the same pitch (or warmth, brightness, or any other sensory attribute) must be given the same numerical value on its scale of sensation magnitude. Psychophysics should agree with sensory physics.

The criterion of internal consistency might appear almost too trivial and obvious to require statement. The criterion was implicit, if not explicit, in the treatment and exposition of the compositional, temporal, spatial, and qualitative parameters involved in sensation. One might ask could any of the attempts to measure sensory magnitudes—to obtain scales of sensation—have failed to meet so elemental a requirement? In fact, some have. The most salient examples of failure occurred in scales of summated JNDs. Recall that a concept central to Fechner's system was that JNDs measured at all intensity levels are subjectively equal. Thus, the JND became a unit of sensory magnitude. In Fechner's scheme, the equality of JNDs was an assumption. As mentioned above, others (Cobb, 1932; Luce & Galanter, 1963) pointed out that it might, instead, be taken as a postulate.

In either case, it is possible to examine whether JNDs provide a constant, consistent unit of sensory magnitude; such examination becomes possible when two or more stimulus parameters are varied. For instance, assume that there are 10 JNDs between the absolute threshold of hearing and some suprathreshold level for frequency A, and 10 JNDs between the threshold and some suprathreshold level for frequency B. Now a simple question to ask is whether the two suprathreshold levels appear equally loud. Newman (1933) examined such data for loudness (summated JND scales and loudness matches across frequencies) and for brightness (JND scales and brightness matches across wavelengths) and concluded that JND scales lead to systematic violations of equality. Equalities predicted by summated JND scales are not substantiated by direct matches.

Results of a number of other studies have led to the same conclusion. Lemberger (1908) started with equally sweet solutions of sucrose and saccharin, then increased the concentrations of both by the same number of JNDs. The new concentration of sucrose tasted sweeter than the new concentration of saccharin. That result comes about because differential sensitivity to sucrose and saccharin are nearly the same. The exponent of the psychophysical function that governs sweetness of sucrose is, however, about three times the size of the exponent that governs sweetness of saccharin (Moskowitz, 1970b).

Piéron (1952), apparently independently of Newman, decided that JND scales violated cross-frequency matches of loudness. Piéron (1934) also found that brightness matches obtained for red light presented to the peripheral and foveal retina failed to agree with matches predicted by counting off JNDs. Again in the realm of vision, Durup and Piéron (1933) reported that foveal brightness matches for different wavelengths failed to agree with summated JNDs. Perilhou and Piéron (1942) came to the equivalent conclusion for tactile stimulation of different body loci.

In a few instances, support for the Fechnerian hypothesis did obtain. Bujas (1937) reported that solutions of sodium chloride and sucrose that tasted equally strong remained equally strong when increased by equal numbers of JNDs. The positive result of Bujas contrasts with the negative result of Lemberger described above. A reason for the difference might be that sodium chloride and sucrose have similar power-function exponents, whereas saccharin and sucrose have dissimilar exponents.

Heinemann (1961) examined a generalization of the Fechnerian argument, namely that differential sensitivity is a function of the rate at which sensory magnitude changes with stimulus intensity, regardless of how the sensory magnitude is produced. In particular, Heinemann looked at sensitivity to changes in luminance for stimuli viewed under simultaneous brightness contrast (cf., Chapter 5). The presence of a high-luminance surround diminished the brightness of the test field and also modified differential sensitivity to changes in its luminance. Heinemann's conclusion was that differential sensitivity was the same whenever the rate of change of brightness with luminance ($\delta B/\delta L$) was the same.

A similar question was asked by Nachmias and Steinman (1965) with respect to the brightness and discriminability of short and long flashes of light. Brightness increases more rapidly with increase in luminance when duration is short than when it is long (see Chapter 4), and, concomitantly, discriminability is greater for short flashes. However, Nachmias and Steinman found the increase in discriminability to be smaller than the increase in rate of growth of brightness. They concluded that JNDs are not all subjectively equal when measured at the same level of brightness.

Over all, the evidence appears conclusive: scales of summated JNDs do not usually predict equal-sensation functions, at least for some sensory continua. And in fact, there is really no good reason to expect that they would. Realistically, although this may be considered arguing ad hoc, there is neither reason to believe that JNDs should always reflect some constant increment in sensation, nor that the rule "equals added to equals are equal" will obtain when the "added equals" are JNDs. JNDs reflect discriminability, and discriminability is not always so intimately related to sensory magnitude as some of us might wish. The fact that a large number of JNDs separates two levels of loudness at one sound frequency, for example, but a smaller number of JNDs separates the same loudnesses at another frequency, means that discriminability over equally loud portions of the loudness scale is, in some sense, better at the first frequency than at the second.

A number of simple hypotheses could account for the distinction between sensory intensity and discriminability. For example, under the assumption of a quantal theory of discrimination, such as that espoused by Stevens, Morgan, and Volkmann (1941), the subjective size of a sensory quantum might be larger for the second frequency than for the first when measured at equally loud levels. On the other hand, if the discrimination task is viewed as one of detecting signal in noise (e.g., Swets, Tanner, & Birdsall, 1961), or of setting a decision criterion in the face of variability in sensory response, then an increase in variability or noise at the second frequency would produce a larger physical JND. By means of either, or both, of these hypotheses we could predict failure of summated JND scales to agree with direct matches. A similar point was made by Stevens:

> Some people claim to enjoy a compelling intuition to the effect that, when the psychophysical function gets steeper, the resolving power must get better. This, I suppose, is a tacit admission that the person in question has not made the proper distinction between a differential and a standard deviation [1961c, p. 83].

To the extent that consistency fails to hold between summated JNDs and intrasensory matches, then so too must fail any of a number of simple hypotheses concerning the subjective size of the JND. Fechner proposed that JNDs correspond to constant unit changes in sensory magnitude. Teghtsoonian (1971) has proposed that JNDs correspond to constant ratio changes in sensory magnitude. From data collected on a large number of sensory modalities, Teghtsoonian estimated that JNDs typically reflect constant changes of about 3% in sensory intensity. Such a suggestion must be regarded with skepticism, in the light of the analyses performed by Newman and by Piéron (see above).

It is possible, one might argue, to avoid the inconsistency between summated JND scales and direct matches by means of some rescaling procedure. That is, by combining the results of matching with the measurement of JNDs, it might be possible to assign, in accordance with both sets of data, scale values to sensation levels that remove the inconsistency but maintain the same general psychophysical relationships between sensory scale values and stimulus intensity. In other words, the summated JND scale might be left unchanged at one sound frequency, for example, but transformed at other frequencies in order to maintain the appropriate equalities of loudness. Although such a procedure could provide the basis for producing consistency, to the present author such an approach appears as useless as some arbitrary assignment of numbers to stand for sensory magnitudes, with the only restriction that the same number always stands for all stimuli judged equal with regard to the sensory attribute in question.

It is reasonable to ask whether other processed scales have better success at meeting a criterion of consistency. Perhaps better agreement might obtain for scales derived by means of other types of discriminability-scaling procedure, e.g., the method of successive intervals. A successive-intervals analysis of ratings of loudness for different combinations of intensity and frequency, for example, might produce scales that do not violate direct loudness matches. The present author knows of no attempts at such tests of consistency using other discriminability-scaling procedures.

Surely, however, there could be no question that this simple measure of consistency is fulfilled for scales derived by the various direct procedures, e.g., category rating, magnitude estimation, etc. Indeed, the supposition has been made that, regardless of how idiosyncratic a person's use of available categories or numbers might be, to any given sensory magnitude he will always assign the same category or number (on the average), no matter how that magnitude is brought about. [See, for example, Marks (1970).] In a few experiments (e.g., Hellman & Zwislocki, 1964;

Marks, 1966), results from magnitude estimation and direct matching have been compared, and the predicted agreement has been satisfactorily observed. Schemes, such as Stevens's (1956a, 1961a, 1972) methods for calculating the loudness of complex noise, entail such consistency, and the accuracy of the predicted matches support consistency. Consistency is evident also in results of cross-modality matching (e.g., J. C. Stevens *et al.*, 1960; see Chapter 1). A formal model of magnitude estimation and cross-modality matching, based on internal consistency, is given by Krantz (1972).

Strangely enough, at least two experiments exist whose results, based on direct judgments, seem to disagree with observations determined by matching. Raab, Fehrer, and Hershenson (1961) and Lewis (1965) obtained category ratings of brightness for visual stimuli that varied with respect to luminance and duration. The results of both experiments failed to show the otherwise universally observed Broca-Sulzer peak, which was discussed in Chapter 4. The Broca-Sulzer effect has been verified by direct matches (Aiba & Stevens, 1964; Katz, 1964) and by magnitude estimation (Raab, 1962; J. C. Stevens & Hall, 1966). Lewis's data also failed to show the intensity-dependent change in the critical duration for temporal summation. It would appear, then, that rating procedures that limit the number of responses available to subjects may be constraining enough, at least in some instances, to camouflage important sensory phenomena (see Stevens, 1966a). Results of Raab *et al.* and Lewis serve to cast suspicion on the procedure of category rating as a general means for obtaining valid scales of sensory magnitude and suggest caution in the use of constraining rating procedures to obtain information concerning sensory processes and functioning.

It is quite possible that some of the difficulties inherent in the use of rating procedures can be avoided if care is taken to eliminate response biases as much as possible. Stevens and Galanter (1957) and Pollack (1965a & b) describe procedures of experimental iteration that may remove many of the biases (such as the tendency of some subjects to use all categories equally often). In experiments where an understanding of sensory-physical phenomena is of prime importance, and when it becomes particularly important to be able to determine as accurately as possible curves of constant sensitivity (equal-sensation relations), there is much to be gained by utilizing procedures whereby the subject varies one or more of the stimulus parameters. Category production and magnitude production are examples of procedures that make it possible to eliminate many of the numerical response biases and also to obtain directly measures of the values of various stimuli that produce the same sensory effect.

With production procedures it is never necessary, as it almost always is necessary with category rating and magnitude estimation, to interpolate between psychological values in order to derive equal-sensation functions; instead, the sensory-physical relations are obtained directly from the stimulus settings. In fact, the production procedures may be looked upon as complex matching tasks where the several matching levels are identified by categories or by numbers.

Scales derived by direct ratio procedures give us the most extensive evidence concerning internal consistency. That evidence is positive. However, consistency is still a far cry from validity. In fact, it is imperative to bear in mind that any monotonic transformation of direct ratio scales, e.g., logarithmic or power transformation, will leave the degree of consistency intact. In this sense, consistency is indeed a weak measure of validity.

How Valid Are Numerical Estimates of Sensation?

A number of investigators (Attneave, 1962; Ekman, 1964; Garner, 1958) have suggested that subjects may use numbers in a decidedly non-linear manner in the course of judging the strength of sensory stimuli, as by the method of magnitude estimation. In one sense, this suggestion can hardly be disputed. Sizeable differences are typically found among psychophysical functions obtained from various subjects, and these differences almost always arise from differences in subjects' use of numbers. However, there is another sense in which, if taken as a general statement, we can reject that notion. At least insofar as the pitch continuum is concerned, for example, subjects appear to be perfectly capable of using numbers in a linear fashion, though, of course, the suggestion could be correct in the context of some other modalities. Ekman's (1964) proposal is formally similar, or identical, to those of MacKay (1963) and of Treisman (1964); it retains Fechner's law relating stimulus and (underlying) sensation magnitudes, but predicts a power relation between stimulus intensity and numerical estimates because of a presumed logarithmic function relating numbers to their subjective values.

The suppositions of Attneave (1962) and of Garner (1958) are, essentially, that subjects' conceptions of numerical ratios may deviate systematically and markedly from nominal numerical values. Garner (1954b), for instance, generated a scale for the loudness of a 1000-Hz tone on the basis of both fractionation and equisection data. The fractionation results were used only to indicate equality of the nominal values of fractional loudness, but the judged fractions themselves were not used.

Rather, the results of equisection experiments were employed in order to generate equal-appearing loudness intervals and thereby to determine the subjective size of the loudness fractions. The scale that resulted was termed by Garner the lambda scale in order to distinguish it from the sone scale. (Chapter 5 describes use of the lambda scale for analysis of masking and of binaural summation.) Loudness in lambda units increases as a power function of stimulus energy, but with an exponent (about .13) much smaller than the exponent (about .3) that describes increase of loudness in sones. In fact, the fractionation data could probably have been dispensed with, if one were to make the assumption that the equisection data should be consistent with a power function.

The results of interval-scaling procedures, such as equisection or category rating, can often be described in terms of a power function whose exponent is smaller than that of the corresponding magnitude function. Stevens and Guirao (1962), for example, reported equisections of loudness that are consistent with a power relation, but, unlike Garner's results, with a substantially larger exponent—one nearly the same as that of the sone function (see Marks, 1968b).

Attneave (1962) pointed out that subjects may use numbers in a markedly nonlinear manner when asked to estimate sensory magnitudes. Instead of a logarithmic relation between "numbers" and their "psychological magnitudes," as Ekman suggested, Attneave indicated a power function. Such a formulation parallels Garner's conception that fractional judgments may reflect underlying sensory ratios that differ from the nominal fractions reported by the subject or given by the experimenter. A subject might, for instance, when asked to set one sound half the loudness of another, always set it to be two-thirds as loud. This proposal implies that so-called ratio-scaling procedures fail to produce ratio scales of sensory magnitude; instead, they would produce logarithmic interval scales. There is no question that some subjects can and do, at least some of the time, use numbers in a nonlinear manner. The "centering tendency" or "regression," in which the subject constricts the range of whatever stimulus variable is under his control, is an example of an omnipresent effect that occurs in numberical judgment and in matching in general (see Stevens & Greenbaum, 1966). Attneave's proposal, which goes beyond this, is that there are regular and systematic deviations from proportionality between numbers and the underlying sensory magnitudes they are presumed to reflect.

Several papers have elaborated Attneave's suggestion into a model and equation for judgment of sensory and perceptual magnitudes (e.g., Curtis, Attneave, & Harrington, 1968). By asking subjects to judge sensory intervals rather than merely sensory magnitudes, they hoped to parcel out the input (sensory) effects from output (nonlinear use of number) effects. In the case of estimation of magnitudes, it is assumed that in the typically observed power relation

$$\psi = k\phi^{\beta} \tag{7.4}$$

the size of β is equal to the product of δ, the exponent that governs the sensory process, and ϵ, the exponent that governs the nonlinear use of numbers. For the judgment of intervals (ψ_{int}) the expected relation becomes

$$\psi_{int} = k(\phi_1^{\delta} - \phi_2^{\delta})^{\epsilon} \tag{7.5}$$

where ϕ_1 and ϕ_2 are the two stimuli that comprise each pair to be judged. Support for the two-process model expressed by Eq. (7.5) was obtained for judgments of magnitudes of, and intervals between, heaviness of lifted weights (Curtis et al., 1968), brightness of lights (Curtis, 1970), and sums of heavinesses of weights (Curtis & Fox, 1969). Of special interest was the report by Curtis et al. (1968) that the value of δ, which they interpreted to reflect only a sensory process, was practically invariant from one subject to another. The intersubject differences in the size of the exponent β for lifted weights could be accounted for merely in terms of the parameter ϵ, which Curtis et al. interpreted as judgmental in nature and origin. For most of the subjects, the value of β fell close to the product, $\delta \cdot \epsilon$. Unfortunately, this simplicity did not obtain between results for judgments of brightness and of brightness intervals (Curtis, 1970); in that experiment the size of δ varied considerably among subjects, and the average size of the magnitude-estimation exponent β was 20% larger than the product, $\delta \cdot \epsilon$. A major difficulty with that study, however, was the failure to control for simultaneous contrast (spatial inhibition) in the estimation of intervals. The degree of contrast could potentially vary greatly from stimulus pair to stimulus pair. [The product, $\delta \cdot \epsilon$ did predict β for judgments of lightness of grays when surfaces were viewed in a surround of high reflectance, but failed to predict β when surfaces were viewed in a surround of low reflectance (Curtis & Rule, 1972).]

One of the difficulties in interpreting the results of these studies stems from the fact that the exponents β, δ, and ϵ were estimated from generalizations of Eqs. (7.4) and (7.5) that included additive constants. Additive constants have frequently been used to rectify psychophysical functions in the vicinity of the absolute threshold. However, in the present examples, the additive constants were used as curve-fitting devices, that is, solely in order to increase the goodness of fit of functions to data. All of the exponents—β of the magnitude functions, and δ and ϵ of the interval functions—are sensitive to the size of any additive constant. It is worthwhile to point out that the use of an additive constant solely to increase goodness of fit has been seen to help create intransitivities among cross-modality matching functions (see Stevens, 1969a).

In another experiment (Marks & Cain, 1972), subjects gave magnitude estimates of intervals of heaviness of lifted weights, apparent area of circles, and apparent roughness of sandpapers. The results for all three modalities agreed in suggesting that the two-process Eq. (7.5), although capable of describing the experimental results fairly well, is of little use in predicting the exponent β of the magnitude functions. The variation in the size of the exponent δ was as great as the variation in β or in ϵ, and the product of $\delta \cdot \epsilon$ was consistently smaller than the size of β. It appears that, in general, the product of $\delta \cdot \epsilon$ may fall below the value of the magnitude-estimation exponent β. Above it was pointed out that power functions can often describe the results of interval-scaling experiments, but typically with exponents smaller than those found by ratio-scaling procedures. The procedure whereby subjects estimate intervals is, by its nature, an interval-scaling procedure, and it is not surprising that the observed exponents are relatively small. In fact, with the same subjects who did magnitude estimation and estimation of intervals, Marks and Cain obtained category ratings for the same three modalities. Power functions usually described the results satisfactorily, and the exponents were smaller than the exponents for the corresponding magnitude functions. For two of the three modalities investigated, average scale values derived from the estimates of intervals were linearly related to category ratings.

The two-process Eq. (7.5) may well be appropriate to account for data on judgments of intervals, but there is no compelling evidence that its application sheds much light on the processes involved in judgment of magnitudes. In particular, the view that the exponent δ reflects sensory processing only is most likely false. Rather, the size of δ obtained from interval estimation probably depends as much, if not more on judgmental factors than does the size of β obtained from magnitude estimation. This can be seen most clearly in terms of bisection and equisection experiments, where, the two-stage model would predict, the results should reflect only the value of δ. In these tasks, stimuli are adjusted in order to make sensory intervals appear equal; under this condition, the value of ϵ (which supposedly reflects judgmental nonlinearity) becomes irrelevant.

Bisection and equisection points would presumably remain the same under any transformation of ϵ; results should reflect, therefore, only sensory processing. It should be clear from the earlier discussion that results from any particular bisection or equisection experiments are determined in large measure by non-sensory factors. The fundamental scale underlying interval-type judgments may depend as much on similarity as on magnitude per se.

IMPLICATIONS FOR SENSORY-PHYSICAL LAWS

These digressions to consider the hypotheses of Attneave and of Garner can serve to put some problems into appropriate perspective. Let us consider what sorts of changes would be entailed by the assumption that subjects use numbers in a nonlinear manner when they set about to estimate sensory magnitudes or sensory ratios. As was mentioned at the beginning of the present discussion, any such nonlinearity would have no effect at all on the prediction of matches for subjective equality, i.e., on sensory physics. Subjects could square or cube all of their numerical responses, could take logarithms, could exponentiate, could add a constant, without changing the fact that to the same sensory magnitude they would always assign the same number, at least on the average. Thus, sensory-physical laws, laws which relate physical variables to one another in order to describe sensory invariances, would be totally unaffected.

IMPLICATIONS FOR PSYCHOSENSORY LAWS

Some types of sensory law might be affected, however, by nonlinearity in subjects' use of numbers. These types of laws are psychosensory laws, laws which relate psychological or sensory variables to one another, without regard to any physical magnitudes. The nature of any change in a psychosensory law depends upon the mathematical form of the law and the nature of the nonlinear transformation. For example, Stevens *et al.* (1965) reported that when subjects are asked to estimate numerically the magnitudes of three auditory attributes—loudness, volume, and density—it turns out the loudness (L) is proportional to the product of volume (V) and density (D) over a wide range of sound frequencies and intensities. Thus

$$L = V \cdot D \tag{7.6}$$

This is a curious and interesting result (whose implications for auditory processes remain obscure). Note, however, what happens if we assume nonlinear behavior on the part of the subjects; for instance, if the subjects used ratios in some "improper" manner, but did so consistently, then we might wish to recompute all of the scale values by raising them to some (constant) power. But this transformation would have no effect on the nature of the psychosensory law: loudness would still be proportional to density times volume, in accordance with Eq. (7.6). On the other hand, if logarithmic transformation were necessary, then the nature of the psychosensory law would change: after logarithmic transformation, loudness would equal the sum of volume plus density.

It is of interest to consider Garner's (1954b) lambda scale in this context. As was described earlier, the lambda scale can be described as a power transformation of the sone scale. This relation implies that some psychosensory laws utilizing the two scales may differ. And they do. Let us consider the predicted loudness of complex sounds. When tones of different frequencies are played simultaneously (assuming they are far enough apart in frequency to preclude significant masking) measurements made utilizing the sone scale suggest that the total loudness for the tonal complex (L_t) is the linear sum of the loudnesses of the individual components (L) (Howes, 1950; Zwislocki, 1965), as Fletcher (1940) claimed

$$L_t = \Sigma L \qquad (7.7)$$

Measurements utilizing the lambda scale suggest that total loudness equals the square root of the sum of the squares of the loudnesses of the individual components (Garner, 1959)

$$L_t = (\Sigma L^2)^{1/2} \qquad (7.8)$$

A similar outcome results from considering stimulation of the two ears: evidence (Hellman & Zwislocki, 1963; Reynolds & Stevens, 1960; Scharf & Fishken, 1970) suggests that if the sone scale is used, then to a first approximation loudnesses sum linearly; but if the lambda scale is used, they sum vectorally, i.e., as the square root of the sum of squares (Garner, 1959).

VALIDITY RELATED TO THEORIES OF SENSORY PROCESSES

In the present context it is worthwhile to mention Anderson's (1970) argument in favor of "functional" measurement. Anderson regards psychophysical measurement and scaling to be coextensive with determination of behavioral—and, presumably, sensory—laws. That is, scaling and the determination of substantive relations among variables go hand in hand, particularly with regard to the validation of scales. Rescaling of response values may, for example, be necessary in order to correlate the scales with behavioral laws.

Consider as one example the interrelations among three auditory attributes: loudness, volume, and density. We have seen that loudness when measured by magnitude estimation equals the *product* of volume and density. If, however, the psychosensory law were to give loudness as the linear *sum* of volume and density, it would be necessary to rescale (by logarithmic transformation) all three perceptual variables. At present, there is neither empirical nor theoretical basis from which to choose between multiplicative and additive laws. As another example, if loudness is converted from units of sones to lambda units, the rule of binaural summation changes from linear addition to root-mean-square addition. In the present context, results obtained by Levelt *et al.* (1972) support a law of linear addition of loudnesses by the two ears and, by implication, the sone scale of loudness.

Scales, therefore, are determined as much by the nature of the behavioral laws as the laws are determined by the nature of the scales. (Anderson has tended to apply the functional measurement model primarily to the examination of judgmental—"information integration"—tasks, but it can just as well apply to the examination of sensory functions.) A general polynomial transformation has been formulated by Bogartz and Wackwitz (1971), which is applicable, for example, to results of bisection experiments. It is not completely certain, however, that validational bases always can be guaranteed. As we just saw, response rescaling may affect certain laws (psychophysical or psychosensory), but leave others unchanged. In the latter case, what may be required are theories that demand one formulation or another.

In a fundamental sense, Anderson's position is quite close to that taken in this book. It is in the context of the study of sensory processes that direct ratio procedures show their utility and importance. And it is in the study of sensory behavior that the possibility for validating sensory scales resides: not just from sensory laws, but also from theories of sensory function. To take an example, there are both neurophysiological data and psychophysical evidence to suggest an intimate relationship (inverse proportionality) between sensitivity in the visual system (e.g., the quantity of light needed to reach some physiological or sensory response criterion) and response latency or critical duration (the time for physiological response to reach its maximum, or the maximum stimulus duration over which luminous energy is integrated). A model of brightness vision, described in Chapters 4 and 5, extends the notion of an inverse proportionality between sensitivity and critical duration to the realm of brightness magnitude. The data are quite compatible with the model; the evidence that the critical duration for temporal brightness summation is inversely proportional to the cube root of intensity lends some validational support for the bril scale of brightness, which is related to luminance also by a cube-root relation. Another example is the derivation of a power law for loudness from data and theory on discrimination of pure tones (McGill & Goldberg, 1968). Sensory laws and theories of these types, if they can be developed in other modalities, may pave the way for at least tentative validation of scales of sensory magnitude. If sensory scales are to play no substantive role in theories of sensory function, then any scale type is as good as any other, as long as it meets the requirements of internal consistency. In fact, it is only through such theoretical formulations that the nature of sensory scales truly becomes relevant.

Scales of Sensation: Theoretical Interpretations

SENSATION AND NEUROPHYSIOLOGY

Allusions have already been made to several theoretical accounts of the nature of direct scales of sensory magnitude. One prominent interpretation is that such sensory scales reflect directly some aspect of sensory, neural activity. This is the position taken by Stevens (1967, 1970, 1971b), who has referred to psychophysical power functions obtained by direct ratio procedures as "transducer functions."

In a similar vein, Luce and Green (1972) interpret sensory magnitude scales in terms of time intervals between neural impulses, and they attempt thereby to account for some correlations between judged sensory magnitudes and reaction times. This type of theoretical viewpoint also underlies the various models, described in Chapters 4 and 5, to account for psychophysical and sensory-physical relations, as well as some others that have been propounded. A model for audition proposed by Stewart (1963) predicts phenomena as diverse as the psychophysical power law, Weber's law, masking, and diplacusis. Stewart's model implies a direct relation between scaled loudness and neural firing rate. An often unfortunate byproduct of this type of viewpoint has been the search for neurophysiological correlates to the power-law behavior observed psychophysically.

Some discussion of neurophysiological relations is worthwhile, if for no other reason than to demonstrate the vagueness of attempts to correlate neural results with certain psychophysical relations. In the early twentieth

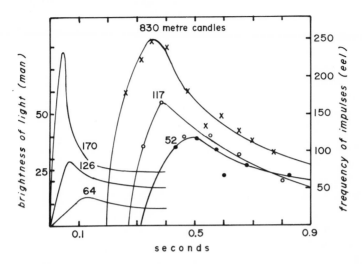

Figure 7.7. Brightness to man and frequency of neural impulses in eel, as functions of the duration of the flashes of light. The parameter is the illuminance of the light in meter-candles. [From Adrian (1928). Reproduced by permission of the author and Chatto and Windus.]

century, evidence began to accumulate that suggested that frequency of neural firing could provide a measure of and correlate to sensory intensity. Figure 7.7 shows neural responses obtained from flashes of light of different durations presented to the eye of an eel. It is difficult to avoid comparison of the rise and overshoot in response rate to temporal summation of brightness and the Broca-Sulzer effect described in Chapter 4. In fact, even in the nineteenth century it was noted that physiological responses (e.g., activity in the optic nerve) increase in a strongly nonlinear manner with increases in light intensity (Waller, 1895). Such observations were taken as evidence to give credence to Fechner's claim that sensation magnitude grows as the logarithm of stimulus intensity. Logarithmic stimulus–response functions have been reported for physiological responses in several afferent systems—e.g., the visual (Hartline & Graham, 1932), auditory (Tasaki, 1954), and muscle afferent (Matthews, 1931) systems. There is probably basis for agreement with Rosner and Goff's (1967) suggestion that logarithmic relations were observed at least in part because they were looked for. It is not infrequent to observe that the same data may be fitted as well by power functions as by logarithmic functions (Rosner & Goff, 1967; Stevens, 1970, 1971b), and, in general, the past few years has seen a marked increase in the number of power functions reported in the neurophysiological literature.

In many, if not all, sensory systems there appears to be a simple hyperbolic equation that can describe how various physiological responses χ (magnitude of receptor potential, frequency of neural spikes) depend on stimulus intensity ϕ

$$X = \frac{a\phi}{1 + b\phi} \tag{7.9}$$

An interesting feature of Eq. (7.9) is that, over different portions of the range of stimulus intensity ϕ, it defines a stimulus–response relation that can vary from linear to logarithmic (Lipetz, 1971). When intensity is low, the stimulus–response relation is linear; at higher intensities, it can be approximated by power functions with exponents less than 1; at still higher intensities it can be approximated by a logarithmic relation.

Rosner and Goff (1967) and Stevens (1970, 1971b) gathered results from a large number of studies that purport to demonstrate power-law behavior at the physiological level. Some of the evidence consists of data, such as those of Hartline and Graham (1932) and Matthews (1931), that were originally reported to support a logarithmic relation. We shall not make any attempt to review again all of these studies. To summarize, however, power functions have been shown to describe adequately relations between stimulus intensity and physiological responses in the visual, auditory, tactile, and proprioceptive systems. A large number of different types of physiological measure have been used: these include electrical potentials in peripheral neurons [for example, Stevens's (1970) analysis of data of Fuortes & Hodgkin (1964)], peripheral neural action potentials (for example, Boudreau, 1965), central neural responses (for example, Mountcastle, Poggio, & Werner, 1963; Rosner & Goff, 1967). Cohen and Little (1971) described a simple model, based on a mechanism of addition and subtraction among responses from several neurons, which can produce neural power functions between stimulus and response with variable exponent.

Unfortunately, exponents obtained from physiological experiments do not always coincide with exponents typically obtained from psychological experiments run under comparable conditions. Keidel and Spreng (1965) showed that the voltage of the cortical evoked auditory potential grows as a power function of sound energy, but the size of the exponent depends on the point in time the potential is measured. At a point 100 msec after stimulus onset, the exponent equaled .07, but 60 msec later, it equaled .36. (It is of considerable interest to note Rosner and Goff's observation that the variation in the size of the exponent from one neural sample to another is quite large. The variation is as large or larger than the variation of exponents typically observed psychophysically from one subject to another.)

The problem of relating sensory responses to physiological measures is especially thorny. Rosner and Goff's review makes it quite clear that as far as peripheral sensory responses are concerned, a large number of different stimulus–response functions may be observed. The most typical stimulus–response relation is one that saturates (i.e., the response reaches a maximum) at high levels of stimulus intensity, although at lower intensities the stimulus-response relation may appear to be a power or a logarithmic function. Equation (7.9) can often describe such relations. But which of the several responses (different from one neural unit to another), if any, is relevant for sensory perception? It would seem unlikely, for example, that sensory magnitude depends on the response of a single neuron. Relatively large-scale activity seems more probable. Some investigators (e.g., Borg, Diamant, Ström, & Zotterman, 1967) have looked at "total activity" in a sensory nerve; Borg et al. reported that responses of the chorda tympani nerve were proportional to simultaneously measured magnitude estimates of taste. Others (e.g., Rosner & Goff, 1967) have examined cortical evoked potentials, which appear to reflect activity of many neurons. But should we expect such integrated activities to be indicative merely of sensory magnitude? Might not integrated activity also reflect other sensory dimensions, such as quality? There is good reason for maintaining a large degree of caution with regard to possible similarities between neural and psychophysical responses. It is important to remain as sensitive as possible to the several different means by which sensory information might be coded at any particular level in the nervous system (see Uttal, 1969).

It is an understatement that the relation between sensory perceptions and underlying neural responses is exceedingly complex. It is probably a mistake to focus simply on how magnitudes of neural responses depend on stimulus intensity when examining physiological data, just as it is too narrow to focus simply on how sensory magnitudes depend on intensity when examining psychophysical data. Instead, it is important to look at these stimulus–response functions in the context of measurements that involve several stimulus parameters. In this regard, studies of special interest were conducted by Drake et al. (1969) and by Døving (1966). Drake et al. compared magnitude estimates of odor intensity to electro-olfactograms (EOGs) taken from olfactory epithelia of frogs. They employed several different odorants. Not only did power functions describe stimulus–response relations both for magnitude estimation and for EOG, but for the most part, there were good correlations between frog electro-physiological and human psychophysical responses. Døving (1966) found that patterns of response from frog olfactory bulbs were most simi-lar the closer in structure were the odorants in terms of carbon-chain length. Stimuli that were most similar physiologically to frogs were also most similar psychologically to humans [as measured by Engen (1964)].

Perhaps the most important question to ask is: Are there neural re-sponses that display the same invariances displayed by psychophysical re-sponses? Just as the validation of psychophysical scales of sensory magni-tude appears to be possible only, if at all, through systematic investigation of sensory behavior, so an understanding of the physiological mechanisms underlying sensation magnitudes will probably come about only through systematic examination of interrelationships among stimulus variables with regard to their effects on neural responses.

It can be only an hypothesis, therefore, albeit a seemingly reasonable one, that psychological measures of sensory magnitude primarily reflect, or directly relate to, some simple underlying sensory-physiological process or event. In the case of the sensory attribute pitch, the hypothesis gains cre-dence because of the linear relation between pitch in mels and locus of maximal stimulation on the basilar membrane. Any constant number of mels seems to correspond approximately to a constant distance along the basilar membrane. It is unlikely, however, that other sensory attributes will turn out to show such simple correlations to anatomical structures or physiological processes. The mechanism or mechanisms by which pitch is neurally coded remains still in doubt (see Raab, 1971).

Because sensory-physiological data are far from clear cut, there exist several alternative ways to relate psychological magnitudes to physiologi-cal measures. The alternative that is selected may depend in part on which data are emphasized or what function is used to describe a particu-lar set of data. Stevens's (1970) examination of data from *Limulus* [frequencies of neural impulses obtained by Hartline and Graham (1932) and electrical potentials recorded by Fuortes and Hodgkin (1964)] shows functions roughly consistent with a cube-root relation be-tween output and input. If it is assumed that similar responses occur in vertebrate vision, then such an outcome suggests that peripheral receptor function may comprise most, if not all, of the basis for the nonlinearity between stimulus intensity and psychological response. That is, all of the nonlinear transformation might occur at the earliest stages of sensory re-ception, with linear transmission of signals thereafter. Mountcastle has made a similar suggestion with regard to somesthetic responses (Mount-castle, 1966; Werner & Mountcastle, 1965).

Alternatively, one might consider that the relation between stimulus intensity and physiological response is better described in some instances by a logarithmic relation. It then becomes necessary to assume some sort of additional nonlinear process, such as the matching process suggested by MacKay (1963), to transform the logarithmic relation at the psychophysiological level into a power function at the psychophysical level. This would appear to be the position taken by Treisman (1963, 1964), who suggested that subjects perform exponential transformations when they judge sensory magnitudes by procedures such as magnitude estimation. Treisman (1965) later modified this view, on the basis of some measurements of physiological and psychophysical responses to weak visual stimuli; he concluded that the receptor transformation process is linear, but some more central process is nonlinear and logarithmic.

SENSATION AND LEARNING

A third alternative (though not completely distinct from the last one) is to consider as totally unknown the nature of the physiological responses which underlie the perception of sensory magnitudes. In that case, the behavior that subjects emit in sensory scaling experiments may primarily reflect learning and experiential factors.

The view that we *learn* to judge and, therefore, that we judge *stimuli,* not *sensations,* was put forth by Fullerton and Cattell (1892) and by Ebbinghaus (1902). It is also the position taken by Warren (see Warren, 1958, 1969; Warren & Warren, 1958), who proposes a "physical correlate theory" of psychophysical judgment. This theory assumes that judgments of sensory magnitude are based on available physical correlates of stimulation, correlates which vary monotonically with sensory magnitude and which can readily be quantified by subjects on the basis of their past experience. Warren and his co-workers have concerned themselves to a great extent with brightness and loudness. For both of those attributes, the physical correlate assumed to operate is distance. Thus, in the case where subjects are called upon to judge relative loudness, the theory states that they judge the relative distances to sound sources that would yield the sensory experiences. Judging the loudness of one sound to be half that of another, according to this hypothesis, is based on the judgment that it appears to come from a source twice as far away. Because the sound intensity at any point varies approximately as the inverse square of the distance from its source, the physical correlate theory predicts a square-root relation between loudness and sound energy. A similar outcome is predicted for brightness, where, however, the distance involved is that between the object judged and the source of illumination, rather than the distance from stimulus object to the subject. On the basis of our experience with sound sources at various distances from our ears and with light sources at various distances from the objects they illuminate, Warren argues that we learn relations that we then apply when asked to judge sensory magnitudes.

One of the major difficulties with Warren's hypothesis is that much of the data fails to support the expected square-root relations. The clearest example is in the case of loudness, where typical exponents lie nearer to one-third than to one-half (e.g., J. C. Stevens & Guirao, 1964; Stevens, 1972). The cube-root result is also most typical for brightness (e.g., Marks & J. C. Stevens, 1966), but it might be argued that the appropriate condition in vision to test Warren's hypothesis is the perception of a complex pattern in the steady state, since that condition most closely simulates our typical visual experiences. For a simple visual field, the "equilibrium brightness function" determined by J. C. Stevens and Stevens (1963) is more nearly related to luminance by a logarithmic, than by a power, function. On the other hand, estimates of the brightness of objects in a complex field may not even conform to a power equation (Bartleson & Breneman, 1967).

Other questions are raised by Warren's hypothesis, such as why we are not able to learn to "correct" the steepening in psychophysical functions that occurs at weak stimulus levels, or why the scale for autophonic loudness (the judgment of self-produced sound) is the same for congenitally deaf subjects as for normals (Lane, 1963). The distinction between Warren's view (that subjects estimate physical correlates) and Stevens's view (that they estimate sensations) was reviewed by Natsoulas (1967), who put the controversy into a larger framework, namely whether subjects' reports in perceptual experiments should be interpreted as phenomenal or cognitive. Natsoulas suggested a possible resolution of the Warren–Stevens controversy in terms of Treisman's hypothesis that percepts mediate between stimuli and responses. Thus, the relation of stimulus to percept is psychological, the relation of percept to response, cognitive.

It would unfortunately require too much detail to examine Warren's points in greater depth. Although his approach is of interest, the empirical evidence would seem to weigh against any general applicability of the physical correlate theory.

Ranges of Sensory Magnitude and Stimulus Intensity

A close relative to Warren's theory is the "range hypothesis," put forth by Jones and Woskow (1962) and later by Poulton (1967, 1968). Poulton took note of the fact, observed and reported previously by several investigators (e.g., Jones & Woskow, 1962; Stevens & J. C. Stevens, 1960), that the power-function exponent determined in any particular scaling experiment depends to some extent on the range of the physical intensities of stimuli. (Additional references and discussion can be found in Chapter 2.) Within a single sensory modality, there is some tendency for larger exponents to obtain in experiments in which the stimulus range is relatively small. Across different modalities, those in which large stimulus ranges are employed (e.g., loudness and brightness) tend to give smaller exponents than do modalities in which smaller stimulus ranges are employed. The possibility arises, therefore, that the size of the exponent may be to some extent an experimental artifact of the stimulus range. In other words, subjects might use a relatively constant response range (i.e., a constant range of numbers in a magnitude-estimation experiment) regardless of the stimulus range to which they are exposed. Then, if a large stimulus range is presented, the resulting exponent will be lower than if a small range is presented. Poulton estimated that about one-third of the variance among reported exponents could be accounted for in terms of variations in stimulus range.

There can be little doubt that the range of stimuli used in a scaling experiment to some extent influences the value of the resultant exponent. To consider first the relation between stimulus range and exponent within a modality (where the correlation *is* an indicant of the effect of range on the exponent), the influence of stimulus range seems to be a second-order effect: the size of the exponent appears to depend much more significantly on the nature and parametric values of the stimulus. The slopes (exponents) of the brightness function under contrast and of the loudness function under masking depend strongly on, and vary continuously with, the intensity of the inhibiting stimulus (Stevens, 1966e). The size of the exponent relating odor intensity to the concentration of aliphatic alcohol varies monotonically with the length of carbon chain (Cain, 1969); one study in olfaction (Henion, 1971c) found an approximately 15-to-1 range of exponents for different odorous substances, in spite of the fact that the stimulus range was the same for all substances.

Next for consideration is the relation between stimulus range and exponent across different sensory modalities. In the present case, of course, a correlation between range and exponent may not at all reflect any *influence* of range on exponent. Rosenblith (1959) pointed out some evidence that the product of stimulus range and psychophysical exponent is approximately constant across sensory modalities. Such an invariance could reflect either something about the way subjects use numbers or else a fundamental property of sensory systems. Teghtsoonian (1971) suggested that Poulton actually underestimated the extent to which stimulus range and size of exponent are correlated across modalities, but gave an explanation for that correlation different from Poulton's. He proposed:

> that the ratio of the greatest to the smallest possible sensory magnitude is approximately constant for all perceptual continua, and that variation in power law exponents among continua reflects variation in the ratio of the greatest to the smallest stimulus intensity to which S is responsive (a ratio which is sometimes called the dynamic range). In short, widely varying dynamic ranges may all be mapped into the same sensory range; and if some nearly constant proportion of those dynamic ranges is presented, the result will still be a nearly constant sensory range . . . [Teghtsoonian, 1971, p. 72].

Under this interpretation, the correlation between exponent and range is the result of the fact that different sensory systems respond over different dynamic ranges, but produce approximately the same range of sensory (and, presumably, underlying neural) response. The use of different stimulus ranges for the study of different sensory systems is also a reflection of the nature of the system's psychophysical behavior. Experimenters select stimulus ranges appropriate to the sensory modalities studied. Teghtsoonian's hypothesis is consistent with the notions that the sizes of exponents reflect the nature of transformations produced by sensory systems, and that there is markedly similarity from one sense to another in the nature and magnitudes of the neural responses produced by stimuli, responses that underlie magnitudes of sensation.

Appendix

UNITS OF MEASUREMENT

The physical units of measurement used throughout this book are the metric units of the SI (Système Internationale). These include the following fundamental units:

meter (m): length
kilogram (kg): mass
second (sec): duration
candela (cd): luminous intensity

In addition, there are derived units, such as:

newton (N): force
pascal (P): pressure

Permissible multiples and fractions of these units consist of all multiples by the factor $10^{\pm 3n}$. Thus, the gram is 10^{-3} kilogram; the nanometer is 10^{-9} meter; the megasecond is 10^6 second. The only exception to this limitation on units is the use throughout the book of the centimeter (10^{-2} meter), a unit whose size is especially convenient.

Special mention should be made of the employment of *decibel scales* of sound and light intensity. In a fundamental sense, decibel scales are superordinate to the units described above, since the reference levels of decibel scales are arbitrary. Decibels may be used with any units, whether SI metric, non-SI metric, or nonmetric.

The decibel is a measure of relative flow of energy (power), i.e., of energy relative to some reference. Number of decibels (dB) is defined as

$$dB = 10 \log \frac{E}{E_{ref}}$$

Used as an acoustical measure, i.e., applied to sound pressure, the formula is:

$$dB = 20 \log \frac{P}{P_{ref}}$$

(The change from multiplication by 10 to multiplication by 20 derives from the fact that energy or power is proportional to the square of pressure).

Three decibel scales are common in psychoacoustics. One is sound pressure level (SPL). Here the reference level is .00002 pascal, where a pascal equals one newton per square meter ($P = N/m^2$). Thus

$$dB\ SPL = 20 \log \frac{P}{2 \times 10^{-5}}$$

A second common scale is sensation level (SL). Here the reference pressure is the sound pressure at threshold.

$$dB\ SL = 20 \log \frac{P}{P_{threshold}}$$

The third scale is loudness level (LL), a somewhat different creature. The loudness level of a test sound is defined as the SPL of a sound of 1000 Hz that appears to be of the same loudness as the test sound. When the test sound is a 1000-Hz tone, LL and SPL are, of course, identical.

Decibel scales are not limited to sound energies, but can be used with any type of energy. (In fact, decibels were first used to describe losses of electrical power.) A frequent application of decibels is to relative light energy, in particular to luminance. The fundamental unit of luminance in SI is the candela per square meter (cd/m^2). The decibel scale of luminance used here takes as the reference $10^{-6}\ cd/m^2$ (a value that is itself a permissible unit of SI—the microcandela per square meter). Thus,

$$dB = 10 \log \frac{cd/m^2}{10^{-6}}$$

This reference level—10^{-6} cd/m²—is slightly different from a reference level that has sometimes been used elsewhere. The older reference is $1/\pi$ · 10^{-6} cd/m². The multiplicative factor $1/\pi$ arises from the fact that the reference level was originally stated in terms of the lambert, which is a non-SI unit equal to $(1/\pi)$ · $(10^4$ cd/m²$)$. Transformation from the "old" to the "new" decibel scale of luminance can be effected quite simply: just subtract 5 dB. Thus 1 cd/m² was 65 dB on the old scale, but is 60 dB on the new decibel scale. Whenever possible, units of luminance used in studies described in the present work have been transformed into decibels *re* 10^{-6} cd/m².

REFERENCES

Abraham, O. Zur psychologischen Akustik von Wellenlänge und Schwingungszahl. *Zeitschrift für Sinnesphysiologie,* 1920, **51**, 121–152.

Abrahams, H., Krakauer, D., & Dallenbach, K. M. Gustatory adaptation to salt. *American Journal of Psychology,* 1937, **49**, 462–469.

Adair, E. R., Stevens, J. C., & Marks, L. E. Thermally induced pain, the dol scale, and the psychophysical power law. *American Journal of Psychology,* 1968, **81**, 147–164.

Adams, E., & Messick, S. An axiomatic formulation and generalization of successive intervals scaling. *Psychometrika,* 1958, **23**, 355–368.

Adrian, E. D. *The basis of sensation.* London: Christophers, 1928.

Aiba, T. S., & Stevens, S. S. Relation of brightness to duration and luminance under light- and dark-adaptation. *Vision Research,* 1964, **4**, 391–401.

Amoore, J. E. Psychophysics of odor. *Cold Springs Harbor Symposia on Quantitative Biology,* 1965, **30**, 623–637.

Amoore, J. E. A plan to identify most of the primary odors. In C. Pfaffmann (Ed.), *Olfaction and taste III.* New York: Rockefeller Univ. Press, 1969. Pp. 158–171.

Anderson, N. H. Functional measurement and psychophysical judgment. *Psychological Review,* 1970, **77**, 153–170.

Angell, F. Untersuchungen über die Schätzung von Schallintensitäten nach der Methode der mittleren Abstufungen. *Philosophische Studien,* 1891, **7**, 414–468.

Anglin, J. W., & Mansfield, R. J. W. On the brightness of short and long flashes. *Perception & Psychophysics,* 1968, **4**, 161–162.

Attneave, F. A method of graded dichotomies for the scaling of judgments. *Psychological Review,* 1949, **56**, 334–340.

Attneave, F. Perception and related areas. In S. Koch (Ed.), *Psychology: A study of a science.* Vol. 4. New York: McGraw-Hill, 1962. Pp. 619–659.

Attneave, F., & Olson, R. K. Pitch as a medium: A new approach to psychophysical scaling. *American Journal of Psychology,* 1971, **84**, 147–165.

Bain, A. *Mental science: A compendium of psychology.* New York: Appleton, 1868.

Baird, J. C. A cognitive theory of psychophysics. II. Fechner's law and Stevens' law. *Scandinavian Journal of Psychology,* 1970, **11**, 89–102.

Barlow, H. B. Temporal and spatial summation in human vision at different background intensities. *Journal of Physiology,* 1958, **141**, 337–350.

Bartleson, C. J., & Breneman, E. J. Brightness perception in complex fields. *Journal of the Optical Society of America,* 1967, **57**, 953–957.

Bartley, S. H. Subjective brightness in relation to flash rate and light-dark ratio. *Journal of Experimental Psychology,* 1938, **23**, 313–319.

Bartoshuk, L. M. Water taste in man. *Perception & Psychophysics,* 1968, **3**, 69–72.

Baumgardt, E. On direct scaling methods. *Vision Research,* 1967, **7**, 679–681.

Baumgardt, E., & Hillman, B. Duration and size as determinants of peripheral retinal response. *Journal of the Opical Society of America,* 1961, **51**, 340–344.

Beck, J., & Shaw, W. A. The scaling of pitch by the method of magnitude estimation. *American Journal of Psychology,* 1961, **74**, 242–251.

Beck, J., & Shaw, W. A. Magnitude estimations of pitch. *Journal of the Acoustical Society of America,* 1962, **34**, 92–98.

Beck, J., & Shaw, W. A. Single estimates of pitch magnitude. *Journal of the Acoustical Society of America,* 1963, **35**, 1722–1724.

Beck, J., & Shaw, W. A. Magnitude of the standard, numerical value of the standard, and stimulus spacing in the estimation of loudness. *Perceptual and Motor Skills,* 1965, **21**, 151–156.

Beck, J., & Shaw, W. A. Ratio-estimations of loudness-intervals. *American Journal of Psychology,* 1967, **80**, 59–65.

Bedford, R. E., & Wyszecki, G. W. Luminosity functions for various field sizes and levels of retinal illuminance. *Journal of the Optical Society of America,* 1958, **48**, 406–411.

Beebe-Center, J. G. *The psychology of pleasantness and unpleasantness.* Princeton, New Jersey: Van Nostrand, 1932.

Beebe-Center, J. G., Rogers, M. S., Atkinson, W., & O'Connell, D. N. Sweetness and saltiness of compound solutions of sucrose and NaCl as a function of concentration of solutes. *Journal of Experimental Psychology.* 1959, **57**, 231–234.

Beebe-Center, J. G., & Waddell, D. A general psychological scale of taste. *Journal of Psychology,* 1948, **26**, 517–524.

Békésy, G. von. On the resonance curve and the decay period at various points on the cochlear partition. *Journal of the Acoustical Society of America,* 1949, **21**, 245–254. (a)

Békésy, G. von. The vibration of the cochlear partition in anatomical preparations and in models of the inner ear. *Journal of the Acoustical Society of America,* 1949, **21**, 233–245. (b)

Békésy, G. von. Human skin perception of traveling waves similar to those on the cochlea. *Journal of the Acoustical Society of America,* 1955, **27**, 830–841.

Békésy, G. von. Funneling in the nervous system and its role in loudness and sensory intensity on the skin. *Journal of the Acoustical Society of America,* 1958, **30**, 399–412.

Békésy, G. von. Similarities between hearing and skin sensations. *Psychological Review,* 1959, **66**, 1–22.

Békésy, G. von. Lateral inhibition of heat sensations on the skin. *Journal of Applied Physiology,* 1962, **17**, 1003–1008.

Békésy, G. von. *Sensory inhibition.* Princeton, New Jersey: Princeton Univ. Press, 1967.

Békésy, G. von. Brightness distribution across the Mach bands measured with flicker photometry, and the linearity of sensory nervous interaction. *Journal of the Optical Society of America,* 1968, **58**, 1–8. (a)

Békésy, G. von. Mach- and Hering-type lateral inhibition in vision. *Vision Research,* 1968, **8**, 1483–1499. (b)

Bénéze, G. Notes sur la loi de Fechner. *Revue Philosophique,* 1929, **108**, 429–432.

Berglund, B., & Berglund, U. A further study of the temporal integration of loudness. *Report from the Psychological Laboratory, University of Stockholm,* 1967, No. 229.

Berglund, B., Berglund, U., & Ekman, G. Temporal integration of vibrotactile stimulation. *Perceptual and Motor Skills,* 1967, **25,** 549–560.

Berglund, B., Berglund, U., & Lindvall, T. On the principle of odor interaction. *Acta Psychologica,* 1971, **35,** 255–268.

Berglund, B., Berglund, U., Lindvall, T., & Svennson, L. T. A quantitative principle of perceived intensity summation in odor mixtures. *Report from the Psychological Laboratory, University of Stockholm,* 1972, No. 346.

Berglund, U., & Berglund, B. Adaptation and recovery in vibrotactile perception. *Perceptual and Motor Skills,* 1970, **30,** 843–853.

Berglund, U., Berglund, B., Ekman, G., & Engen, T. Individual psychophysical functions for 28 odorants. *Perception & Psychophysics,* 1971, **9,** 379–384.

Berlyne, D. E. *Aesthetics and psychobiology.* New York: Appleton, 1971.

Biersdorf, W. R. Critical duration in visual brightness discrimination for retinal areas of various sizes. *Journal of the Optical Society of America,* 1955, **45,** 920–925.

Björkman, M., & Strangert, B. The relation between ratio estimates and stimulus dispersion. *Report from the Psychological Laboratory, University of Stockholm,* 1960, No. 81.

Bogartz, R. S., & Wackwitz, J. W. Polynomial response scaling and functional measurement. *Journal of Mathematical Psychology,* 1971, **8,** 418–443.

Borg, G. A. V. *Physical performance and perceived exertion.* Lund: Gleerup, 1962.

Borg, G., Diamant, H., Ström, L., & Zotterman, Y. The relation between neural and perceptual intensity. A comparative study on the neural and psychophysical response to taste stimuli. *Journal of Physiology,* 1967, **192,** 13–20.

Borg, G., Edström, C.-G., & Marklund, G. A new method to determine the exponent for perceived force in physical work. *Report from the Institute of Applied Psychology, University of Stockholm,* 1970, No. 4.

Boring, E. G., & Stevens, S. S. The nature of tonal brightness. *Proceedings of the National Academy of Sciences,* 1936, **22,** 514–521.

Bornstein, M. H., & Marks, L. E. Photopic luminosity measured by the method of critical frequency. *Vision Research,* 1972, **12,** 2023–2033.

Boudreau, J. C. Stimulus correlates of wave activity in the superior-olivary complex of the cat. *Journal of Acoustical Society of America.* 1965, **37,** 779–785.

Boynton, R. M., & Gordon, J. Bezold-Brücke hue shift measured by color-naming technique. *Journal of the Optical Society of America,* 1965, **55,** 78–86.

Boynton, R. M., & Kaiser, P. K. Vision: The additivity law made to work for heterochromatic photometry with bipartite fields. *Science,* 1968, **161,** 366–368.

Boynton, R. M., Schafer, W., & Neun, M. E. Hue-wavelength relation measured by color-naming method for 3 retinal locations. *Science,* 1964, **146,** 666–668.

Brentano, F. *Psychologie vom empirischen Standpunkte.* Leipzig: Duncker & Humblot, 1874.

Breton, P. Mesure des sensations lumineuses, en fonction des quantités de lumière. *Comptes Rendus de l'Académie des Sciences (Paris),* 1887, **103,** 426–429.

Brindley, G. S. The summation areas of human color-receptive mechanisms at increment threshold. *Journal of Physiology,* 1954, **124,** 400–408.

Brindley, G. S. *Physiology of the retina and visual pathway.* London: Arnold, 1960.

Broca, A. Etudes théoretiques et expérimentales sur les sensations visuelles et la photométrie. *Journal de Physique et le Radium,* 1894, **3,** 206–218.

Broca, A., & Sulzer, D. La sensation lumineuse en fonction du temps. *Comptes Rendus de l'Académie des Sciences (Paris)*, 1902, **134**, 831–834. (a)

Broca, A., & Sulzer, D. La sensation lumineuse en fonction du temps. *Comptes Rendus de l'Académie des Sciences (Paris)*, 1902, **137**, 944–946, 977–979, 1046–1049. (b)

Brown, J. H. Magnitude estimation of angular velocity during passive rotation. *Journal of Experimental Psychology*, 1966, **72**, 169–172.

Brown, J. H. Cross-modal estimation of angular velocity. *Perception & Psychophysics*, 1968, **3**, 115–117.

Bryngdahl, O. Characteristics of the visual system: Psychophysical measurement of the response to spatial sine-wave stimuli in the mesopic region. *Journal of the Optical Society of America*, 1964, **54**, 1152–1160.

Bryngdahl, O. Characteristics of the visual system: Psychophysical measurements of the response to spatial sine-wave stimuli in the photopic region. *Journal of the Optical Society of America*, 1966, **56**, 811–821.

Bujas, Z. La mesure de la sensibilité différentielle dans le domaine gustatif. *Acta Instituti Psychologici, Universitatis Zagrebensis*, 1937, **2**, 3–18.

Bujas, Z. L'adaptation gustative et son méchanisme. *Acta Instituti Psychologici, Universitatis Zagrebensis*, 1953, **No. 17**, 1–10.

Bujas, Z., & Ostojčić, A. L'évolution de la sensation gustative en fonction du temps d'excitation. *Acta Instituti Psychologici, Universitatis Zagrebensis*, 1939, **3**, 3–24.

Bujas, Z., Ostojčić, A. La sensibilité gustative en fonction de la surface excitée. *Acta Instituti Psychologici, Universitatis Zagrebensis*, 1941, **No. 13**, 1–20.

Cabanac, M. Plaisir ou déplaisir de la sensation thermique et homeothermie. *Physiology and Behavior*, 1969, **4**, 359–364.

Cabanac, M. Physiological role of pleasure. *Science*, 1971, **173**, 1103–1107.

Cabanac, M., Minaire, Y., & Adair, E. R. Influence of internal factors on the pleasantness of a gustative sweet sensation. *Communications in Behavioral Biology*, 1968, **1A**, 77–82.

Cain, W. S. An analysis of the odor properties of the linear aliphatic alcohols by direct psychophysical methods. Masters thesis, Brown University, 1966.

Cain, W. S. Odor intensity: Differences in the exponent of the psychophysical function. *Perception & Psychophysics*, 1969, **6**, 349–354.

Cain, W. S. Odor intensity after self-adaptation and cross-adaptation. *Perception & Psychophysics*, 1970, **7**, 271–275.

Cain, W. S. Physicochemical characteristics and supraliminal odor intensity: Reply to Mitchell. *Perception & Psychophysics*, 1971, **9**, 478–479.

Cain, W. S., and Engen, T. Olfactory adaptation and the scaling of odor intensity. In C. Pfaffmann (Ed.), *Olfaction and taste III*. New York: Rockefeller Univ. Press, 1969. Pp. 127–141.

Cain, W. S., & Stevens, J. C. Effort in sustained and phasic handgrip contractions. *American Journal of Psychology*, 1971, **84**, 52–65.

Cameron, A. T. The taste sense and the relative sweetness of sugars and other sweet substances. *Sugar Research Foundation, Scientific Report Series, No. 9*, 1947.

Carterette, E. C., Friedman, M. P., & Lovell, J. D. Mach bands in hearing. *Journal of the Acoustical Society of America*, 1969, **45**, 986–998.

Carvellas, T., & Schneider, B. Direct estimation of multidimensional tonal dissimilarity. *Journal of the Acoustical Society of America*, 1972, **51**, 1839–1848.

Causée, R., & Chavasse, P. Recherches sur le seuil de l'audition binauriculaire comparé au seuil monauriculaire. *Comptes Rendus de la Societé de Biologie (Paris)*, 1941, **85**, 1272–1275.

Causée, R., & Chavasse, R. Différences entre le seuil de l'audition binauriculaire et le seuil monauriculaire en fonction de la fréquence. *Comptes Rendus de la Societé de Biologie (Paris)*, 1942, **86**, 301–302.

Chocholle, R. Les effets des interactions interaurales dans l'audition. *Journal de Psychologie Normale et Pathologique*, 1962, **59**, 255–282.

Chocholle, R., & Greenbaum, H. B. La sonie de sons purs partiellement masqués. Etude comparative par une méthode d'égalisation et par la méthode des temps de réaction. *Journal de Psychologie Normale et Pathologique*, 1966, **63**, 387–414.

Churcher, B. G. A loudness scale for industrial noise measurement. *Journal of the Acoustical Society of America*, 1935, **6**, 216–226.

Clark, B., & Stewart, J. D. Magnitude estimates of rotational velocity during and following prolonged increasing, constant, and zero angular acceleration. *Journal of Experimental Psychology*, 1968, **78**, 329–339.

Cobb, P. W. Weber's law and the Fechnerian muddle. *Psychological Review*, 1932, **39**, 533–551.

Cohen, J., & Little, W., IV. Neural model for the Stevens power law. *Perception & Psychophysics*, 1971, **10**, 269–270.

Comrey, A. L. A proposed method for absolute ratio scaling. *Psychometrika*, 1950, **15**, 317–325.

Conklin, H. C. Hanunóo color categories. *Southwestern Journal of Anthropology*, 1955, **11**, 339–344.

Cornsweet, T. N. Changes in the appearance of stimuli of very high luminance. *Psychological Review*, 1962, **69**, 257–273.

Corso, J. F. A theoretico-historical review of the threshold concept. *Psychological Bulletin*, 1963, **60**, 356–370.

Cowdrick, M. The Weber-Fechner law and Sanford's weight experiment. *American Journal of Psychology*, 1917, **28**, 585–588.

Craig, J. C. Vibrotactile loudness addition. *Perception & Psychophysics*, 1966, **1**, 185–190.

Craig, J. C., & Sherrick, C. E. The role of skin coupling in the determination of vibrotactile spatial summation. *Perception & Psychophysics*, 1969, **6**, 97–101.

Craik, K. J. W. The effect of adaptation on subjective brightness. *Proceedings of the Royal Society (London)*, 1940, **128B**, 232–247.

Crocker, E. C., & Henderson, L. F. Analysis and classification of odors. *American Perfumery*, 1927, **22**, 325–327, 356.

Cross, D. W. An application of mean value theory to psychophysical measurement. Unpublished manuscript, University of Michigan, 1965.

Curtis, D. W. Magnitude estimations and category judgments of brightness and brightness intervals: A two-stage interpretation. *Journal of Experimental Psychology*, 1970, **83**, 201–208.

Curtis, D. W., Attneave, F., & Harrington, T. L. A test of a two-stage model for magnitude estimation. *Perception & Psychophysics*, 1968, **3**, 25–31.

Curtis, D. W., & Fox, B. E. Direct quantitative judgments of sums and a two-stage model for psychophysical judgments. *Perception & Psychophysics*, 1969, **5**, 89–93.

Curtis, D. W., & Rule, S. J. Magnitude judgments of brightness and brightness difference as a function of background reflectance. *Journal of Experimental Psychology*, 1972, **95**, 215–222.

Davies, J. T., & Taylor, F. H. The role of absorption and molecular morphology in olfaction: The calculation of olfactory thresholds. *Biological Bulletin*, 1959, **117**, 222–238.

Dawson, W. E. Magnitude estimation of apparent sums and differences. *Perception & Psychophysics*, 1971, **9**, 368–374.

Decker, T. A. The effect of the level of retinal illumination upon the foveal spectral luminosity function. Doctoral dissertation, Brown University, 1967.

de Lange, H., dzn. Eye's response at flicker fusion to square-wave modulation of a test field surrounded by a large steady field of equal mean luminance. *Journal of the Optical Society of America*, 1961, **51**, 415–421.

Delboeuf, J. R. L. Etude psychophysique: Recherches théoretiques et expérimentales sur la mesure des sensations, et spécialement des sensations de lumière et de fatigue. *Mémoires de l'Académie Royale de Belgique*, 1873, **23**, No. 5.

de Vries, H., & Stuiver, M. The absolute sensitivity of the human sense of smell. In W. A. Rosenblith (Ed.), *Sensory communication*. New York: Wiley, 1961. Pp. 159–167.

Diamond, A. L. A theory of depression and enhancement in the brightness response. *Psychological Review*, 1960, **67**, 168–199.

Diamond, A. L. Brightness of a field as a function of its size. *Journal of the Optical Society of America*, 1962, **52**, 700–706.

Docq, A.-J. Recherches physico-physiologiques sur la fonction collective des deux organes de l'appareil auditif. *Mémoires Couronnés de l'Académie Royale de Belgique*, 1870, **34**, 1–39.

Døving, K. B. An electrophysiological study of odour stimilarities of homologous substances. *Journal of Physiology*, 1966, **186**, 97–109.

Drake, B., Johannson, B., von Sydow, E., & Døving, K. B. Quantitative psychophysical and electrophysiological data on some odorous compounds. *Scandinavian Journal of Psychology*, 1969, **10**, 89–96.

Dravnieks, A., & Laffort, P. Odor intensity function in adult humans: Physicochemical correlates of slope for single odorants. Manuscript, IIT Research Institute, 1970.

Duclaux, R., & Cabanac, M. Effets d'une ingestion de glucose sur la sensation et la perception d'un stimulus olfactif, alimentaire. *Comptes Rendus de l'Académie des Sciences (Paris)*, 1970, **270**, 1006–1009.

Durup, G., & Piéron, H. Recherches au sujet de l'interpretation du phénomème de Purkinje par des différences dans les courbes de sensation des recepteurs chromatiques. *L'Année Psychologique*, 1933, **33**, 57–83.

Ebbinghaus, H. *Grundzüge der Psychologie*. Leipzig: Verlag von Veit, 1902.

Egan, J. P. Perstimulatory fatigue as measured by heterophonic loudness balance. *Journal of the Acoustical Society of America*, 1955, **27**, 111–120.

Egan, J. P., & Thwing, E. J. Further studies on perstimulatory fatigue. *Journal of the Acoustical Society of America*, 1955, **27**, 1225–1226.

Eisler, H. Empirical test of a model relating magnitude and category scales. *Scandinavian Journal of Psychology*, 1962, **3**, 88–96. (a)

Eisler, H. On the problem of category scales in psychophysics. *Scandinavian Journal of Psychology*, 1962, **3**, 81–87. (b)

Eisler, H. How prothetic is the continuum of smell? *Scandinavian Journal of Psychology*, 1963, **4**, 29–32. (a)

Eisler, H. Magnitude scales, category scales, and Fechnerian integration. *Psychological Review* 1963, **70**, 243–253. (b)

Eisler, H. The ceiling of psychophysical power functions. *American Journal of Psychology*, 1965, **78**, 506–509.

Eisler, H., & Ekman, G. A mechanism of subjective similarity. *Acta Psychologica*, 1959, **16**, 1–10.

Ekman, G. Subjective power functions and the method of fractionation. *Report from the Psychological Laboratory, University of Stockholm*, 1956, No. 34.

Ekman, G. Methodological note on scales of gustatory intensity. *Scandinavian Journal of Psychology*, 1961, **2**, 185–190.

Ekman, G. A direct method for multidimensional ratio scaling. *Psychometrika*, 1963, **28**, 33–41.

Ekman, G. Is the power law a special case of Fechner's law? *Perceptual and Motor Skills*, 1964, **19**, 730.

Ekman, G. Temporal integration of brightness. *Vision Research*, 1966, **6**, 683–688.

Ekman, G. Comparative studies on multidimensional scaling and related techniques. *Report from the Psychological Laboratory, University of Stockholm*, 1970, Supplement 3.

Ekman, G., & Åkesson, C. Saltness, sweetness, and preference. *Scandinavian Journal of Psychology*, 1965, **6**, 241–253.

Ekman, G., Berglund, B., & Berglund, U. Loudness as a function of the duration of auditory stimulation. *Scandinavian Journal of Psychology*, 1966, **7**, 201–208.

Ekman, G., Berglund, B., Berglund, U., & Lindvall, T. Perceived intensity of odor as a function of time of adaptation. *Scandinavian Journal of Psychology*, 1967, **8**, 177–186.

Ekman, G., Eisler, H., & Künnapas, T. Brightness of monochromatic light as measured by the method of magnitude production. *Acta Psychologica*, 1960, **17**, 392–397. (a)

Ekman, G., Eisler, H., & Künnapas, T. Brightness scales for monochromatic light. *Scandinavian Journal of Psychology*, 1960, **1**, 41–48. (b)

Ekman, G., Engen, T., Künnapas, T., & Lindman, R. A quantitative principle of qualitative similarity. *Journal of Experimental Psychology*, 1964, **68**, 530–534.

Ekman, G., Frankenhaeuser, M., Levander, S., & Mellis, I. Scales of unpleasantness of electrical stimulation. *Scandinavian Journal of Psychology*, 1964, **5**, 257–261.

Ekman, G., Frankenhaeuser, M., Levander, S., & Mellis, I. The influence of intensity and duration of electrical stimulation on subjective variables. *Scandinavian Journal of Psychology*, 1966, **7**, 58–64.

Ekman, G., Fröberg, J., & Frankenhaeuser, M. Temperal integration of perceptual response to supraliminal electrical stimulation. *Scandinavian Journal of Psychology*, 1968, **9**, 83–88.

Ekman, G., Goude, G., & Waern, Y. Subjective similarity in two perceptual continua. *Journal of Experimental Psychology*, 1961, **61**, 222–227.

Ekman, G., & Gustaffson, U. Threshold values and the psychophysical function in brightness vision. *Vision Research*, 1968, **8**, 747–758.

Ekman, G., Hosman, B., Lindman, R., Ljungberg, L., & Åkesson, C. A. Interindividual differences in scaling performance. *Perceptual and Motor Skills*, 1968, **26**, 815–823.

Ekman, G., Hosman, J., & Berglund, U. Perceived brightness as a function of duration of dark-adaptation. *Perceptual and Motor Skills*, 1966, **23**, 931–943.

Ekman, G., & Künnapas, T. Ratio scales and category scales for brightness of monochromatic light. *Report from the Psychological Laboratory, University of Stockholm*, 1960, No. 86.

Ekman, G., & Künnapas, T. Brightness of monochromatic light in scotopic and photopic vision. *Journal of Psychology*, 1962, **53**, 319–327.

Engel, G. R. The autocorrelation function and binocular brightness mixing. *Vision Research*, 1969, **9**, 1111–1130.

Engen, T. An evaluation of a method for developing ratio-scales. *American Journal of Psychology*, 1956, **69**, 92–95.

Engen, T. The psychophysical similarity of the odors of aliphatic alcohols. *Report from the Psychological Laboratory, University of Stockholm,* 1962, No. 127.

Engen, T. Psychophysical scaling of odor intensity and quality. *Annals of the New York Academy of Sciences,* 1964, **116,** 504–516.

Engen, T. Psychophysical analysis of the odor intensity of homologous alcohols. *Journal of Experimental Psychology,* 1965, **70,** 611–616.

Engen, T., Cain, W. S., & Rovee, C. K. Direct scaling of olfaction in the newborn infant and the adult human observer. In N. Tanyolaç (Ed.) , *Theories of odors and odor measurement.* Istanbul: Robert College, 1968, Pp. 271–294.

Engen, T., & Levy, N. The influence of standards on psychophysical judgments. *Perceptual and Motor Skills,* 1955, **5,** 193–197.

Engen, T., & Levy, N. The influence of context on constant-sum loudness-judgments. *American Journal of Psychology,* 1958, **71,** 731–736.

Engen, T., & Lindström, C.-O. Psychophysical scales of the odor intensity of amyl acetate. *Scandinavian Journal of Psychology,* 1963, **4,** 23–28.

Engen, T., & McBurney, D. H. Magnitude and category scales of the pleasantness of odors. *Journal of Experimental Psychology,* 1964, **68,** 435–440.

Engen, T., & Ross, B. M. Effect of reference number on magnitude estimation. *Perception & Psychophysics,* 1966, **1,** 74–76.

Engen, T., & Tulunay, Ü. Some sources of error in half-heaviness judgments. *Journal of Experimental Psychology,* 1957, **54,** 208–212.

Eskildsen, P. R. Rate of change constancy as a method for psychophysical scaling. Doctoral dissertation, University of Oregon, 1966.

Eyman, R. K., & Kim, P. J. A model for partitioning judgment error in psychophysics. *Psychological Bulletin,* 1970, **74,** 35–46.

Fabian, F. W., & Blum, H. B. Relative taste intensity of some basic food constituents and their competitive and compensatory action. *Food Research,* 1943, **8,** 179–193.

Fagot, R. F. On the psychophysical law and estimation procedures in psychophysical scaling. *Psychometrika,* 1963, **28,** 145–160.

Fagot, R. F. Alternative power laws for ratio scaling. *Psychometrika,* 1966, **31,** 201–214.

Fagot, R. F., & Stewart, M. R. Test of a response bias model of bisection. *Perception & Psychophysics,* 1970, **7,** 257–262.

Falmagne, J. C. The generalized Fechner problem and discrimination. *Journal of Mathematical Psychology,* 1971, **8,** 22–43.

Feallock, J. B. Estimated magnitudes of taste under improved conditions of stimulus control. Doctoral dissertation, University of Virginia, 1965.

Fechner, G. T. *Elemente der Psychophysik.* Leipzig: Breitkopf und Härtel, 1860.

Fletcher, H. Auditory patterns. *Review of Modern Physics,* 1940, **12,** 47–65.

Fletcher, H., & Munson, W. A. Loudness, its definition, measurement and calculation. *Journal of the Acoustical Society of America,* 1933, **5,** 82–108.

Franzén, O. The dependence of vibrotactile threshold and magnitude functions on stimulation frequency and signal level. *Scandinavian Journal of Psychology,* 1969, **10,** 289–298. (a)

Franzén, O. On spatial summation in the tactual sense. *Scandinavian Journal of Psychology,* 1969, **10,** 193–208. (b)

Fraser, W. D., Petty, J. W., & Elliott, D. N. Adaptation: Central or peripheral? *Journal of the Acoustical Society of America,* 1970, **47,** 1016–1021.

Frederiksen, N., & Gulliksen, H. (Eds.) *Contributions to mathematical psychology.* New York: Holt, 1964.

Freeman, R. F. Contrast interpretation of brightness constancy. *Psychological Bulletin,* 1967, **67,** 165–187.

Fullerton, G. S., & Cattell, J. McK. *On the perception of small differences.* Philadelphia: Univ. of Pennsylvania Press, 1892.

Fuortes, M. G. F., & Hodgkin, A. L. Changes in time scale and sensitivity in the ommatidia of *Limulus. Journal of Physiology,* 1964, **172,** 239–263.

Gage, F. H. An experimental investigation of the measurability of auditory sensation. *Proceedings of the Royal Society (London),* 1934, **116B,** 103–122. (a)

Gage, F. H. An experimental investigation of the measurability of visual sensation. *Proceedings of the Royal Society (London),* 1934, **116B,** 123–138. (b)

Galanter, E. Contemporary psychophysics. In *New directions in psychology.* New York: Holt, 1962. Pp. 87–156.

Galanter, E., & Messick, S. The relation between category and magnitude scales of loudness. *Psychological Review,* 1961, **68,** 363–372.

Galifret, Y. Les psychophysiques de la saturation chromatique. *L'Année Psychologique,* 1959, **59,** 35–46.

Garner, W. R. The loudness of repeated short tones. *Journal of the Acoustical Society of America,* 1948, **20,** 513–527.

Garner, W. R. An equal discriminability scale for loudness judgments. *Journal of Experimental Psychology,* 1952, **43,** 232–238.

Garner, W. R. Context effects and the validity of loudness scales. *Journal of Experimental Psychology,* 1954, **48,** 218–224. (a)

Garnerf, W. R. A technique and a scale for loudness measurement. *Journal of the Acoustical Society of America,* 1954, **26,** 73–88. (b)

Garner, W. R. Advantages of the discriminability criterion for a loudness scale. *Journal of the Acoustical Society of America,* 1958, **30,** 1005–1012.

Garner, W. R. On the lambda loudness function, masking, and the loudness of multi-component tones. *Journal of the Acoustical Society of America,* 1959, **31,** 602–607.

Garner, W. R., & Hake, H. W. The amount of information in absolute judgments. *Psychological Review,* 1951, **58,** 446–459.

Gässler, G. Über die Hörschwelle für Schallereignisse mit verschieden breitem Frequenzspektrum. *Acustica,* 1954, **4,** 408–414.

Geblewicz, E. La sommation spatiale des excitations thermiques. *L' Année Psychologique,* 1938, **39,** 199–217.

Geiger, P. H., & Firestone, F. A. The estimation of fractional loudness. *Journal of the Acoustical Society of America,* 1933, **5,** 25–30.

Geldard, F. A. The measurement of retinal fatigue to achromatic stimulation. II. *Journal of General Psychology,* 1928, **1,** 578–590.

Gescheider, G. A., & Wright, J. H. Effects of sensory adaptation on the form of the psychophysical magnitude function for cutaneous vibration. *Journal of Experimental Psychology,* 1968, **77,** 308–313.

Gescheider, G. A., & Wright, J. H. Effects of vibrotactile adaptation on the perception of stimuli of varied intensity. *Journal of Expermental Psychology,* 1969, **81,** 449–453.

Glasser, L. G., McKenney, A. H., Reilly, A. H., & Schnelle, P. D. Cube-root color coordinate system. *Journal of the Optical Society of America,* 1958, **48,** 736–740.

Goff, G. D. Differential discrimination of frequency of cutaneous mechanical vibration. *Journal of Experimental Psychology,* 1967, **74,** 294–299.

Graham, C. H., Brown, R. H., & Mote, F. A., Jr. The relation of size of stimulus and intensity in the human eye. I. Intensity thresholds for white light. *Journal of Experimental Psychology,* 1939, **24,** 555–573.

Graham, C. H., & Kemp, E. H. Brightness discrimination as a function of the duration of the increment in intensity. *Journal of General Physiology*, 1938, 21, 635–650.

Graham, C. H., & Margaria, R. Area and the intensity-time relation in the peripheral retina. *American Journal of Physiology*, 1935, 113, 299–305.

Green, D. G. Sinusoidal flicker characteristics of the color-sensitive mechanisms of the eye. *Vision Research*, 1969, 9, 591–602.

Greene, L. C., & Hardy, J. D. Spatial summation of pain. *Journal of Applied Physiology*, 1958, 13, 457–464.

Greene, L. C., & Hardy, J. D. Adaptation to thermal pain in the skin. *Journal of Applied Physiology*, 1962, 17, 693–696.

Greenwood, D. D. Auditory masking and the critical band. *Journal of the Acoustical Society of America*, 1961, 33, 484–502.

Gregson, R.-A. M. Representation of taste mixture cross-modal matching on a Minkowski R-metric. *Australian Journal of Psychology*, 1965, 17, 195–204.

Gregson, R. A. M. Theoretical and empirical multidimensional scalings of taste mixture matchings. *British Journal of Mathematical and Statistical Psychology*, 1966, 19, 59–75.

Griffin, D. R., Hubbard, R., & Wald, G. The sensitivity of human rod and cone vision to infrared radiation. *Journal of the Optical Society of America*, 1947, 37, 546–554.

Guilford, J. P. A generalized psychophysical law. *Psychological Review*, 1932, 39, 73–85.

Guilford, J. P. System in the relationship of affective value to frequency and intensity of auditory stimuli. *American Journal of Psychology*, 1954, 67, 691–695.

Guilford, J. P., & Dingman, H. F. A validation study of ratio-judgment methods. *American Journal of Psychology*, 1954, 67, 395–410.

Guillot, M. Anosmies partielles et odeurs fondamentales. *Comptes Rendus de l'Académie des Sciences (Paris)*, 1948, 226, 1307–1309.

Guirao, M., & Stevens, S. S. Measurement of auditory density. *Journal of the Acoustical Society of America* 1964, 36, 1176–1182.

Guth, S. L., Donley, N. J., & Marrocco, R. T. On luminance additivity and related topics. *Vision Research*, 1969, 9, 537–576.

Haber, R. N. Discrepancy from adaptation level as a source of affect. *Journal of Experimental Psychology*, 1958, 56, 370–375.

Hahn, J. F. The unfinished chapter. In G. R. Hawkes (Ed.), *Symposium on cutaneous sensitivity*. Fort Knox: U. S. Army Medical Research Laboratory, 1960. Pp. 131–142.

Hahn, J. F. Vibrotactile adaptation and recovery measured by two methods. *Journal of Experimental Psychology*, 1966, 71, 655–658.

Ham, L. B., & Parkinson, J. S. Loudness and intensity relations. *Journal of the Acoustical Society of America*, 1932, 3, 511–534.

Hamilton, P. M. Noise masked thresholds as a function of tonal duration and masking noise bandwidth. *Journal of the Acoustical Society of America*, 1957, 29, 506–511.

Hanes, R. M. The construction of subjective brightness scales from fractionation data: A validation. *Journal of Experimental Psychology*, 1949, 39, 719–728. (a)

Hanes, R. M. A scale of subjective brightness. *Journal of Experimental Psychology*, 1949, 39, 438–452. (b)

Hanes, R. M. Suprathreshold area brightness relationships. *Journal of the Optical Society of America*, 1951, 41, 28–31.

Hardy, J. D. The nature of pain. *Journal of Chronic Diseases*, 1956, 4, 22–51.

Hardy, J. D., & Oppel, T. W. Studies on temperature sensation. III. The sensitivity of the body to heat and the spatial summation of the end organ responses. *Journal of Clinical Investigation*, 1937, 16, 533–540.

Hardy, J. D., & Oppel, T. W. Studies in temperature sensation. IV. The stimulation of cold sensation by radiation. *Journal of Clinical Investigation*, 1938, **17**, 771–778.

Hardy, J. D., Wolff, H. G., & Goodell, H. Studies on pain: A new method for measuring pain threshold: Observations on spatial summation of pain. *Journal of Clinical Investigation*, 1940, **19**, 649–657.

Hardy, J. D., Wolff, H. G., & Goodell, H. Studies on pain: Discrimination of differences in pain as a basis of a scale of pain intensity. *Journal of Clinical Investigation*, 1947, **26**, 1152–1158.

Harlow, H. F., Harlow, M. K., & Suomi, S. J. From thought to therapy: Lessons from a primate laboratory. *American Scientist*, 1971, **59**, 538–549.

Harper, R., Bate-Smith, E. C., Land, D. G., & Griffiths, N. M. A glossary of odour stimuli and their qualities. *Perfumery and Essential Oil Record*, 1968, **59**, 22–37.

Harper, R. S., & Stevens, S. S. A psychological scale of weight and a formula for its derivation. *American Journal of Psychology*, 1948, **61**, 343–351.

Harris, J. D., & Pikler, A. G. The stability of a standard of loudness as measured by compensatory tracking. *American Journal of Psychology*, 1960, **73**, 573–580.

Hartline, H. K. Inhibition of activity of visual receptors by illuminating nearby retinal areas in the Limulus eye. *Federation Proceedings*, 1949, **8**, 69.

Hartline, H. K., & Graham, C. H. Nerve impulses from single receptors in the eye. *Journal of Cellular and Comparative Physiology*, 1932, **1**, 277–295.

Hartshorne, C. *The philosophy and psychology of sensation*. Chicago, Illinois: Univ. of Chicago Press, 1934.

Heinemann, E. G. Simultaneous brightness induction as a function of inducing- and test-field luminances. *Journal of Experimental Psychology*, 1955, **50**, 89–96.

Heinemann, E. G. The relation of apparent brightness to the threshold for differences in luminance. *Journal of Experimental Psychology*, 1961, **61**, 389–399.

Hellman, R. P. Effect of noise bandwidth on the loudness of a 1000-Hz tone. *Journal of the Acoustical Society of America*, 1970, **48**, 500–504.

Hellman, R. P. Asymmetry of masking between noise and tone. *Perception & Psychophysics*, 1972, **11**, 241–246.

Hellman, R. P., & Zwislocki, J. J. Some factors affecting the estimation of loudness. *Journal of the Acoustical Society of America*, 1961, **33**, 687–694.

Hellman, R. P., & Zwislocki, J. J. Monaural loudness function at 1000 cps and interaural summation. *Journal of the Acoustical Society of America*, 1963, **35**, 856–865.

Hellman, R. P., & Zwislocki, J. J. Loudness function of a 1000-cps tone in the presence of a masking noise. *Journal of the Acoustical Society of America*, 1964, **36**, 1618–1627.

Hellman, R. P., & Zwislocki, J. J. Loudness determination at low sound frequencies. *Journal of the Acoustical Society of America*, 1968, **43**, 60–64.

Helm, C. E. Multidimensional ratio scaling analysis of perceived color relations. *Journal of the Optical Society of America*, 1964, **54**, 256–262.

Helm, C. E., Messick, S., & Tucker, L. R. Psychological models for relating discrimination and magnitude estimation scales. *Psychological Review*, 1961, **68**, 167–177.

Helmholtz, H. von. *Handbuch der physiologischen Optik*. Leipzig: Voss, 1866.

Helson, H., & Lansford, T. The role of spectral energy of source and background color in pleasantness of object colors. *Applied Optics*, 1970, **9**, 1513–1562.

Henion, K. E. Direct psychophysical scaling of the odor intensity of undiluted n-aliphatic alcohols. *Journal of Experimental Psychology*, 1970, **85**, 300–304. (a)

Henion, K. E. Psychophysical scales of the odor oiliness of homologous alcohols. *Perception & Psychophysics*, 1970, **7**, 351–353. (b)

Henion, K. E. Direct psychophysical scaling of the olfactory pleasantness of diluted n-amyl acetate. *Perception & Psychophysics*, 1971, **10**, 158–160. (a)

Henion, K. E. Odor pleasantness and intensity: A single dimension? *Journal of Experimental Psychology*, 1971, **90**, 275–279. (b)

Henion, K. E. Olfactory intensity of diluted n-alphatic alcohols. *Psychonomic Science*, 1971, **22**, 213–214. (c)

Henion, K. E. Psychophysical scales of the olfactory pleasantness of homologous alcohols. *Perception & Psychophysics*, 1971, **9**, 234–236. (d)

Henning, H. *Der Geruch*. Leipzig: Barth, 1916. (a)

Henning, H. Die Qualitätenreihe des Geschmacks. *Zeitschrift für Psychologie*, 1916, **74**, 203–219. (b)

Hensel, H., & Kenshalo, D. R. Warm receptors in the nasal region of cats. *Journal of Physiology*, 1969, **204**, 99–112.

Herget, C. M., Granath, L. P., & Hardy, J. D. Warmth sense in relation to the area of skin stimulated. *American Journal of Physiology*, 1941, **135**, 20–26.

Hering, E. *Zur Lehre vom Lichtsinne*. Vienna: C. Gerold's Sohn, 1878.

Herrick, R. M. Foveal luminance discrimination as a function of the duration of the decrement or increment in luminance. *Journal of Comparative and Physiological Psychology*, 1956, **49**, 437–443.

Holladay, L. L. The fundamentals of glare and visibility. *Journal of the Optical Society of America*, 1926, **12**, 271–319.

Hollingworth, H. L. The inaccuracy of movement. *Archives of Psychology*, 1909, No. 13.

Hood, J. D. Studies in auditory fatigue and adaptation. *Acta Oto-Laryngologica*, Supplement, 1950, **92**, 1–57.

Horeman, H. W. Inductive brightness depression as influenced by configurational conditions. *Vision Research*, 1963, **3**, 121–135.

Horeman, H. W. Relations between brightness and luminance under induction. *Vision Research*, 1965, **5**, 331–340.

Houston, R. A. New observations on the Weber-Fechner law. *Report on a discussion in vision*. London: The Physical Society, 1932. Pp. 167–181.

Houtsma, A. J. M., & Goldstein, J. L. The central origin of the pitch of complex tones: Evidence from musical interval recognition. *Journal of the Acoustical Society of America*, 1972, **51**, 520–529.

Howes, D. H. The loudness of multicomponent tones. *American Journal of Psychology*, 1950, **63**, 1–30.

Hughes, J. W. The monaural threshold: Effect of a subliminal contralateral stimulus. *Proceedings of the Royal Society (London)*, 1938, **124B**, 406–420.

Hurvich, L. M., & Jameson, D. An opponent-process theory of color vision. *Psychological Review*, 1957, **64**, 384–404.

Indow, T. Saturation scales for red. *Vision Research*, 1967, **7**, 481–495.

Indow, T., & Kanazawa, K. Multidimensional mapping of Munsell colors varying in hue, chroma, and value. *Journal of Experimental Psychology*, 1960, **59**, 330–336.

Indow, T., & Stevens, S. S. Scaling of saturation and hue. *Perception & Psychophysics*, 1966, **1**, 253–271.

Indow, T., & Uchizono, T. Multidimensional mapping of Munsell colors varying in hue and chroma. *Journal of Experimental Psychology*, 1960, **59**, 321–329.

Irwin, R. J., & Corballis, M. C. On the general form of Stevens' law for loudness and softness. *Perception & Psychophysics*, 1968, **3**, 137–143.

Irwin, R. J., & Zwislocki, J. J. Loudness effects in pairs of tone bursts. *Perception & Psychophysics*, 1971, **10**, 189–192.

Ishak, I. G. H., Bouma, H., & van Bussel, H. J. J. Subjective estimates of colour attributes for surface colours. *Vision Research*, 1970, **10**, 489–500.

Jacobs, G. H. Saturation estimates and chromatic adaptation. *Perception & Psychophysics*, 1967, **2**, 271–274.

Jacobs, G. H., & Gaylord, H. A. Effects of chromatic adaptation on color naming. *Vision Research*, 1967, **7**, 643–654.

James, W. *Principles of psychology*. New York: Holt, 1890.

James, W. *Psychology: Briefer course*. New York: Holt, 1892.

Jameson, D. Threshold and supra-threshold relations in vision. In *Proceedings of the International Colour Meeting: Lucerne*, Vol. 1. Göttingen: Musterschmidt-Verlag, 1965. Pp. 128–136.

Jameson, D., & Hurvich, L. Perceived color and its dependence on focal, surrounding, and preceding stimulus variables. *Journal of the Optical Society of America*, 1959, **49**, 890–898.

Jameson, D., & Hurvich, L. Theory of brightness and color contrast in human vision. *Vision Research*, 1964, **4**, 135–154.

Jeddi, E. Confort du contact et thermoregulation comportementale. *Physiology and Behavior*, 1970, **5**, 1487–1493.

Johnson, D. M. The central tendency of judgment as a regression phenomenon. *American Psychologist*, 1952, **7**, 281.

Jones, F. N. Scales of subjective intensity for odors of diverse chemical nature. *American Journal of Psychology*, 1958, **71**, 305–310. (a)

Jones, F. N. Subjective scales of intensity for three odors. *American Journal of Psychology*, 1958, **71**, 423–425. (b)

Jones, F. N., & Marcus, M. J. The subject effect in judgments of subjective magnitude. *Journal of Experimental Psychology*, 1961, **61**, 40–44.

Jones, F. N., & Woskow, M. H. On the relationship between estimates of loudness and pitch. *American Journal of Psychology*, 1962, **75**, 669–671.

Jones, F. N., & Woskow, M. H. On the intensity of odor mixtures. *Annals of the New York Academy of Sciences*, 1964, **116**, 484–494.

Jones, F. N., & Woskow, M. H. Some effects of context on the slope in magnitude estimation. *Journal of Experimental Psychology*, 1966, **71**, 177–180.

Judd, D. B. Chromaticity sensibility to stimulus differences. *Journal of the Optical Society of America*, 1932, **22**, 72–108.

Judd, D. B. Report of the U. S. Secretariat Commission on Colorimetry and Artificial Daylight. *CIE Proceedings*, 1951, **Part 7.**

Junge, K. The Garner-Attneave theory of ratio scaling. *Scandinavian Journal of Psychology*, 1967, **8**, 7–10.

Kaiser, P. Color names of very small fields varying in duration and luminance. *Journal of the Optical Society of America*, 1968, **58**, 849–852.

Kamen, J. M., Pilgrim, F. J., Guttman, N. J., & Kroll, B. J. Interactions of suprathreshold taste stimuli. *Journal of Experimental Psychology*, 1961, **62**, 348–356.

Kameoka, A., & Kuriyagawa, M. Consonance theory part I: Consonance of dyads. *Journal of the Acoustical Society of America*, 1969, **45**, 1451–1459. (a)

Kameoka, A., & Kuriyagawa, M. Consonance theory part II: Consonance of complex tones and its calculation method. *Journal of the Acoustical Society of America*, 1969, **45**, 1460–1469. (b)

Karn, H. W. Area and the intensity-time relation in the fovea. *Journal of General Psychology*, 1936, **14**, 360–369.

Katz, M. S. Brief flash brightness. *Vision Research*, 1964, **4**, 361–373.

Keidel, W. D., & Spreng, M. Neurophysiological evidence for the Stevens power function in man. *Journal of the Acoustical Society of America*, 1965, **38**, 191–195.

Keller, M. The relation between the critical duration and intensity in brightness discrimination. *Journal of Experimental Psychology*, 1941, **28**, 407–418.

Kelly, D. H. Visual responses to time-dependent stimuli. I. Amplitude sensitivity measurements. *Journal of the Optical Society of America*, 1961, **51**, 422–429. (a)

Kelly, D. H. Visual responses to time-dependent stimuli. II. Single-channel model of the photopic visual system. *Journal of the Optical Society of America*, 1961, **51**, 747–754. (b)

Kenshalo, D. R. Behavioral and electrophysiological responses of cats to thermal stimuli. In D. R. Kenshalo (Ed.), *The skin senses*. Springfield, Illinois: Thomas, 1968. Pp. 400–416.

Kenshalo, D. R., Decker, T., & Hamilton, A. Spatial summation on the forehead, forearm, and back produced by radiant and conducted heat. *Journal of Comparative and Physiological Psychology*, 1967, **63**, 510–515.

Keys, J. W. Binaural vs. monaural hearing. *Journal of the Acoustical Society of America*, 1947, **19**, 629–631.

Kiesow, F. Beiträge zur physiologischen Psychologie des Geschmackssinnes. *Philosophische Studien*, 1896, **12**, 255–278.

Kingsbury, B. A. A direct comparison of the loudness of pure tones. *Physical Review*, 1927, **29**, 588–600.

Kocher, E. C., & Fisher, G. L. Subjective intensity and taste preference. *Perceptual and Motor Skills*, 1969, **28**, 735–740.

Körber, C. A. *Versuch einer Ausmessung menschlicher Seelen und aller endlichen Dinge überhaupt, Wie solche der inneren Beschaffenheit derselbn gemäss ins Werck zu richten ist, Wenn man ihre Kräfte, Vermögen, und Würkungen recht will kennen lernen*. Halle: Hemmerde, 1746.

Krakauer, D., & Dallenbach, K. M. Gustatory adaptation to sweet, sour, and bitter. *American Journal of Psychology*, 1937, **49**, 469–475.

Krantz, D. Integration of just-noticeable differences. *Journal of Mathematical Psychology*, 1971, **8**, 591–599.

Krantz, D. H. A theory of magnitude estimation and cross-modality matching. *Journal of Mathematical Psychology*, 1972, **9**, 168–199.

Krüger, J. G. *Naturlehre*. Halle-Magdeburg: Hemmerde, 1743.

Kruger, L., Feldzamen, A. N., & Miles, W. R. Comparative olfactory intensities of the aliphatic alcohols in man. *American Journal of Psychology*, 1955, **68**, 386–395.

Künnapas, T. M. Scales for subjective distance. *Scandinavian Journal of Psychology*, 1960, **1**, 187–192.

Künnapas, T., & Künnapas, U. On the mechanism of subjective similarity for unidimensional continua. *Report from the Psychological Laboratory, University of Stockholm*, 1971, No. 342.

Ladefoged, P., & McKinney, N. P. Loudness, sound pressure, and subglottal pressure in speech. *Journal of the Acoustical Society of Amerca*, 1963, **35**, 454–460.

Laffort, P. The prediction of olfactory thresholds and slopes of odor intensity. Manuscript, IIT Research Institute, 1970.

Laird, D. A., Taylor, E., & Wille, H. H., Jr. The apparent reduction of loudness. *Journal of the Acoustical Society of America*, 1932, **3**, 393–401.

Lane, H. The autophonic scale of voice level for congenitally deaf subjects. *Journal of Experimental Psychology*, 1963, **66**, 328–331.

Lane, H. L., Catania, A. C., & Stevens, S. S. Voice level: Autophonic scale, perceived loudness, and effect of side tone. *Journal of the Acoustical Society of America*, 1961, **33**, 160–167.

Le Grand, Y. *Light, colour and vision*. London: Chapman & Hall, 1957.

Lemberger, F. Psychophysische Untersuchungen über den Geschmack von Zucker and Saccharin (Saccharose und Krystallose). *Pflügers Archiv für die gesamte Physiologie*, 1908, **123**, 293–311.

Levelt, W. J. M. *On binocular rivalry*. The Hague: Mouton, 1968.

Levelt, W. J. M., Riemersma, J. B., & Bunt, A. A. Binaural additivity of loudness. *British Journal of Mathematical and Statistical Psychology*, 1972, **25**, 51–68.

Lewis, D. R. Psychological scales of taste. *Journal of Psychology*, 1948, **26**, 437–446.

Lewis, M. F. Category judgments as functions of flash luminance and duration. *Journal of the Optical Society ofAmerica*, 1965, **55**, 1655–1660.

Lipetz, L. E. The relation of physiological and psychological aspects of sensory intensity. In W. R. Loewenstein (Ed.), *Handbook of sensory physiology*, Vol. I. Berlin and New York: Springer-Verlag, 1971. Pp. 191–225.

Lochner, J. P. A., & Burger, J. F. Form of the loudness function in the presence of masking noise. *Journal of the Acoustical Society of America*. 1961, **33**, 1705–1707.

Lochner, J. P. A., & Burger, J. F. Pure-tone loudness relations. *Journal of the Acoustical Society of America*, 1962, **34**, 576–581.

Luce, R. D. On the possible psychophysical laws. *Psychological Review*, 1959, **66**, 81–95.

Luce, R. D. What sort of measurement is psychophysical measurement? *American Psychologist*, 1972, **27**, 96–106.

Luce, R. D., & Edwards, W. The derivation of subjective scales from just noticeable differences. *Psychological Review*, 1958, **65**, 222–237.

Luce, R. D., & Galanter, E. Discrimination. In R. D. Luce, R. R. Bush, and E. Galanter (Eds.), *Handbook of mathematical psychology*, Vol. I. New York: Wiley, 1963. Pp. 191–243.

Luce, R. D., & Green, D. M. A neural timing theory for response times and the psychophysics of intensity. *Psychological Review*, 1972, **79**, 14–57.

Luce, R. D., & Mo, S. S. Magnitude estimation of heaviness and loudness by individual subjects: A test of a probabilistic response theory. *British Journal of Mathematical and Statistical Psychology*, 1965, **18**, 159–174.

Luce, R. D., & Tukey, J. Simultaneous conjoint measurement: A new type of fundamental measurement. *Journal of Mathematical Psychology*, 1964, **1**, 1–27.

MacAdam, D. L. Loci of constant hue and brightness determined with various surrounding colors. *Journal of the Optical Society of America*, 1950, **40**, 589–595.

Mach, E. Über die physiologische Wirkung räumliche verheilter Lichtreise. *Sitzungsberichte der mathematisch-naturwissenschaftlichen Classe der Kaiserlichen Akadamie der Wissenschaften, Wien*, 1868, **57**, 11–19.

MacKay, D. M. Psychophysics of perceived intensity: A theoretical basis for Fechner's and Stevens' laws. *Science*, 1963, **139**, 1213–1216.

MacLeod, S. A construction and attempted validation of sensory sweetness scales. *Journal of Experimental Psychology*, 1952, **44**, 316–323.

Mansfield, R. J. W. Intensity relations in vision: Analysis and synthesis in a non-linear sensory system. Doctoral dissertation, Harvard University, 1970.

Marks, L. E. Brightness as a function of retinal locus. *Perceptions & Psychophysics*, 1966, **1**, 335–341.

Marks, L. E. Brightness as a function of retinal locus in the light-adapted eye. *Vision Research*, 1968, **8**, 525–535. (a)

Marks, L. E. Stimulus-range, number of categories, and form of the category-scale. *American Journal of Psychology*, 1968, 81, 467–479. (b)

Marks, L. E. Apparent depth of modulation as a function of frequency and amplitude of temporal modulations of luminance. *Journal of the Optical Society of America*, 1970, 60, 970–977.

Marks, L. E. Brightness and retinal locus: Effects of target size and spectral composition. *Perception & Psychophysics*, 1971, 9, 26–30. (a)

Marks, L. E. Spatial summation in relation to the dynamics of warmth sensation. *International Journal of Biometeorology*, 1971, 15, 106–110. (b)

Marks, L. E. Visual brightness: Some applications of a model. *Vision Research*, 1972, 12, 1409–1421.

Marks, L. E. Brightness and equivalent intensity of intrinsic light. *Vision Research*, 1973, 13, 371–382.

Marks, L. E., & Cain, W. S. Perception of intervals and ratios for three prothetic continua. *Journal of Experimental Psychology*, 1972, 94, 6–17.

Marks, L. E., & Slawson, A. W. Direct test of the power function for loudness. *Science*, 1966, 154, 1036–1037.

Marks, L. E., & Stevens, J. C. Individual brightness functions. *Perception & Psychophysics*, 1966, 1, 17–24.

Marks, L. E., & Stevens, J. C. The form of the psychophysical function near threshold. *Perception & Psychophysics*, 1968, 4, 315–318. (a)

Marks, L. E., & Stevens, J. C. Perceived warmth and skin temperature as functions of the duration and level of thermal irradiation. *Perception & Psychophysics*, 1968, 4, 220–228. (b)

Marks, L. E., & Stevens, J. C. Perceived cold and skin temperature as functions of stimulation level and duration. *American Journal of Psychology*, 1972, 85, 407–419.

Marks, L. E., & Stevens, J. C. Spatial summation of warmth: Influence of duration and configuration of the stimulus. *American Journal of Psychology*, 1973 86, in press.

Marks, L. E., & Stevens, J. C. Temporal summation related to the nature of the proximal stimulus for warmth. *Perception & Psychophysics*, in press.

Marsden, A. M. An elemental theory of induction. *Vision Research*, 1969, 9, 653–663.

Mashhour, M., & Hosman, J. On the new "psychophysical law": A validation study. *Perception & Psychophysics*, 1968, 3, 367–375.

Matin, L. Binocular summation at the absolute threshold of perpheral vision. *Journal of the Optical Society of America*, 1962, 52, 1276–1286.

Matin, L. Critical duration, the differential luminance threshold, critical flicker frequency and visual adaptation: A theoretical treatment. *Journal of the Optical Society of America*, 1968, 58, 404–415.

Matthews, B. H. C. The response of a single end organ. *Journal of Physiology*, 1931, 71, 64–110.

Maxwell, R. S. The quantitative estimation of the sensation of colour. *British Journal of Psychology*, 1929, 20, 181–189.

McBurney, D. H. Magnitude estimation of the taste of sodium chloride after adaptation to sodium chloride. *Journal of Experimental Psychology*, 1966, 77, 869–873.

McBurney, D. H. Gustatory cross adaptation between sweet-tasting compounds. *Perception & Psychophysics*, 1972, 11, 225–227.

McBurney, D. H., & Bartoshuk, L. M. Water taste in mammals. In D. Schneider (Ed.), *Olfaction and taste IV*. Stuttgart: Wissenschaftliche Verlagsgesellschaft, 1972. Pp. 329–335.

McBurney, D. H., & Lucas, J. A. Gustatory cross adaptation between salts. *Psychonomic Science*, 1966, 4, 301–302.

McBurney, D. H., & Shick, T. R. Taste and water taste of twenty-six compounds for man. *Perception & Psychophysics*, 1971, 10, 249–252.

McBurney, D. H., Smith, D. V., & Shick, T. R. Gustatory cross adaptation: Sourness and bitterness. *Perception & Psychophysics*, 1972, 11, 228–232.

McGill, W. J. The slope of the loudness function: A puzzle. In H. Gulliksen and S. Messick (Eds.), *Psychological scaling: Theory and applications*. New York: Wiley, 1960. Pp. 67–81.

McGill, W. J., & Goldberg, J. P. Pure-tone intensity discrimination and energy detection. *Journal of the Acoustical Society of America*, 1968, 44, 576–581.

Meiselman, H. L. Adaptation and cross-adaptation of the four gustatory qualities. *Perception & Psychophysics*, 1968, 4, 368–372. (a)

Meiselman, H. L. Magnitude estimations of the course of gustatory adaptation. *Perception & Psychophysics*, 1968, 4, 193–196. (b)

Meiselman, H. L. Effect of presentation procedure on taste intensity functions. *Perception & Psychophysics*, 1971, 10, 15–18.

Meiselman, H. L., Bose, H. E., & Nykvist, W. F. Magnitude production and magnitude estimation of taste intensity. *Perception & Psychophysics*, 1972, 12, 249–252.

Mendel, M. I., Sussman, H. M., Merson, R. M. Naeser, M. A., & Minifie, F. D. Loudness judgments of speech and nonspeech stimuli. *Journal of the Acoustical Society of America*, 1969, 46, 1556–1561.

Merkel, J. Die Abhängigkeit zwischen Reiz und Empfindung. *Philosophische Studien*, 1888, 4, 541–594.

Merkel, J. Die Abhängigkeit zwischen Reiz und Empfindung. *Philosophische Studien*, 1889, 5, 245–291.

Merkel, J. Die Abhängigkeit zwischen Reiz und Empfindung. *Philosophische Studien*, 1894, 10, 203–248.

Metfessel, M. A proposal for quantitative reporting of comparative judgments. *Journal of Psychology*, 1947, 24, 229–235.

Michels, W. C., & Helson, H. A reformulation of the Fechner law in terms of adaptation-level applied to rating-scale data. *American Journal of Psychology*, 1949, 62, 355–368.

Miller, G. A. Sensitivity to changes in the intensity of white noise and its relation to loudness and masking. *Journal of the Acoustical Society of America*, 1947, 19, 609–619.

Mitchell, M. J., & McBride, R. I. Effects of propanol masking odor on the olfactory intensity scaling of eugenol. *Journal of Experimental Psychology*, 1971, 87, 309–313.

Moncrieff, R. W. *The chemical senses*. London: Leonard Hall, 1944.

Moskowitz, H. R. Ratio scales of sugar sweetness. *Perception & Psychophysics*, 1970, 7, 315–320. (a)

Moskowitz, H. R. Sweetness and intensity of artificial sweeteners. *Perception & Psychophysics*, 1970, 8, 40–42. (b)

Moskowitz, H. R. Intensity scales for pure tastes and for taste mixtures. *Perception & Psychophysics*, 1971, 9, 51–57. (a)

Moskowitz, H. R. Ratio scales of acid sourness. *Perception & Psychophysics*, 1971, 9, 371–374. (b)

Moskowitz, H. R. The sweetness and pleasantness of sugars. *American Journal of Psychology*, 1971, 84, 387–405. (c)

Mountcastle, V. B. Discussion. In A. V. S. de Reuck and J. Knight (Eds.), *Touch, heat, pain*. Boston, Massachusetts: Little, Brown, 1966. Pp. 21–22.

Mountcastle, V. B., Poggio, G. F., & Werner, G. The relation of thalamic cell response to peripheral stimuli varied over an intensive continuum. *Journal of Neurophysiology*, 1963, **26**, 807–834.

Münsterberg, H. *Beiträge zur experimentellen Psychologie, III* Freiburg: Mohr, 1890.

Murgatroyd, D. Spatial summation of pain for large body areas. *Defense Atomic Support Agency Report*, 1964.

Nachmias, J., & Steinman, R. M. Brightness and discriminability of light flashes. *Vision Research*, 1965, **5**, 545–557.

Natsoulas, T. What are perceptual reports about? *Psychological Bulletin*, 1967, **67**, 249–272.

Newhall, S. M. The ratio method in the review of the Munsell colors. *American Journal of Psychology*, 1939, **52**, 394–405.

Newhall, S. M., Nickerson, D., & Judd, D. B. Final report of the O.S.A. Subcommittee on the spacing of the Munsell colors. *Journal of the Optical Society of America*, 1943, **33**, 385–418.

Newman, E. B. The validity of the just noticeable difference as a unit of psychological magnitude. *Transactions of the Kansas Academy of Science*, 1933, **36**, 172–175.

Newman, E. B., Volkmann, J., & Stevens, S. S. On the method of bisection and its relation to a loudness scale. *American Journal of Psychology*, 1937, **49**, 134–137.

Ogawa, T., Kozaki, T., Takano, Y., & Okayama, K. Effect of area on apparent brightness. *Report from the Psychological Laboratory, Keio University*, 1966.

Onley, J. W. Brightness scaling of white and colored stimuli. *Science*, 1960, **132**, 1668–1670.

Onley, J. W. Light adaptation and the brightness of brief foveal stimuli. *Journal of the Optical Society of America*, 1961, **51**, 667–673.

Onley, J. W., Klingberg, C. L., Dainoff, M. J., & Rollman, G. B. Quantitative estimates of saturation. *Journal of the Optical Society of America*, 1963, **53**, 487–493.

Østerberg, G. Topography of the layer of rods and cones in the human retina. *Acta Ophthalmologica*, Supplement, 1935, **61**, 1–102.

Padgham, C. A. The direct estimation of the luminosity of coloured light sources. *Vision Research*, 1971, **11**, 577–590.

Padgham, C. A., & Saunders, J. E. Scales of apparent brightness. *Transactions of the Illumination Engineering Society (London)*, 1966, **31**, 122–142.

Panek, D. W., & Stevens, S. S. Saturation of red: A prothetic continuum. *Perception & Psychophysics*, 1966, **1**, 59–66.

Pangborn, R. M. Taste interrelationships. *Food Research*, 1960, **25**, 245–256.

Périlhou, P., & Piéron, H. Quelques charactéristiques des sensations vibratoires. *Comptes Rendus de la Societé de Biologie (Paris)*, 1942, **136**, 448–449.

Petty, J. W., Fraser, W. D., & Elliott, D. N. Adaptation and loudness decrement: A reconsideration. *Journal of the Acoustical Society of America*, 1970, **47**, 1074–1081.

Pfaffmann, C. Taste and smell. In S. S. Stevens (Ed.), *Handbook of experimental psychology*. New York: Wiley, 1951. Pp. 1143–1171.

Pfaffmann, C. The pleasures of sensation. *Psychological Review*, 1960, **67**, 253–268.

Pfanzagl, J. A general theory of measurement: Applications to utility. *Naval Research Logistics Quarterly*, 1959, **6**, 283–294.

Piéron, H. Le problème du mechanisme physiologique impliqué par l'échelon différentiel de sensation. *L'Année Psychologique*, 1934, **34**, 217–236.

Piéron, H. Recherches sur la validité de la loi d'Abney impliquant l'addition intégrale des valences lumineuses élémentaires dans les flux composities. *L'Année Psychologique*, 1939, **40**, 52–83.

Piéron, H. *The sensations: Their functions, processes and mechanisms*. Translated by M. H. Pirenne and B. C. Abbot. New Haven, Connecticut: Yale Univ. Press, 1952.

Piéron, H. Les échelles subjectives peuvent-elles fournir la base d'une nouvelle loi psychophysique? *L'Année Psychologique*, 1959, **59**, 1–34.

Pikler, J. *Schriften zur Anpassungstheorie des Empfindungsvorganges. IV. Theorie der Empfindungsqualität als Abbild des Reizes*. Leipzig: Barth, 1922.

Pinegrin, N. I. Absolute photopic sensitivity of the eye in the ultraviolet and in the visible spectrum. *Nature*, 1944, **154**, 770.

Pinegrin, N. I. Absolute scotopic sensitivity of the eye in the ultraviolet and in the visible spectrum. *Nature*, 1945, **155**, 20–21.

Pitz, G. F. Magnitude scales of line length. *Psychonomic Science*, 1965, **2**, 213–214.

Plateau, J. A. F. Sur la mesure des sensations physiques, et sur la loi qui lie l'intensité de ces sensations à l'intensité de la cause excitante. *Bulletins de l'Académie Royale des Sciences, des Lettres, et des Beaux-Arts de Belqique*, 1872, **33**, 376–388.

Plomp, R. Pitch of complex tones. *Journal of the Acoustical Society of America*, 1967, **41**, 1526–1533.

Plomp, R., & Levelt, W. J. M. Tonal consonance and critical bandwidth. *Journal of the Acoustical Society of America*, 1965, **38**, 548–560.

Pollack, I. The effect of white noise on the loudness of speech of assigned average level. *Journal of the Acoustical Society of America*, 1949, **21**, 255–258.

Pollack, I. Iterative techniques for unbiased rating scales. *Quarterly Journal of Experimental Psychology*, 1965, **17**, 139–148. (a)

Pollack, I. Neutralization of stimulus bias in the rating of grays. *Journal of Experimental Psychology*, 1965, **69**, 564–578. (b)

Pollack, I. The apparent pitch of short tones. *American Journal of Psychology*, 1968, **81**, 165–169.

Port, E. Über die Lautstärke einzelner kurzer Schallimpulse. *Acustica*, 1963, **13**, 212–223.

Poulton, E. C. Population norms of top sensory magnitudes and S. S. Stevens' exponents. *Perception & Psychophysics*, 1967, **2**, 312–316.

Poulton, E. C. The new psychophysics: Six models for magnitude estimation. *Psychological Bulletin*, 1968, **69**, 1–19.

Poulton, E. C., & Simmonds, D. C. V. Value of standard and very first variable in judgments of reflectance of grays with various ranges of available numbers. *Journal of Experimental Psychology*, 1963, **65**, 297–304.

Pradhan, P. L., & Hoffman, P. J. Effect of spacing and range of stimuli on magnitude estimation. *Journal of Experimental Psychology*, 1963, **66**, 533–541.

Pryor, G. T., Steinmetz, G., & Stone, H. Changes in absolute detection threshold and subjective intensity of suprathreshold stimuli during olfactory adaptation and recovery. *Perception & Psychophysics*, 1970, **8**, 331–335.

Purdy, D. M. Spectral hue as a function of intensity. *American Journal of Psychology*, 1931, **43**, 541–559.

Pütter, A Studien zur Theorie der Reizvorgänge. *Pflügers Archiv für die gesamte Physiologie*, 1918, **171**, 201–261.

Raab, D. H. Magnitude estimation of the brightness of brief foveal stimuli. *Science*, 1962, **135**, 42–44.

Raab, D. H. Audition. *Annual Review of Psychology*, 1971, **22**, 95–118.

Raab, D., Fehrer, E., & Hershenson, M. Visual reaction time and the Broca-Sulzer phenomenon. *Journal of Experimental Psychology*, 1961, **61**, 193–199.

Raab, D. H., & Osman, E. Effect of temporal overlap on brightness matching of adjacent flashes. *Journal of the Optical Society of America*, 1962, **52**, 1174–1178.

Rabelo, C., & Grüsser, O.-J. Die Abhängigkeit der subjektiven Helligkeit intermittierender Lichtreize von der Flimmerfrequenz (Brücke-Effekt, "brightness enhancement") : Untersuchungen bei verschiedener Leuchtdichte und Feldgrösse. *Psychologische Forschung*, 1961, **26**, 299–312.

Rainbolt, H., & Small, A. M. Mach bands in auditory masking: An attempted replication. *Journal of the Acoustical Society of America*, 1972, **51**, 567–574.

Ramsay, J. O. Economical method of analyzing perceived color differences. *Journal of the Optical Society of America*, 1968, **58**, 19–22.

Ratliff, F., & Hartline, H. K. The response of limulus optic nerve fibers to patterns of illumination on the receptor mosaic. *Journal of General Physiology*, 1959, **42**, 1241–1255.

Reese, T. S., & Stevens, S. S. Subjective intensity of coffee odor. *American Journal of Psychology*, 1960, **73**, 424–428.

Reese, T. W. The application of the theory of physical measurement to the measurement of psychological magnitudes, with three experimental examples. *Psychologial Monographs*, 1943, **55**, Whole No. 251.

Reynolds, G. S., & Stevens, S. S. Binaural summation of loudness. *Journal of the Acoustical Society of America*, 1960, **32**, 1337–1344.

Riccò, A. Relazione fra il minimo angolo visuale e l'intensità luminosa. *Annali d'Ottalmologia*, 1877, **6**, 373–479.

Rich, G. J. A study of tonal attributes. *American Journal of Psychology*, 1919, **30**, 121–164.

Richards, A. M. Monaural loudness functions under masking. *Journal of the Acoustical Society of America*, 1968, **44**, 599–604.

Richardson, L. F. Imagery, conation, and cerebral conductance. *Journal of General Psychology*, 1928, **2**, 324–352.

Richardson, L. F. Quantitative mental estimates of light and colour. *British Journal of Psychology*, 1929, **20**, 27–37.

Richardson, L. F., & Ross, J. S. Loudness and telephone current. *Journal of General Psychology*, 1930, **3**, 288–306.

Riesz, R. R. Differential intensity sensitivity of the ear for pure tones. *Physical Review*, 1928, **31**, 867–875.

Riggs, L. A., Ratliff, F., Cornsweet, J. C., & Cornsweet, T. N. The disappearance of steadily fixated test objects. *Journal of the Optical Society of America*, 1953, **43**, 495–501.

Robinson, D. W., & Dadson, R. S. A re-determination of the equal-loudness relations for pure tones. *British Journal of Applied Physics*, 1956, **7**, 166–181.

Rosenblith, W. A. Some quantifiable aspects of the electrical activity of the nervous system (with emphasis upon responses to sensory systems). *Review of Modern Physics*, 1959, **31**, 532–545.

Rosner, B. S., & Goff, W. R. Electrical responses of the nervous system and subjective scales of intensity. In W. D. Neff (Ed.), *Contributions to sensory physiology*, Vol. II. New York: Academic Press, 1967. Pp. 169–221.

Ross, J. L., & di Lollo, V. A vector model for psychophysical judgment. *Journal of Experimental Psychology*, 1968, **77** (No. 3, Pt. 2, Suppl.)

Ross, S. Matching functions and equal-sensation contours for loudness. *Journal of the Acoustical Society of America*, 1967, **42**, 778–793.

Rowley, R. R., & Studebaker, G. A. Loudness-intensity relations under various levels of contralateral noise. *Journal of the Acoustical Society of America*, 1971, **49**, 499–504.

Rschevkin, S. N., & Rabinovitch, A. V. Sur le problème de l'estimation quantitative de la force d'un son. *Revue d'Acoustique*, 1936, **5**, 183–200.

Rule, S. J. Equal discriminability scale of number. *Journal of Experimental Psychology*, 1969, **79**, 35–38.

Rule, S. J., Curtis, D. W., & Markley, R. P. Input and output transformations from magnitude estimation. *Journal of Experimental Psychology*, 1970, **86**, 343–349.

Rushton, W. A. H. Visual adaptation. *Proceedings of the Royal Society (London)*, 1965, **162B**, 20–46.

Rushton, W. A. H. Review lecture. Pigments and signals in colour vision. *Journal of Physiology*, 1972, **220**, 1–31P.

Rushton, W. A. H., & Gubisch, R. W. Glare: Its measurement by cone thresholds and by the bleaching of cone pigments. *Journal of the Optical Society of America*, 1966, **56**, 104–110.

Saffir, M. A. A comparative study of scales constructed by three psychophysical methods. *Psychometrika*, 1937, **2**, 179–198.

Saunderson, J. L., & Milner, B. I. A further study of ω space. *Journal of the Optical Society of America*, 1944, **34**, 167–173.

Scharf, B. Critical bands and the loudness of complex sounds near threshold. *Journal of the Acoustical Society of America*, 1959, **31**, 365–370. (a)

Scharf, B. Loudness of complex sounds as a function of the number of components. *Journal of the Acoustical Society of America*, 1959, **31**, 783–785. (b)

Scharf, B. Dichotic summation of loudness. *Journal of the Acoustical Society of America*, 1969, **45**, 1193–1205.

Scharf, B. Critical bands, In J. V. Tobias (Ed.), *Foundations of modern auditory theory*, Vol. I. New York: Academic Press, 1970. Pp. 157–202.

Scharf, B. Fundamentals of auditory masking. *Audiology*, 1971, **10**, 30–40.

Scharf, B., & Fishken, D. Binaural summation of loudness: Reconsidered. *Journal of Experimental Psychology*, 1970, **86**, 374–379.

Schiffman, S. S., & Erickson, R. P. A psychophysical model for gustatory quality. *Physiology and Behavior*, 1971, **7**, 617–633.

Schjelderup, H. K. Über die Abhängigkeit zwischen Empfindung und Reiz. *Zeitschrift für Psychologie*, 1918, **80**, 226–243.

Schneider, B., & Lane, H. Ratio scales, category scales, and variability in the production of loudness and softness. *Journal of Acoustical Society of America*, 1963, **35**, 1953–1961.

Schneider, B., Wright, A. A., Edelheit, W., Hock, P., & Humphrey, C. Equal loudness contours derived from sensory magnitude judgments. *Journal of the Acoustical Society of America*, 1972, **51**, 1951–1959.

Schneider, C. W., & Bartley, S. H. Changes in sensory phenomena and observer criteria at low rates of intermittent photic stimulation. *Journal of Psychology*, 1966, **63**, 53–66.

Schouten, J. F. The perception of subjective tones. *Nederlandse Akademie van Wetenschappen (Amsterdam)*, 1938, **41**, 1086–1093.

Schouten, J. F., & Ornstein, L. S. Measurements on direct and indirect adaptation by means of a binocular method. *Journal of the Optical Society of America*, 1939, **29**, 168–182.

Shaw, W. A., Newman, E. B., & Hirsh, I. J. The difference between monaural and binaural thresholds. *Journal of Experimental Psychology*, 1947, **37**, 229–242.

Shepard, R. Attention and the metric structure of the stimulus space. *Journal of Mathematical Psychology*, 1964, **1**, 54–87.

Shepard, R. Metric structures in ordinal data. *Journal of Mathematical Psychology*, 1966, **3**, 287–315.

Sherrick, C. E., Jr. Variables affecting sensitivity of the human skin to mechanical vibration. *Journal of Experimental Psychology*, 1953, **45**, 273–282.

Sherrick, C. E., Jr. Observations relating to some common psychophysical functions as applied to the skin. In G. R. Hawkes (Ed.), *Symposium on cutaneous sensitivity*. Fort Knox: U. S. Army Medical Research Laboratory, 1960. Pp. 147–158.

Sherrington, C. *The integrative action of the nervous system*. London: Constable, 1906.

Shower, E. G., & Biddulph, R. Differential pitch sensitivity of the ear. *Journal of the Acoustical Society of America*, 1931, **3**, 275–287.

Siegel, R. J. A replication of the mel scale of pitch. *American Journal of Psychology*, 1965, **78**, 615–620.

Singer, W. B., & Young, P. T. Studies in affective reaction: II. Dependence of affective ratings upon the stimulus situation. *Journal of General Psychology*, 1941, **24**, 303–325.

Sivian, L. J., & White, S. D. On minimum audible sound fields. *Journal of the Acoustical Society of America*, 1933, **4**, 288–321.

Skramlik, E. von. *Handbuch der Physiologie der niederen Sinne. I. Die Physiologie des Geruchs-und Geschmackssinnes*. Leipzig: Thienne, 1926.

Sloan, L. L. The effect of intensity of light, state of adaptation of the eye, and size of photometric field on the visibility curve. *Psychological Monographs*, 1928, **38** (Whole No. 173).

Small, A. M., Jr. Auditory adaptation. In J. Jerger (Ed.), *Modern developments in audiology*. New York: Academic Press, 1963. Pp. 287–336.

Small, A. M., Jr., Brandt, J. F., & Cox, P. G. Loudness as a function of signal duration. *Journal of the Acoustical Society of America*, 1962, **34**, 513–514.

Smith, D. V. Taste intensity as a function of area and concentration: Differentiation between compounds. *Journal of Experimental Psychology*, 1971, **87**, 163–171.

Smith, D. V., & McBurney, D. H. Gustatory cross-adaptation: Does a single mechanism code the salty taste? *Journal of Experimental Psychology*, 1969, **80**, 101–105.

Snow, W. B. Change of pitch with loudness at low frequencies. *Journal of the Acoustical Society of America*, 1936, **8**, 14–19.

Sperling, G. Model of visual adaptation and contrast detection. *Perception & Psychophysics*, 1970, **8**, 143–157.

Sperling, G., & Sondhi, M. M. Model for visual luminance discrimination and flicker detection. *Journal of the Optical Society of America*, 1968, **58**, 1133–1145.

Sperling, H. G. An experimental investigation of the relationship between colour mixture and luminous efficiency. In *Visual problems of colour*, Vol. I. London: Her Majesty's Stationery Office, 1958. Pp. 253–297.

Stecher, S., Sandberg, M., & Minsky, P. J. Successive luminance difference thresholds and brightness as a function of the interstimulus interval and durations of successive flashes. *Perception & Psychophysics*, 1970, **7**, 79–85.

Steinberg, J. C., & Gardner, M. B. The dependence of hearing impairment on sound intensity. *Journal of the Acoustical Society of America*, 1937, **9**, 11–23.

Sternbach, R. A., & Tursky, B. On the psychophysical power function in electric shock. *Psychonomic Science*, 1964, **1**, 217–218.

Sternheim, C., & Boynton, R. M. Uniqueness of perceived hues investigated with a continuous judgmental technique. *Journal of Experimental Psychology*, 1966, **72**, 770–776.

Stevens, J. C. A comparison of ratio scales for the loudness of white noise and the brightness of white light. Doctoral dissertation, Harvard University, 1957.

Stevens, J. C. Stimulus spacing and the judgment of loudness. *Journal of Experimental Psychology*, 1958, **56**, 246–250.

Stevens, J. C. Brightness function: Binocular versus monocular stimulation. *Perception & Psychophysics*, 1967, **2**, 451–454. (a)

Stevens, J. C. Brightness inhibition re size of surround. *Perception & Psychophysics*, 1967, **2**, 189–192. (b)

Stevens, J. C., & Cain, W. S. Effort in muscular contractions related to force level and duration. *Perception & Psychophysics*, 1970, **8**, 240–244.

Stevens, J. C., & Guirao, M. Individual loudness functions. *Journal of the Acoustical Society of America*, 1964, **36**, 2210–2213.

Stevens, J. C., & Hall, J. W. Brightness and loudness as functions of stimulus duration. *Perception & Psychophysics*, 1966, **1**, 319–327.

Stevens, J. C., & Mack, J. D. Scales of apparent force. *Journal of Experimental Psychology*, 1959, **58**, 405–413.

Stevens, J. C., Mack, J. D., & Stevens, S. S. Growth of sensation on seven continua as measured by force of handgrip. *Journal of Experimental Psychology*, 1960, **59**, 60–67.

Stevens, J. C., & Marks, L. E. Cross-modality matching of brightness and loudness. *Proceedings of the National Academy of Sciences*, 1965, **54**, 407–411.

Stevens, J. C., & Marks, L. E. Spatial summation and the dynamics of warmth sensation. *Perception & Psychophysics*, 1971, **9**, 291–298.

Stevens, J. C., Marks, L. E., & Gagge, A. P. The quantitative assessment of thermal discomfort. *Environmental Research*, 1969, **2**, 149–165.

Stevens, J. C. & Shickman, G. M. The perception of repetition rate. *Journal of Experimental Psychology*, 1959, **58**, 433–440.

Stevens, J. C., & Stevens, S. S. Brightness function: Effects of adaptation. *Journal of the Optical Society of America*, 1963, **53**, 375–385.

Stevens, J. C., & Tulving, E. Estimations of loudness by a group of untrained observers. *American Journal of Psychology*, 1957, **70**, 600–605.

Stevens, S. S. Are tones spatial? *American Journal of Psychology*, 1934, **46**, 145–147.

Stevens, S. S. The relation of pitch to intensity. *Journal of the Acoustical Society of America*, 1935, **6**, 150–154.

Stevens, S. S. A scale for the measurement of a psychological magnitude: Loudness. *Psychological Review*, 1936, **43**, 405–416.

Stevens, S. S. On the theory of scales of measurement. *Science*, 1946, **103**, 677–680.

Stevens, S. S. Mathematics, measurement, and psychophysics. In S. S. Stevens (Ed.), *Handbook of experimental psychology*. New York: Wiley, 1951. Pp. 1–49.

Stevens, S. S. The measurement of loudness. *Journal of the Acoustical Society of America*, 1955, **27**, 815–829.

Stevens, S. S. Calculation of the loudness of complex noise. *Journal of the Acoustical Society of America*, 1956, **28**, 807–832. (a)

Stevens, S. S. The direct estimation of sensory magnitudes—loudness. *American Journal of Psychology*, 1956, **69**, 1–25. (b)

Stevens, S. S. On the psychophysical law. *Psychological Review*, 1957, **64**, 153–181.

Stevens, S. S. Problems and methods of psychophysics. *Psychological Bulletin,* 1958, **55,** 177–196.

Stevens, S. S. Cross-modality validation of subjective scales for loudness, vibration, and electric shock. *Journal of Experimental Psychology,* 1959, **57,** 201–209. (a)

Stevens, S. S. Tactile vibration: Dynamics of sensory intensity. *Journal of Experimental Psychology,* 1959, **57,** 210–218. (b)

Stevens, S. S. Ratio scales, partition scales and confusion scales. In H. Gulliksen and S. Messick (Eds.), *Psychological scaling: Theory and applications.* New York: Wiley, 1960. Pp. 49–66b.

Stevens, S. S. Procedure for calculating loudness: Mark VI. *Journal of the Acoustical Society of America,* 1961, **33,** 1577–1585. (a)

Stevens, S. S. The psychophysics of sensory function. In W. A. Rosenblith (Ed.), *Sensory communication.* New York: Wiley, 1961. Pp. 1–33. (b)

Stevens, S. S. To honor Fechner and repeal his law. *Science,* 1961, **133,** 80–86. (c)

Stevens, S. S. Concerning the measurement of brightness. *Journal of the Optical Society of America,* 1966, **56,** 1135–1136. (a)

Stevens, S. S. Duration, luminance, and the brightness exponent. *Perception & Psychophysics,* 1966, **1,** 96–100. (b)

Stevens, S. S. Matching functions between loudness and ten other continua. *Perception & Psychophysics,* 1966, **1,** 5–8. (c)

Stevens, S. S. On the operation known as judgment. *American Scientist,* 1966, **54,** 385–401. (d)

Stevens, S. S. Power-group transformations under glare, masking, and recruitment. *Journal of the Acoustical Society of America,* 1966, **39,** 725–735. (e)

Stevens, S. S. Intensity functions in sensory systems. *International Journal of Neurology,* 1967, **6,** 202–209.

Stevens, S. S. Tactile vibration: Change of exponent with frequency. *Perception & Psychophysics,* 1968, **3,** 223–228.

Stevens, S. S. On predicting exponents for cross-modality matches. *Perception & Psychophysics,* 1969, **6,** 251–256. (a)

Stevens, S. S. Sensory scales of taste intensity. *Perception & Psychophysics,* 1969, **6,** 302–308. (b)

Stevens, S. S. Neural events and the psychophysical law. *Science,* 1970, **170,** 1043–1050.

Stevens, S. S. Issues in psychophysical measurement. *Psychological Review,* 1971, **78,** 426–450. (a)

Stevens, S. S. Sensory power functions and neural events. In W. R. Loewenstein (Ed.), *Handbook of sensory physiology,* Vol. I. Berlin and New York: Springer-Verlag, 1971. Pp. 226–242. (b)

Stevens, S. S. Perceived level of noise by Mark VII and decibels (E). *Journal of the Acoustical Society of America,* 1972, **51,** 575–601.

Stevens, S. S., & Diamond, A. L. Effect of glare angle on the brightness function for a small target. *Vision Research,* 1965, **5,** 649–659.

Stevens, S. S., & Galanter, E. Ratio scales and category scales for a dozen perceptual continua. *Journal of Experimental Psychology,* 1957, **54,** 377–411.

Stevens, S. S., & Greenbaum, H. B. Regression effect in psychophysical judgment. *Perception & Psychophysics,* 1966, **1,** 439–446.

Stevens, S. S., & Guirao, M. Loudness, reciprocality, and partition scales. *Journal of the Acoustical Society of America,* 1962, **34,** 1466–1471.

Stevens, S. S., & Guirao, M. Loudness functions under inhibition. *Perception & Psychophysics*, 1967, **2**, 459–465.

Stevens, S. S., Guirao, M., & Slawson, A. W. Loudness, a product of volume times density. *Journal of Experimental Psychology*, 1965, **69**, 503–510.

Stevens, S. S., Morgan, C. T., & Volkmann, J. Theory of the neural quantum in the discrimination of loudness and pitch. *American Journal of Psychology*, 1941, **54**, 315–335.

Stevens, S. S., & Poulton, E. C. The estimation of loudness by unpracticed observers. *Journal of Experimental Psychology*, 1956, **51**, 71–78.

Stevens, S. S., & Stevens, J. C. *The Dynamics of visual brightness.* Psychophysical Project Report PPR–246, Harvard University, 1960.

Stevens, S. S., & Volkmann, J. The relation of pitch to frequency: A revised scale. *American Journal of Psychology*, 1940, **53**, 329–353.

Stevens, S. S., Volkmann, J., & Newman, E. B. A scale for the measurement of the psychological magnitude pitch. *Journal of the Acoustical Society of America*, 1937, **8**, 185–190.

Stewart, J. L. Quantitative laws for sensory perception. *Psychological Review*, 1963, **70**, 180–192.

Stewart, M. R., Fagot, R. F., & Eskildsen, P. R. Invariance tests for bisection and fractionation scaling. *Perception & Psychophysics*, 1967, **2**, 323–327.

Stiles, W. S., & Crawford, B. H. The effect of a glaring light source on extrafoveal vision. *Proceedings of the Royal Society, (London)*, 1937, **122B**, 255–280.

Stokinger, T. E., Cooper, W. A., Jr., & Meissner, W. A. Influence of binaural interaction on the measurement of perstimulatory loudness adaptation. *Journal of the Acoustical Society of America*, 1972, **51**, 602–607.

Stokinger, T. E., Cooper, W. A. Jr., Meissner, W. A., & Jones, K. O. Intensity, frequency, and duration effects in the measurement of monaural perstimulatory loudness adaptation. *Journal of the Acoustical Society of America*, 1972, **51**, 608–616.

Stokinger, T. E., & Studebaker, G. A. Measurement of perstimulatory loudness adaptation. *Journal of the Acoustical Society of America*, 1968, **44**, 250–256.

Stolwijk, J. A. J., & Hardy, J. D. Skin and subcutaneous temperature changes during exposure to intense thermal radiation. *Journal of Applied Physiology*, 1965, **20**, 1006–1013.

Stout, G. F. *A manual of psychology.* New York: Hinds, Nobel, & Eldredge, 1915.

Strangert, B. A validation study of the method of ratio estimation. *Report from the Psychological Laboratory, University of Stockholm*, 1961, No. 95.

Suppes, P., & Zinnes, J. L. Basic measurement theory. In R. D. Luce, R. R. Bush, and E. Galanter (Eds.), *Handbook of mathematical psychology*, Vol. I. New York: Wiley, 1963. Pp. 1–76.

Swets, J. A. Is there a sensory threshold? *Science*, 1961, **134**, 168–177.

Swets, J. A., Tanner, W. P., & Birdsall, T. G. Decision processes in perception. *Psychological Review*, 1961, **68**, 301–340.

Tasaki, I. Nerve impulses in individual auditory nerve fibers of guinea pig. *Journal of Neurophysiology*, 1954, **17**, 97–122.

Teghtsoonian, M., & Teghtsoonian, R. Seen and felt length. *Psychonomic Science*, 1965, **3**, 465–466.

Teghtsoonian, M., & Teghtsoonian, R. How repeatable are Stevens' power law exponents for individual subjects? *Perception & Psychophysics*, 1971, **10**, 147–149.

Teghtsoonian, R. On the exponents in Stevens' law and the constant in Ekman's law. *Psychological Review*, 1971, **78**, 71–80.

Terrace, H. S., & Stevens, S. S. The quantification of tonal volume. *American Journal of Psychology*, 1962, **75**, 596–604.

Thomas, G. J. Equal-volume judgments of tones. *American Journal of Psychology*, 1949, **62**, 182–201.

Thomas, G. J. Volume and loudness of noise. *American Journal of Psychology*, 1952, **65**, 588–593.

Thomson, L. C. The effect of change of brightness level upon the foveal luminosity curve measured with small fields. *Journal of Physiology*, 1947, **106**, 368–377.

Thurlow, W. R., & Melamed, L. E. Some new hypotheses on the mediation of loudness judgments. *Perception & Psychophysics*, 1967, **2**, 77–80.

Thurstone, L. L. A law of comparative judgment. *Psychological Review*, 1927, **34**, 273–286. (a)

Thurstone, L. L. Psychophysical analysis. *American Journal of Psychology*, 1927, **38**, 368–389. (b)

Titchener, E. B. *Experimental psychology: A manual of laboratory practice*, Vol. II, Quantitative. 2, Instructor's Manual. New York: MacMillan, 1905.

Titchener, E. B. The psychophysics of climate. *American Joural of Psychology*, 1909, **20**, 1–14.

Torgerson, W. S. A law of categorical judgment. *American Psychologist*, 1954, **7**, 408.

Torgerson, W. S. Distances and ratios in psychological scaling. *Acta Psychologica*, 1961, **19**, 201–205.

Treisman, M. Laws of sensory magnitude. *Nature*, 1963, **198**, 914–915.

Treisman, M. Sensory scaling and the psychophysical law. *Quarterly Journal of Experimental Psychology*, 1964, **16**, 11–22.

Treisman, M. Signal detection theory and Crozier's law: Derivation of a new sensory scaling procedure. *Journal of Mathematical Psychology*, 1965, **2**, 205–218.

Treisman, M. Brightness contrast and the perceptual scale. *British Journal of Mathematical and Statistical Psychology*, 1970, **23**, 205–224.

Treisman, M., & Irwin, R. J. Auditory intensity and discrimination scale I. Evidence derived from binaural intensity summation, *Journal of the Acoustical Society of America*, 1967, **42**, 586–592.

Troland, L. T. *The fundamentals of human motivation*. Princeton, New Jersey: Van Nostrand, 1928.

Troland, L. T. *Principles of psychophysiology*, Vol. 2. Sensation. Princeton, New Jersey: Van Nostrand, 1930.

Tversky, A., & Krantz, D. H. The dimensional representation and the metric structure of similarity data. *Journal of Mathematical Psychology*, 1970, **7**, 572–596.

Tyler, W. F. A scheme for the comparison of climates. *Journal of Balneology and Climatology*, 1904, **8**, 17–44.

Tyler, W. F. *The psycho-physical aspect of climate with a theory concerning intensities of sensation*. London: Bale, Sons, & Danielsson, 1907.

Uttal, W. Emerging principles of sensory coding. *Perspectives in Biology and Medicine*, 1969, **12**, 344–368.

van den Brink, G. Subjective brightness during dark-adaptation. *Vision Research*, 1962, **2**, 495–502.

Verrillo, R. T. Investigation of some parameters of the cutaneous threshold for vibration. *Journal of the Acoustical Society of America*, 1962, **34**, 1768–1773.

Verrillo, R. T. Effect of contactor area on the vibrotactile threshold. *Journal of the Acoustical Society of America*, 1963, **35**, 1962–1966.

Verrillo, R. T. Temporal summation in vibrotactile sensitivity. *Journal of the Acoustical Society of America*, 1965, **37**, 843–846.

Verrillo, R. T. Effect of spatial parameters on the vibrotactile threshold. *Journal of Experimental Psychology*, 1966, **71**, 570–575. (a)

Verrillo, R. T. Specificity of a cutaneous receptor. *Perception & Psychophysics*, 1966, **1**, 149–152. (b)

Verrillo, R. T. Vibrotactile thresholds for hairy skin. *Journal of Experimental Psychology*, 1966, **72**, 47–50. (c)

Verrillo, R. T., & Chamberlain, S. C. The effect of neural density and contactor surround on vibrotactile sensation magnitude. *Perception & Psychophysics*, 1972, **11**, 117–120.

Verrillo, R. T., Fraoli, A. J., & Smith, R. L. Sensory magnitude of vibrotactile stimuli. *Perception & Psychophysics*, 1969, **6**, 366–372.

Vitz, P. C. Preference for tones as a function of frequency (hertz) and intensity (decibels). *Perception & Psychophysics*, 1972, **11**, 84–88.

von Neumann, J., & Morgenstern, O. *Theory of games and economic behavior*. Princeton, New Jersey: Princeton Univ. Press, 1947.

Wald, G. Human vision and the spectrum, *Science*, 1945, **101**, 653–658.

Wald, G. Blue-blindness in the normal fovea. *Journal of the Optical Society of America*, 1967, **57**, 1289–1301.

Wallace, S. R., Jr. Studies in binocular interdependence: I. Bioncular relations in macular adaptation. *Journal of General Psychology*, 1937, **17**, 307–322.

Waller, A. D. Points relating to the Weber-Fechner law. Retina; muscle; nerve. *Brain*, 1895, **18**, 200–216.

Walters, H. V., & Wright, W. D. The spectral sensitivity of the fovea and the extrafovea in the Purkinje range. *Proceedings of the Royal Society (London)*, 1943, **131B**, 340–361.

Warren, R. M. A basis for judgments of sensory intensity. *American Journal of Psychology*, 1958, **71**, 675–687.

Warren, R. M. Quantitative judgments of color: The square root rule. *Perception & Psychophysics*, 1967, **2**, 448–452.

Warren, R. M. Visual intensity judgments: An empirical rule and a theory. *Psychological Review*, 1969, **76**, 16–30.

Warren, R. M., & Poulton, E. C. Ratio- and partition-judgments. *American Journal of Psychology*, 1962, **75**, 109–114.

Warren, R. M., & Poulton, E. C. Lightness of grays: Effects of background luminance. *Perception & Psychophysics*, 1966, **1**, 145–148.

Warren, R. M., & Warren, R. P. Basis for judgments of relative brightness. *Journal of the Optical Society of America*, 1958, **48**, 445–450.

Warren, R. M., & Warren, R. P. A critique of S. S. Stevens' "new psychophysics". *Perceptual and Motor Skills*, 1963, **16**, 797–810.

Wasserman, G. S. Brightness enhancement in intermittent light: Variation of luminance and light-dark ratio. *Journal of the Optical Society of America*, 1966, **56**, 242–250.

Watanabe, A., Mori, T., Nagata, S., & Hiwatashi, K. Spatial sine-wave responses of the human visual system. *Vision Research*, 1968, **8**, 1245–1263.

Weaver, K. S. A provisional standard observer for low level photometry. *Journal of the Optical Society of America*, 1949, **39**, 278–291.

Weber, E. H. *De pulsu, resorptione, auditu et tactu: Annotationes anatomicae et physiologicae*. Leipzig: Koehler, 1834.

Wegel, R. L., & Lane, C. E. The auditory masking of one pure tone by another and its probable relation to the dynamics of the inner ear. *Physical Review*, 1924, **23**, 266–285.

Werner, G., & Mountcastle, V. B. Neural activity in mechanoreceptive cutaneous afferents: Stimulus-response relations, Weber functions, and information transmission. *Journal of Neurophysiology*, 1965, **28**, 359–397.

Whittle, P., & Challands, P. D. C. The effect of background luminance on the brightness of flashes. *Vision Research*, 1969, **9**, 1095–1110.

Willmer, E. V. Subjective brightness and size of field in the central fovea. *Journal of Physiology*, 1954, **123**, 315–323.

Wilson, B. C. An experimental examination of the spectral luminosity construct. Doctoral dissertation, New York University, 1964.

Winslow, C.-E. A., Herrington, L. P., & Gagge, A. P. Physiological reactions of the human body to varying environmental temperatures. *American Journal of Physiology*, 1937, **120**, 1–22.

Wolff, W. Versuche zur Lautstärkeempfindung. *Zeitschrift für Psychologie*, 1935, **136**, 324–340.

Woskow, M. H. Multidimensional scaling of odors. In N. Tanyolaç (Ed.), *Theories of odors and odor measurement*. Istanbul: Robert College, 1968. Pp. 147–188.

Wright, H. N. Loudness as a function of duration. *Journal of the Acoustical Society of America*, 1965, **37**, 1174.

Wright, W. D. *Researches on normal and defective colour vision*. St. Louis, Missouri: Mosby, 1947.

Wundt, W. *Grundzüge der physiologischen Psychologie*. Leipzig: Engelmann, 1874.

Wundt, W. *Grundzüge der physiologischen Psychologie*. Leipzig: Engelmann, 1902.

Wundt, W. *Outlines of psychology*. Translated by C. H. Judd. Leipzig: Engelmann, 1907.

Yager, D., & Taylor, E. Experimental measures and theoretical account of hue scaling as a function of luminance. *Perception & Psychophysics*, 1970, **7**, 360–364.

Yilmaz, H. Perceptual invariance and the psychophysical law. *Perception & Psychophysics*, 1967, **2**, 533–538.

Yoshida, M. Similarity among different kinds of taste near the threshold concentration. *Japanese Journal of Psychology*, 1963, **34**, 34–35. (English translation of paper in Japanese, pp. 25–34.)

Young, P. T. The role of affective processes in learning and motivation. *Psychological Review*, 1959, **66**, 104–125.

Zinner, E. Die Reizempfindungskurve. *Zeitschrift für Sinnesphysiologie*, 1930–31, **61**, 247–266.

Zinnes, J. L. Scaling. *Annual Review of Psychology*, 1969, **20**, 447–478.

Zwicker, E. Die Verdeckung von Schmalbandgeräuschen durch Sinustöne. *Acustica*, 1954, **4**, 415–420.

Zwicker, E. Über psychologische und methodische Grundlagen der Lautheit. *Acustica*, 1958, **8**, 237–258.

Zwicker, E. Ein Beitrag zur Lautstärkemessung impulshaltiger Schalle. *Acustica*, 1966, **17**, 11–22.

Zwicker, E., & Feltkeller, R. Über die Lautstärke von gleichformigen Geräuschen. *Acustica*, 1955, **5**, 303–316.

Zwicker, E., Flottorp, G., & Stevens, S. S. Critical band width in loudness summation. *Journal of the Acoustical Society of America*, 1957, **29**, 548–557.

Zwicker, E., & Scharf, B. A model of loudness summation. *Psychological Review*, 1970, **72**, 3–26.

Zwislocki, J. J. Theory of temporal auditory summation. *Journal of the Acoustical Society of America*, 1960, **32**, 1046–1060.

Zwislocki, J. J. Analysis of some auditory characteristics. In R. D. Luce, R. R. Bush, and E. Galanter (Eds.), *Handbook of mathematical psychology*, Vol. III. New York: Wiley, 1965. Pp. 1–97.

Zwislocki, J. J. Temporal summation of loudness: An analysis. *Journal of the Acoustical Society of America*, 1969, **46**, 431–441.

Zwislocki, J. J., Buining, E., & Glantz, J. Frequency distribution of central masking. *Journal of the Acoustical Society of America*, 1968, **43**, 1267–1271.

NAME INDEX

Numbers in italics refer to the pages on which the complete references are listed.

SUBJECT INDEX